Seaweeds of the Southeast Coast of India

Seaweeds of the Southeast Coast of India:

Their Diversity and a Field Guide for Their Identification

By

M. Ganesan, C.R.K. Reddy,
K. Eswaran and Bhavanath Jha

Cambridge Scholars Publishing

Seaweeds of the Southeast Coast of India:
Their Diversity and a Field Guide for Their Identification

By M. Ganesan, C.R.K. Reddy, K. Eswaran and Bhavanath Jha

This book first published 2023

Cambridge Scholars Publishing

Lady Stephenson Library, Newcastle upon Tyne, NE6 2PA, UK

British Library Cataloguing in Publication Data
A catalogue record for this book is available from the British Library

ISBN (10): 1-5275-1774-8
ISBN (13): 978-1-5275-1774-5

CONTENTS

ABOUT THE AUTHORS

Dr. M. Ganesan

Dr. M. Ganesan is Ex Senior Principal Scientist at Marine Algal Research Station, Mandpam, Tamil Nadu an extension centre of CSIR- Central Salt & Marine Chemicals Research Institute. His current endeavour is commercial scale production of *Kappaphycus alvarezii* seedlings through tissue culture and agarophytes seedling production through spore culture. He published 80 research papers in reputed SCI journals with high impact factor; 1 patent and guided 2 Ph.D students.

Dr. C.R.K. Reddy

Dr C R K Reddy is currently Adjunct Professor, DBT centre for Energy Bioscience, Institute of Chemical Technology, Mumbai. Dr Reddy's core expertise lies in cellular biotechnology of seaweeds for genetic improvement and micro-propagation of economically important seaweeds. He led active academic activities teaching course work to Ph.D students and supervised more than 10 students for Ph.D in addition to guiding several postgraduate students in their dissertation studies. Dr Reddy holds a distinguished record with established scientific credentials in S&T authoring 86 publications in SCI journals of high repute, 18 book chapters, 3 patents and two authored books and one edited book.

Dr. K. Eswaran

Dr. K. Eswaran, is Ex Chief Scientist and Scientist in charge at Marine Algal Research Station, Mandpam, Tamil Nadu an extension centre of CSIR- Central Salt & Marine Chemicals Research Institute He has immensely contributed to develop technology for the cultivation of *Kappaphycus alvarezii* in India and R&D work carried out over many years resulted in development of various processes with potential commercial value. In addition to *Kappaphycus alvarezii,* he developed cultivation technology for *Gracilaria dura* which gives direct agarose. He published more than 45 research papers in peer reviewed journals.

Prof. Bhavanath Jha

Prof. Dr. Bhavanath Jha (b 1958) is Ex- Chief Scientist and Divisional Chair, Marine Biotechnology and Ecology, CSIR- Central Salt & Marine Chemicals Research Institute, Bhavnagar, Gujarat, India. His scientific interest is in the areas of Molecular Biology and Marine Biotechnology, Stress Genomics, Gene Mining & Translational Research. He has supervised 45 Ph. D. students. He has 275 research publications (average IF ~ 4.5) in peer reviewed referred journals of high impact (cumulative citations -11150, h-index– 60 and i10–index- 192); 21 international patents, 2 books and 41 book chapters to his credit. Prof. Jha is elected fellow of the Indian National Science Academy (**FNA**), Fellow of the National Academy of Science, India (**FNASc**) and Fellow of the National Academy of Agricultural Sciences (**FNAAS**).

FOREWORD

Currently, seaweeds are gaining significant attention worldwide due to their unparallel nutritional, industrial as well as ecological values. Seaweed aquaculture is a fast emerging sector in the world and is widely touted as a sustainable source with immense potentials to meet the future needs and demands of humankind. The term seaweed represents an assemblage of a diverse group of photosynthetic aquatic plants that are exceptionally unique in their form, function, structure, and biochemical composition offering novel opportunities for their commercial exploitation. They largely remained as unexplored marine living resources in India in particular despite proven economic values. In Far East Asian countries, seaweeds are popularly utilized in human food preparations besides being used as a source of raw material for extraction of industrially important phycocolloids (agar, agarose, carrageenan, and alginates) and agro-based products. More recently, there has been a growing interest in the application of seaweed ingredients in beauty and novelty food products, nutraceuticals, bioplastics, beverages, etc. besides being a potential source for biofuels.

India has a coastline of more than 8000 km long and an exclusive economic zone (EEZ) of 2.17 million km^2 which equals 66% of that of the mainland area. This area provides substantial arable space for farming various economically important seaweeds and also hosts luxuriant seaweed biodiversity of immense value. Unlike other Asian countries, seaweeds in India are exclusively utilized for the production of typical phycocolloids such as agar, carrageenan and alginates from the raw material harvested from natural beds, particularly from southeast coast of Tamil Nadu. There are several households, particularly the womenfolk, traditionally engaged in seaweed collection from natural beds along the Gulf of Mannar coast, Tamil Nadu. Likewise, there are 32 MSMEs in the country engaged in processing of seaweeds for production of phycocolloids. Therefore, seaweed collection from natural sources has not only become a source of raw material supply for local seaweed industry but also forms a new income stream for the rural coastal communities.

CSIR-Central Salt and Marine Chemicals Research Institute (CSIR-CSMCSRI), a premier research laboratory under the aegis of Council of Scientific and Industrial Research (CSIR), New Delhi, India is pursuing seaweed research actively since 1960. The pioneering research efforts carried out over years helped to the development of sustainable farming as well as downstream technologies for various economically important seaweeds such as *Gracilaria dura, Gracilaria edulis, Gracilaria debilis, Gelidiella acerosa, Kappaphycus alvarezii, Hypnea musciformis, Sarconema filiforme* and *Ulva* (*Enteromorpha*) spp. and few of them successfully commercialised as well.

The southeast coast of India is endowed with rich seaweed diversity in the country. This coast also possesses UNESCO identified Gulf of Mannar Marine National Park. The present book on “Seaweed diversity of southeast coast of India- Field guide for identification” is an excellent source for students and researchers for easy identification of seaweeds species. The authors described 256 seaweed species in a simple language focussing morphological, anatomical and reproductive characters supported with high quality *in situ* seaweed photographs. Taxonomic key characters given for the genus add more value to the book as there was no such exercise made before. The unique features of southeast coast of India were elucidated nicely in the introductory chapter. The authors of the book have made a number of significant contributions advancing seaweed research in the country, and this book is yet another milestone in their achievements. I congratulate all the authors for accomplishing this stupendous task that will have a profound impact on the future research in the country. It gives me an immense pleasure and honour to have written foreword for this book and also assumes that scientific community, policy makers, students, resource manager and industry friends find this as most useful resource. Finally, I would also thank Cambridge Scholars Publishers for publishing this work.

Kannan Srinivasan
Director, CSIR-CSMCRI 08.04.2020, Bhavnagar- 364 002
India

ACKNOWLEDGMENTS

We are highly grateful to the Council of Scientific & Industrial Research (CSIR), Government of India, New Delhi, for funding the National network project on "Marine Biodiversity of India" under which program the present work was carried out. We are grateful to the Department of Forest, Government of Tamil Nadu for granting official permission for estimating seaweed diversity in Gulf of Mannar islands under Marine National Park.

We wish to say our deepest gratitude to Dr. P.K Ghosh, Former Director, CSIR-Central Salt & Marine Chemicals Research Institute (CSMCRI), Bhavnagar, for providing laboratory facilities and extending all moral support till the completion of the present work. We would like to pay our special regards to Dr. S. Kannan, Director, CSIR-CSMCRI for his foreword message to this book.

We express our profound gratitude to Prof. M. Umamaheswara Rao (deceased), Former Professor, Department of Botany, Andhra University, Visakhapatnam and Prof. V. Krishnamurthy (deceased), Krishnamurthy Institute of Algology, Chennai, the experts in seaweed taxonomy; who supported us in species level identification of seaweeds belonging to the genus *Sargassum, Dictyota, Padina, Gracilaria* and *Laurencia*.

We are indebted to Dr. N. Kaliparumal, Former Principal Scientist, CMFRI, Bangalore, Prof. S. Nagaraj, CAS in Botany, University of Madras and Dr. V. Veeragurunathan, CSIR- CSMCRI, Mandapam for correcting few chapters of this book.

We are grateful to Mr. A. Sathakku and Mr. A.S.S. Ahamed skin divers (former employees of CSIR- CSMCRI), and scuba diving group from People Action Development Initiative (PADI), Vembar, India for collecting seaweeds up to 20 meter depth.

We gratefully acknowledge Mr. K. Selvaraj, Mrs. K. Chitra, Mrs. J. Malarvizhi, Mrs. A. Hanna Jeeja, Mr. B. Prince, and Miss, N. Nadukkattu Nayaki Research Scholars of our laboratory, for assisting in taxonomic key preparation and taking anatomical sections of seaweeds. We also thank Mr. N. Ramalingam, Mr. A. Nanthakumar, Mr. P. Ramachandran,

Mr. S. Uthayakumar, Mr. M. Muthu and Mr. G. Nesan, field staffs of our Institute for helping us in field activities like labelling and preservation of seaweeds and herbarium preparation. We would recognize the invaluable assistance they all provided for this present work.

We thank Mr. R. Sukumaran (deceased), driver, (former employee of CSIR- CSMCRI) and Mr. Ganesan, Ervadi, and Mr. Kumar, Pamban boat drivers, for their support during seaweed diversity assessment work along the islands of Gulf of Mannar.

We thank Tourism Department of Government of Tamil Nadu and Government of Pondicherry for providing guest house facility during the survey work.

We thank Ms. G. Kiruthika, University of Technology, Sydney, Australia for designing the cover page of the book.

Finally we wish to thank all the people whose assistance was a milestone for successful completion of the project work.

The Authors

INTRODUCTION

The Southeast Indian Coast

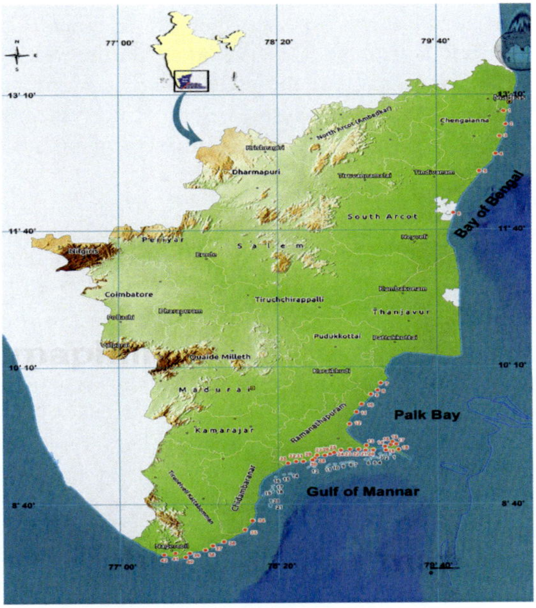

Fig. 1.1. Map showing sampling stations (*red dots*) on the Southeast Indian coast.

Coastline

The southeast Indian coast (Fig. 1.1) is about 1000 km long with diverse and unique ecosystems like mangroves, coral and seagrass and seaweed ecosystems. The world's second longest sandy beach called "Marina beach" is located on the northern tip of this coast (Fig. 1.2A). Boulders and stones were placed parallel to the shore (Fig. 1.2B) to protect the harbours, temples and historic monuments. Pitchavaram mangrove forest spreads over a 1100-hectare area and is the second-largest mangrove cover in India. Muthupet mangrove swamp has thick mangrove vegetation in a 77.20-hectare area. The central part of this coastline is home to the

Cauvery Delta where several rivers and tributaries meet the sea. The Palk Bay shorelines are mostly muddy and extend up to 2-3 km inwards into the sea (Fig. 1.2C). The southern tip of the coastline has a rocky shore (Fig 1.2D), small bays (Fig. 1.2E) and intertidal rock pools (Fig. 1.2F). On the Kanyakumari coast, the sandy coast is replaced with rocks and cliffs of differing height (Fig 1.2G). The several coastal villages located close to the sea were heavily affected by the Tsunami in 2005 (Fig 1.2H). Sand dunes up to a height of 10-25 meters are common along the southern part of this coastline (Figs 1.3A, B). The Gulf of Mannar (GoM), and the UNESCO-identified Biosphere Reserve and Marine National Park are situated on the southern part of the coastline. The GoM has coral reefs (Fig.1.3 C), seagrass meadows, oyster beds, offshore islands (Fig. 1.3D) and lagoons (Figs. 1.3 E, F, and G, H) that form an ideal habitat for seaweeds.

Ports, fishing harbours, nuclear and thermal power plants and a number of refineries, fertilizers and marine chemicals plants are situated on the coastline. A nuclear power plant with 2000 MW (2 × 1000 MW) capacity power generation is present at Kudankulam coast in the southern part of the coastline (Fig. 1.4A). Another nuclear power plant with 2 × 1000 MW capacity power generation is located at Kalpakkam coast, in the northern part of the coastline. A thermal power plant with 1050 MW (5 × 210 MW) capacity power generation is functioning at Tuticorin coast, (Fig. 1.4B) in the southern part of the coastline. A road and railway bridge across the sea with 2.3 km length, connecting Rameswaram Island to the mainland is one of the unique structures of this coastline (Fig. 1.4C). A memorial for Swami Vivekananda and a statue of Tamil poet Thiruvallluvar, constructed on giant offshore rocks in the mid-sea off the Kanyakumari coast attract millions of tourists from all over the world (Fig. 1.4D). The Hindu God Vinayaka's temple in the sea at Kodiyakarai (Fig 4E) and the Swami Vivekananda memorial on the Pamban coastline (Fig. 1.4F) are other tourist attractions on this coast. In addition, a palace constructed by the Pallava King at Tranquebar (Fig. 1.4G), and a tower donated to the British Empire by the Chola King at Manora (Fig 1.4H) are symbols of Indian architecture on the shoreline.

Fig. 1.2. Coast types: A. Long sandy beach (Chennai). B. Big boulders and stones along the shore (Pondicherry) C. Muddy coast with cast-ashore seagrass (Kottaipattinam) D. A long rocky platform edges out the sand (Nochiurani). E. An enclosed bay at Seeniappa Dharga F. Intertidal rock pools at Manapad. G. Massive rocks in the sandy beach at Kanyakumari. H. A coastal village close to the sea.

Fig. 1.3. Coast types: A & B. Beach rocks and intertidal rocks alternating with sand dunes (Manapad). C. A long *Galaxaura* reef (Krusadai Island). D. Two islands (Manoli and Manoli Putti Islands) closely located. E. Lagoon on Valai Island F. Lagoon during lowest low tide on Krusadai Island. G. Huge coastal lagoon at Manapad. H. Sandy beach with country boats for fishing (Keelakarai).

Fig. 1.4. A nuclear power plant (Kudankulam). B. Thermal power station (Tuticorin). C. Pamban bridge connecting Rameswaram Island to the mainland. D. Swami Vivekananda memorial and a giant statue of Tamil poet Thiruvalluvar (Kanyakumari). E. A temple to the God Vinayaka in the sea (Kodiyakarai). F. Swami Vivekananda temple (Pamban). G. Palace of the Pallava king (Tranquebar). H. Monument built by a Chola King at Manora on the seashore of the Adhirampattinam coast.

Diverse Ecosystems along the Southeast Coast of India

Gulf of Mannar Marine Biosphere Reserve

The Gulf of Mannar, the first Marine Biosphere Reserve in India, extends to 140 km in a north-south direction and spreads across an area of 10, 500 km^2. Twenty-one islands parallel to the coastline are located at an average distance of 8 km from the mainland coast. The Gulf is bestowed with a mangrove forest, seaweed beds, seagrass meadows and a coral reef. About

117 species of corals belonging to 37 genera (Figs. 1.5A, B, C, D), 24 mangrove species, (Figs. 1. 5E, F) and 13 seagrass species are reported in the Gulf of Mannar Biosphere Reserve (Figs. 1.5G, H). All three highly productive ecosystems harbour a rich diversity of flora and fauna. The Gulf has 3,600 species of floral and faunal diversity, making it one of the biologically wealthiest coastal ecosystems in India (Global Environment Facility, 1999). Considering its biological wealth, UNESCO declared the Gulf of Mannar as a Biosphere Reserve in 1983. Subsequently, the Union Government of India declared the Gulf of Mannar as a Marine National Park in 1989.

Fig. 1.5. Coral, mangrove and seagrass ecosystem of the Gulf of Mannar, SE Indian coast. A, B, C, D corals; E Mangrove forest on Krusadai Island F. Mangrove on Manoli Island. G. Seagrass *Enhalus acoroides* meadow in Pamban passes. H. Seagrass *Cymodocea rotundata* meadow on Krusadai Island

Pulicat Lake

Fig. 1.6. A. View of Pulicat Lake on the SE Indian coast. B. Birds searching food in Pulicat Lake

Pulicat Lake, the second largest brackish water lake in India is located on the northern tip of this coastline. (Fig. 1.6A). The lake has a 350 sq. km water spread area with an average depth of about 1 m. The lagoon has rich floral and faunal diversity, and supports active commercial fishing. *Chaetomorpha sp., Enteromorpha sp., Cladophora sp., Rhizoclonium sp., Rosenvingea sp., Padina sp., Polysiphonia sp., Gracilaria sp.* and *Hypnea sp.* are the important algal species recorded in Pulicat Lake. Migratory birds, mainly from Central Asia and Eastern Europe, visit this sanctuary (Fig. 1.6B). The lagoon has gained importance through its recognition by the International Union of Conservation of Nature (IUCN).

Pitchavarm Mangrove Forest

Pitchavaram mangrove is the second largest mangrove forest in India. The forest cover is about 1100 hectares and it is separated from the sea by a huge sand bar (Figs. 1. 7A, B). The Pitchavaram mangrove ecosystem has a great wealth of biological diversity. It serves as the best nursery ground for shrimp, lobster, crab and turtle. The seaweeds *Enteromorpha, Cladophora, Chaetomorpha, Padina, Gracilaria* and *Hypnea* and seagrass *Halophila* are reported in this ecosystem. *Avicennia* is the most common mangrove genus, followed by *Rhizophora, Bruguiera* and *Aegiceras.*

Fig. 1.7. A & B. Views of Pitchavaram mangrove forest on the SE Indian coast.

Muthupet Mangrove Swamp

The Muthupet mangrove swamp spreads across an area of 12,500, out of which 77.20 ha. has thickly covered mangrove trees while the remaining area is sparsely covered by the mangrove trees (Figs. 8A, B). *Avicennia marina* is the most abundant species followed by *Excoecaeria agallocha, Aegiceras corniculatum, Acanthus ilicifolius Suaeda maritima* and *Suaeda monoica.* Cutting of the wood for fuel and grazing by cattle have caused great damage to the mangrove forest. Seaweeds like *Chaetomorpha, Enteromorpha, Gracilaria and Hypnea* are found growing in Muthupet swamp.

Fig. 1.8. A & B Views of Muthupet mangrove swamp on the SE Indian coast.

Seasons and Climate

The southeast Indian coast has a typical tropical climate. The climate ranges from dry sub-humid to semi-arid. The maximum temperature occasionally rises to 40° C during the hottest months of April, May and June. The coast is heavily dependent on the monsoonal rains, influenced by the southwest monsoon rain from June to September and the northeast monsoon rain between October and December. The normal annual rainfall is about 945 mm (37.2 in) of which 68% is coming from the northeast monsoon, and 32% from the southwest monsoon.

Currents, Tides, Wave Action, Seawater Temperature and Salinity

The oceanic current changes its direction twice in a year. The "East Indian currents" pass in a northward direction from January to October and, during the remaining period of the year, the "East Indian winter jet current" passes in a southerly direction. The current is very active during September to December, producing severe cyclones that affect the SE Indian coast (Potemra et al. 1991). The tidal amplitude is lowest among the other coastlines of India. The maximum tidal range is about 70 cm. The coast has a semi-diurnal tidal pattern with two high and two low tides in a day. The monsoonal winds influence the wave direction. During the southwest monsoon months, the wave direction is from the southwest to the south and during the northeast monsoon months it is from the northeast to the north. The height of the waves increases from September to January with a peak in November, and then decreases up to August. The seawater temperature ranges between 24 and 32°C. The salinity of the seawater ranges from 26-36 ppt with peak values in May during the summer.

Seaweed Vegetation

The southeast Indian coastline exhibits greater variations in the nature of the substratum in intertidal and subtidal regions. The northern part of the coastline has artificial boulders and stones in the upper intertidal region. It is prone to severe wave surf action and very few algal species which are surf resistant are dominant in this region. The species belonging to the genus *Ulva, Chaetomorpha, Enteromorpha*, *Bryopsis* and *Grateloupia* are common (Figs. 1.9A, B). The Palk Bay coast has muddy intertidal and subtidal regions due to the confluence of many rivers into the sea. Several species of *Gracilaria* like *G. edulis, G. foliifera, G. verrucosa* and *G. salicornia* are dominant in this region. The seaweed-based industries in Tamil Nadu collect *Gracilaria* spp. from this region for agar extraction. The Gulf of Mannar has a rich diversity of all 3 groups of seaweeds. Intertidal and subtidal rocks extending up to 1 meter depth are rich sources of *Sargassum*, *Acanthophora* and *Hypnea* (Figs 1. 9 C, D, and E). The flat coral reef in the subtidal region is bestowed with *Gelidiella acerosa, Turbinaria Saragassum* and *Ulva* vegetation. (Figs. 1.9 F, G). All 20 Gulf of Mannar islands are surrounded by the coral reef. *Asparagopsis taxiformis* is the dominant alga in the deep water around the islands of Gulf of Mannar (Fig. 1.9H).

A
B
C
D

Fig. 1.9. Seaweed vegetation: A. Big boulders with *Grateloupia* population. B. Rock covered with *Bryopsis* population alternates with the sandy coast (Velankanni nagar). C. Intertidal rocks with seaweed vegetation (Seeniappa Dharga). D Lower intertidal rocks with seaweed population (Arockiapuram). E. Rocks in surf zone with rich seaweed diversity (Manapad). F. Intertidal rock pools with dense Sargassum population (Manapad). G. Long flat rocks in the intertidal region with *Ulva fasciata* (Idinthakarai). H. Dense *Asparagopsis taxiformis* population at 10-meter depth (Valimunai Island).

Seaweed Diversity

Sampling of seaweeds

Seaweed sampling was done at 56 sampling stations along the entire southeast coast, of which 42 sampling stations were located on the mainland coast and 14 stations on the Gulf of Mannar islands. The southeast coast has a semi-diurnal tide with two low tides and two high tides in a day. The intertidal region is exposed for four hours per day. The tide starts to recede two hours before the low tide period and floods after two hours. Therefore, the biodiversity assessment and the seaweed sampling were carried out during these 4 hours during which most of the seaweed species were exposed in the intertidal and subtidal regions.

Fig. 1.10. Seaweed Collection: A. Sampling in the intertidal region (0.5 m deep). B, C. Seaweed sampling by snorkelling up to 2-meter depth. D, E, F, G. Fully equipped Scuba diving for sampling beyond 3-meter depth.

All the seaweed collection persons wear rubber safety boots to walk on the uneven rocks, coral surfaces and on muddy areas for the seaweed sampling (Fig. 1.10A). Snorkelling was done to collect the seaweed samples up to 3-meter depth (Figs 1.10B, C). The scuba diving experts were hired for the algal sampling above 3 meters and up to 20 meter-depth (Figs 1.10 D, E, F, G).

Qualitative Assessment

All the seaweeds growing in a particular site, irrespective of quantity, size, as epiphytes or firmly attached on the substratum, either free floating or submerged, are collected. This will ensure having the complete list of the seaweed diversity in that particular coastal locality.

Individual seaweed species were collected in full form, kept in the individual polythene bag with ample seawater, and labelled. The occurrence of the alga either as exposed or submerged in the supra-littoral, intertidal or subtidal region was noted. Global location of the sampling station was recorded with the Global Positioning System (GPS).

The bags with algal specimens were kept in an ice box and brought to the laboratory for further processing.

Preparation of Herbarium

a) The individual alga was transferred to a tray and filled with clean sea water.

b) A thin tin sheet was put in the water and a cardboard sheet corresponding to the size of the specimen was placed over the tin sheet

c) The serial number of the alga given in the field note book, preceded by collection site in three letters (e.g. PUD for Pudumadam), was written in the right down corner of the sheet.

d) The specimen was placed on the sheet, spread in an optimal way by a small brush (Fig.1.11A).

e) A tin sheet along with a cardboard sheet with the specimen were gently lifted out of the water and surplus water was drained off.

f) The specimen was covered with adsorbent cloth (preferably mull cloth) and kept in an air-dryer for some time (Fig.1.11B).

g) The air-dried specimens were kept between the newspapers and arranged evenly in a plant press.

h) The plant press was tightly tied with nylon thread and a weight was placed over it.

i) The plant press was kept under direct sunlight to avoid fungal attack.

j) Newspaper was changed daily until the specimens were completely dried.

k) The specimen was mounted on a standard herbarium sheet with a label in the bottom right corner.

l) Details of the plant name, place of collection, date of collection and the ecology of the collection site have been labelled.

m) Herbarium sheets were stored in a cabinet specially made for this purpose (Fig. 1.11C).

Preservation of Algal Sample

I. *Formalin preservation*

Algal specimens were preserved in the glass bottles with 4% formaldehyde (pure formalin is = 40%, so 1 part of formalin was diluted with 9 part of seawater). The glass bottles were labelled as per the number given on the herbarium sheets (Fig. 1.11D).

II. *Silica- gel preservation*

Algal specimens needing to be subjected to molecular analysis for identification were stored in the silica gel. A small fragment (a few millimetres in length) with an apical part has been cut and cleaned off well with the help of a brush to remove epiphytes and diatoms, before being dried with tissue paper. The fragment was put in the zip lock bag and fine grains of silica gel were added to surround the entire fragment. The zip lock polythene bag was closed tightly, otherwise the silica gel would attract air humidity.

Fig. 1. 11. Preparation of herbarium specimen: A. Specimen is spread over a thick cardboard sheet placed on a tin sheet. B. Specimen is placed over a newspaper on the herbarium press and covered with an adsorbent cloth. C. Herbarium sheets stored in cupboards. D. Racks with preserved seaweed specimens.

The seaweed diversity estimated in 42 stations along the mainland coast and 14 island coasts of the Gulf of Mannar, located on the southeast Indian coastline revealed a total number of 256 seaweed species. Out of these, 71 species belong to green algae (Chlorophyceae), 46 species to brown algae (Phaeophyceae) and 139 species to red algae (Rhodophyceae). Among the green algae, the genus *Caulerpa* represents a maximum number of 18 species; the genus *Codium* represents 7 species, the genus *Halimeda* and the genus *Ulva* represent 6 species each and the genera *Acrosiphonia, Anastomosans, Boergesenia, Dictyosphaeria, Neomeris, Microdictyon, Struvea, Valonia* and *Valoniopsis* are represented by a single species (Table 1.1). In brown algae, the genus *Dictyota* represents 10 species and the genus *Sargassum* represents a maximum number of 9 species; *Padina* represents 5 species. The genera *Turbinaria, Dictyopteris* and *Chnoospora* represent 3 species each and the genera *Hormophysa, Hydroclathrus, Iyengaria, Rosenvingea* and *Zonaria* represent a single species each (Table1.1. 2). In red algae, the genus *Gracilaria* represents a maximum number of 18 species while the genus *Laurencia* represents 11 species. Other major genera in red algae are *Hypnea* (6 species) and *Grateloupia* (5 species). On the other hand, several genera viz., *Asparagopsis Bostrichia, Botryocladia, Chondrococcus, Chondrocanthus, Dasya, Dictyurus, Digenea, Enantiocladia, Griffithsia, Halichrysis, Helminthocladia, Neurymenia, Nitophyllum, Peyssonnelia, Tenaciphyllum,* and *Wrangelia* are represented by a single species only (Table 1.3).

Table 1 -1 List of green algae (Chlorophyta) recorded from the southeast Indian coast

			CHLOROPHYTA		
1.	*Acrosiphonia orientalis*	25	*Caulerpa sertularioides*	49	*Enteromorpha linza*
2.	*Anadyomene stellata*	26	*Caulerpa taxifolia*	50	*Halimeda cuneata*
3.	*Avrainvillea amadelpha*	27	*Caulerpa verticillata*	51	*Halimeda gracilis*
4.	*Avrainvillea erecta*	28	*Caulerpa sp.*	52	*Halimeda macroloba*
5.	*Boergesenia forbesii*	29	*Chaetomorpha aerea*	53	*Halimeda opuntia*
6.	*Boodlea composita*	30	*Chaetomorphaan tennina*	54	*Halimeda simulans*
7	*Bryopsis pennata*	31	*Chaetomorpha crassa*	55	*Halimeda tuna*
8	*Bryopsis pennata var. secunda*	32	*Chaetomorpha indica*	56	*Microdictyon tenuius*
9	*Bryopsis plumosa*	33	*Chaetomorpha linum*	57	*Neomeris annulata*
10	*Bryopsis ramulosa*	34	*Cladophora prolifera*	58	*Penicillus sibogae*
11	*Caulerpa chemnitzia*	35	*Cladophora socialis*	59	*Rhizoclonium hieroglyphicum*
12	*Caulerpa corynephora*	36	*Cladophora vagabunda*	60	*Rhizoclonium riparium*
13	*Caulerpa cupressoides*	37	*Cladophora sp.*	61	*Struvea anastomosans*

14	*Caulerpa fergusonii*	38	*Cladophoropsis sundanensis*	62	*Trichosolen mucronatus*
15	*Caulerpa imbricata*	39	*Codium arabicum*	63	*Udotea caribaea*
16	*Caulerpa laetevirens*	40	*Codium dwarkense*	64	*Udotea flabellum*
17	*Caulerpa lentellifera*	41	*Codium indicum*	65	*Ulva fasciata*
18	*Caulerpa mexicana*	42	*Codium isthmocladum*	66	*Ulva lactuca*
19	*Caulerpa microphysa*	43	*Codium iyengarii*	67	*Ulva reticulata*
20	*Caulerpa peltata*	44	*Codium repens*	68	*Ulva rigida*
21	*Caulerpa racemosa*	45	*Codium tenue*	69	*Ulva covelongensis*
22	*Caulerpa racemosa var. occidentalis*	46	*Dictyosphaeria cavernosa*	70	*Valonia utricularis*
23	*Caulerpa scalpelliformis*	47	*Dictyosphaeria versluysii*	71	*Valoniopsis pachynema*
24	*Caulerpa serrulata*	48	*Enteromorpha compressa*		

Table 1.2 List of brown algae (Phaeophyta) recorded from the southeast Indian coast

		PHAEOPHYTA			
1	*Chnoospora bicanaliculata*	17	*Dictyopteris delicatula*	33	*Sargassum gracile*
2	*Chnoospora implexa*	18	*Dictyopteris woodwardia*	34	*Sargassum odontocarpum*
3	*Chnoospora minima*	19	*Hormophysa cuneiformis*	35	*Sargassum polycystum*
4	*Colpomenia sinuosa*	20	*Hydroclathrus clathratus*	36	*Sargassum swartzii*
5	*Cystoseira trinodis*	21	*Iyengaria stellata*	37	*Sargassum tenerrimum*
6	*Dictyota cervicornis*	22	*Lobophora variegata*	38	*Sargassum wightii*
7	*Dictyota ceylanica*	23	*Lobophora variegata var. Minima*	39	*Spatoglossum asperum*
8	*Dictyota ciliolata*	24	*Padina boergesenii*	40	*Spatoglossum variabile*
9	*Dictyota crispata*	25	*Padina boryana*	41	*Stoechospermum marginatum*
10	*Dictyota dichotoma*	26	*Padina distromatica*	42	*Stoechospermum polydioides*
11	*Dictyota divaricata*	27	*Padina pavonica*	43	*Turbinaria conoides*
12	*Dictyota linearis*	28	*Padina tetrastromatica*	44	*Turbinaria decurrens*
13	*Dictyota maxima*	29	*Rosenvingea sanctae-crucis*	45	*Turbinaria ornata*
14	*Dictyota pinnatifida*	30	*Sargassum acinarium*	46	*Zonaria variegata.*
15	*Dictyota spinulosa*	31	*Sargassum carpophyllum*		
16	*Dictyopteris australis*	32	*Sargassum cinereum*		

Table 1.3 List of red algae (Rhodophyta) recorded from the southeast Indian coast

RHODOPHYTA					
1	*Acanthophora muscoides*	48	*Gelidiopsis variabilis*	95	*Laurencia caraibica*
2	*Acanthophora nayadiformis*	49	*Gigartina acicularis*	96	*Laurencia ceylanica*
3	*Acanthophora spicifera*	50	*Gelidium micropterum*	97	*Laurencia cruciata*
4	*Agardhiella subulata*	51	*Gelidium pusillum*	98	*Laurencia indica*
5	*Ahnfeltiopsis pygmaea*	52	*Gracilaria arcuata*	99	*Laurencia microcladia*
6	*Amphiroa anastomosans*	53	*Gracilaria canaliculata*	100	*Laurencia nidifica*
7	*Amphiroa anceps*	54	*Gracilaria corticata*	101	*Laurencia obtusa*
8	*Amphiroa fragilissima*	55	*Gracilaria corticata var. cylindrica*	102	*Laurencia obtusa* var. *rigidula*
9	*Amphiroa foliacea*	56	*Gracilaria crassa*	103	*Laurencia papillosa*
10	*Amphiroa rigida*	57	*Gracilaria curtissiae*	104	*Laurencia pedicularioides*
11	*Asparagopsis taxiformis*	58	*Gracilaria distiche*	105	*Laurencia rigida*
12	*Bostrychia tenella*	59	*Gracilaria debilis*	106	*Leveillea jungermannioides*
13	*Botryocladia leptopoda*	60	*Gracilaria dura*	107	*Liagora albicans*
14	*Bryocladia thwaitesii*	61	*Gracilaria edulis*	108	*Liagora indica*
15	*Carpopeltis maillardii*	62	*Gracilaria fergusonii*	109	*Liagora mannarensis*
16	*Centroceras clavulatum*	63	*Gracilaria foliifera*	110	*Liagora orientalis*
17	*Ceramium brevizonatum*	64	*Gracilaria opuntia*	111	*Lophocladia lallemandii*
18	*Ceramium cruciatum*	65	*Gracilaria rostrata*	112	*Martensia fragilis*

19	*Ceramium deslongschampii*	66	*Gracilaria pudumadamensis*	113	*Martensia indica*
20	*Ceramium subdichotomum*	67	*Gracilaria salicornia*	114	*Martensia sp.*
21	*Champia compressa*	68	*Gracilaria textorii*	115	*Neurymenia fraxinifolia*
22	*Champia globulifera*	69	*Gracilaria verrucosa*	116	*Nitophyllum punctatum*
23	*Champia parvula*	70	*Grateloupia filicina*	117	*Peyssonnelia obscura*
24	*Champia salicornioides*	71	*Grateloupia filicina* var. *cirrhosa*	118	*Polysiphonia subtilissima*
25	*Champia somalensis*	72	*Grateloupia filicina* var. *horrida*	119	*Polysiphonia tuticorinensis*
26	*Cheilosporum acutilobum*	73	*Grateloupia lithophila*	120	*Portieria hornemanii*
27	*Cheilosporum spectabile*	74	*Grateloupia pectinata*	121	*Portieria spinulosa*
28	*Chondrocanthus acicularis*	75	*Grateloupia wattii*	122	*Pterocladia caloglossoides*
29	*Chondria armata*	76	*Griffithsia opuntioides*	123	*Pterocladia heteroplatos*
30	*Chondria capillaris*	77	*Halichrysis thivyae*	124	*Pterocladia kanyakumariensis*
31	*Chondria dasyphylla*	78	*Haloplegma duperreyi*	125	*Rhodymenia sonderi*
32	*Chondria dasyphylla var. stellata*	79	*Haloplegma preissii*	126	*Sarconema filiforme*
33	*Chondria seticulosa*	80	*Halymenia dilatata*	127	*Scinaia bengalica*
34	*Chondrococcus* sp.	81	*Halymenia venusta*	128	*Solieria filiformis*
35	*Claudea elegans*	82	*Helminthocladia clavadosii f. indica*	129	*Solieria robusta*
36	*Coelarthrum muelleri*	83	*Heterosiphonia muelleri*	130	*Spermothamnion speluncarum*

37	*Coelarthrum opuntia*	84	*Hypnea flagelliformis*	131	*Spyridia fusiformis*
38	*Corynomorpha prismatica*	85	*Hypnea hamulosa*	132	*Spyridia hypnoides*
39	*Cryptonemia coriacea*	86	*Hypnea musciformis*	133	*Tenaciphyllum lobatum*
40	*Cryptonemia lomation*	87	*Hypnea nigrescens*	134	*Tolypiocladia calodictyon*
41	*Dasya flagellifera*	88	*Hypnea pannosa*	135	*Tolypiocladia glomerulata*
42	*Dictyurus purpurascens*	89	*Hypnea valentiae*	136	*Tricleocarpa fragilis*
43	*Digenea simplex*	90	*Jania adhaerens*	137	*Vanvoorstia coccinea.*
44	*Enantiocladia prolifera*	91	*Jania capillacea*	138	*Wrangelia balakrishnanii*
45	*Euptilota articulata*	92	*Jania iyengarii*	139	*Wrangelia penicillata*
46	*Galaxaura rugosa*	93	*Jania rubens*		
47	*Gelidiella acerosa*	94	*Kappaphycus alvarezii*		

Stationwise Diversity of Seaweeds

Along the mainland coast, the Kovalam coast has the maximum number at 84 seaweed species followed by Kanyakumari (83 species), the Manapad coast (73 species), Thonithurai (61 species), Nalupanai (46 species), Pudumadam and Ervadi (41 species), Valinokkam (40 species) and Panchal (36 species). Ennore and Kasimedu coasts have the lowest number at 3 seaweed species. Chlorophycean algae were dominant in Kanyakumari, Kovalam and Thonithurai (22 species in each station) while phaeophycean algae were dominant in Thonithurai (19 species) and Pamban light house area (18 species). Phaeophycean algae were absent in Ennore, Kasimedu, Mahabalipurm, Muttukadu, Pondicherry, Karangadu, Ammapattinam and Kodiakkarai. Rhodophycean algae were dominant in Manapad (53 species), Kovalam (51 species) and Kanyakumari (50 species) (Table 1.4). In general, seaweed diversity is rich along the southern part of the southeast coast and poor along the northern part of the southeast coast.

In the Gulf of Mannar islands, Koswari Island has the maximum number at 92 species followed by 77 species each on Upputhanni Island and Hare Island. Manali Island has the lowest number at 28 species. (Table 4). Chlorophycean algal diversity was high on Hare Island and Valai Island (32 species and 30 species). The phaeophycean algal diversity was rich on Hare Island (26 species) and the rhodphycean diversity was maximum on Koswari Island (47 species).

Table 1. 4 Number of seaweed species recorded at different stations from the southeast Indian coast (mainland and island coasts)

S. No	Places	Latitude and Longitude	Number of Seaweeds Recorded			Total
			Green algae (Chloro-phyceae)	Brown algae (Pheao phycea)	Red algae (Rhodo phyceae)	
Mainland coast						
1	Ennore	13°13.950′N 080°19.926′E	2	0	1	3
2	Kasimedu	13°10.888′N 080°18.962′E	1	0	2	3
3	Muttukadu	12°47.125′N 080°15.295′E	5	0	8	13
4	Kovalam	12°47.900′N 080°14.989′E	7	1	4	12
5	Mahabalipuram	12°11.540′N 080°09.236′E	2	0	3	5
6	Pondicherry	11°55.561′N 079°50.070′E	3	0	3	6
7	Kottaipattinam	09°58.376′N 079°11.956′E	3	1	13	17
8	Arasaneripattinam	09°53.441′N 9079°07.535′E	3	1	14	18
9	Karankadu	09°38.775′N 78°57.953′E	3	0	8	11
10	Ammapattinam	10°00.995′N 079°14.028′E	3	0	8	11
11	Ponnagaram	10°01.642′N 079°14.751′E	1	1	8	10
12	Kodikkarai	10°02.435′N 79°15.748′E	3	0	3	6

13	Munaikadu	09°16.899′N 079°11.349′E	10	5	9	24
14	Pamban light house	09°16.929′N 79°12.511′E	5	18	12	35
15	Nalupanai	09°17.526′N 79°14.408′E	16	11	19	46
16	Villoonditheertham	09°17.717′N 79°16.213′E	14	9	16	39
17	Ariyankundu	09°17.717′N 79°16.212′E	9	8	8	25
18	Rameswaram	09°09.537′N 78°39.437′E	8	3	6	17
19	Pamban	09°19.980′N 079°12.539′E	7	6	9	22
20	Thonithurai	09°16.974′N 079°11.212′E	22	19	20	61
21	Velankanni nagar	09°16.780′N 79°10.318′E	8	3	9	20
22	Seeniappa dharga	09°15.649′N 079°03.981′E	9	5	15	29
23	Nochiurani	09°16.016′N 78°02.043′E	11	6	17	34
24	Pudumadam	09°16.311′N 078°59.921′E	15	6	19	40
25	Thalathoppu	09°12.485′N 078°43.585′E	07	3	11	21
26	Sullimunthal	09°15.221′N 078°52.184′E	3	12	7	22
27	Kanchirankudi	09°14.243′N 78°48.994′E	4	13	14	31
28	Keelakarai light house	09°14.966′N 78°51.238′E	4	14	6	24
29	Keelakarai	09°13.067′N 78°45.533′E	12	13	12	37
30	Bharathi nagar	09°13.067′N 78°45.533′E	9	13	7	29
31	Ervadi	09°12.415′N 078°43.536′E	17	15	9	41
32	Sadaimuniyan valasai	09’11.434’N 078^{0}43.11’E	11	8	7	26
33	Valinokkam	09°09.530′N 078°39.383′E	14	6	20	40
34	Manapad	08°22.313′N 78°03.953′E	14	6	53	73

35	Kuduthalai	08°17.905′N 77°55.684′E	11	7	12	30
36	Idinthakarai	08°10.615′N 077°44.648′E	9	6	6	21
37	Panjal	08°09.514′N 77°39.503′E	10	6	20	36
38	Leepuram	08°06.639′N 077°33.438′E	8	6	12	26
39	Chinnamuttam	08°05.548′N 77°33.845′E	10	5	19	34
40	Vattakottai	08°07.570′N 77°33.963′E	9	7	9	25
41	Kanyakumari	08°04.682′N 77°32.031′E	22	11	50	83
42	Kovalam	0843°04.732′N 77°31.872′E	22	11	51	84
Island coast						
1	Shingle Island	09°14.540′N 79°13.825′E	21	17	16	54
2	Pullivasal Island	09°14.295′N 079°12.388′E	21	17	10	48
3	Krusadai Island	09°14.928′N 079°13.245′E	23	18	20	61
4	Manali Island	09°13.193′N 079°09.091′E	11	9	8	28
5	Hare Island	09°11.656′N 79°03.414′E	32	26	19	77
6	Mulli Island	09°11.590′N 78°59.918′E	18	18	27	63
7	Valai Island	09°11.500’N 79°00.172’E	30	18	23	71
8	Appa Island	09°10.326’N 78°50.172	22	14	23	59
9	Valimunai Island	09°08.939′N 78°43.577′E	20	13	18	51
10	Anaipar Island	09°09.496′N 78°41.812′E	17	18	29	64
11	Salli Island	09°05.917′N 78°32.054′E	21	22	19	62
12	Upputhanni Island	09°05.450′N 78°29.280′E	27	17	33	77
13	Van Island	08°49.686′N 78°13.090′E	22	21	18	61
14	Koswari Island	08°51.192′N 078°14.059′E	24	21	47	92

Seaweed Diversity at Different Depths

About 48% of the green algae were found growing in 0-to-0.5-meter depth with *Caulerpa, Enteromorpha, Cladophora* and *Chaetomorpha* as dominant species. Only 4% of green algae were found occurring in deeper water (5-to-10-meter depth). *Udotea* is the dominant alga occurring in deeper waters.

About 72% of brown seaweeds were found thriving in 0.5-to-2.5-meter depth with *Sargassum, Padina, Dictyota and Lobophora* as dominant seaweeds. Like green algae, only 6% of brown algae were recorded in deeper water (5-to-10-meter depth) with *Dictyopteris* as the dominant species.

Contrary to the green and brown algae, the red algal diversity was abundant in deeper waters. About 44% of the red algae were found growing between 0.5 and 2.5 meter depth with *Gracilaria, Laurencia, Liagora, Galaxaura* and *Spyridia* as the dominant genera and 22% of red algal diversity was registered in 5-to-10-meter depth. *Agardhiella, Coelathrum, Cryptonemia, Neurymenia, Haloplegma and Halichrysis* are the dominant red algae found in 5-10-meter depth (Table 1.5).

Table 1.5 Depth-wise distribution (in percent) and abundance of seaweeds

Group	%	0 to 0.5 meter		0.5 to 2.5 meter		2.5 to 5.0 meter		5.0 to10.0 meter
Green algae (Chlorophyceae)	48%	*Caulerpa* *Enteromorpha* *Cladophora* *Chaetomorpha*	36%	*Valonia* *Ulva* *Halimeda* *Microdictyon*	12%	*Codium*	4%	*Udotea*
Brown algae (Phaeophyceae)	6%	*Stoechospermum* *Hydroclathrus*	72%	*Sargassum* *Dictyota* *Padina* *Lobophora*	16%	*Spathoglossum* *Turbinaria* *Zonaria*	6%	*Dictyopteris*
Red algae (Rhodo phyceae)	23%	*Acanthophora* *Hypnea* *Grateloupia* *Jania*	44%	*Gracilaria,* *Laurencia,* *Liagora,* *Galaxaura,* *Spyridia*	11%	*Asparagopsis* *Halymenia* *Martensia*	22%	*Agardhiella, Coelathrum* *Haloplegma,Neurymenia* *Cryptonema,Halichrysis*

Seaweeds Distributed in Multiple Stations

Caulerpa racemosa were distributed in 35 stations followed by *Enteromorpha compressa* in 34 stations and *Caulerpa taxifolia* in 33 stations. *Sargassum wightii* were distributed in 38 stations and *Dictyota dichotoma* were distributed in 32 stations. *Gelidiella acerosa* and *Hypnea valentiae* were distributed in 33 stations (Table 1.6)

Table 1.6 List of seaweed species distributed in maximum number of stations.

S. No	Group	Seaweed species	Number of stations distributed
1	Green algae	*Caulerpa racemosa*	35
2		*Caulerpa peltata*	26
3		*Caulerpa taxifolia*	33
4		*Caulerpa sertularioides*	25
5		*Enteromorpha compressa*	34
6		*Halimeda gracilis*	20
7		*Halimeda macroloba*	20
8	Brown algae	*Colpomenia sinuosa*	25
9		*Dictyota dichotoma*	32
10		*Lobophora variegata*	25
11		*Sargassum wightii*	38
12		*Stoechospermum marginatum*	28
13		*Turbinaria conoides*	20
14		*Turbinaria ornata*	22
15	Red algae	*Acanthophora spicifera*	28
16		*Gelidium pusillum*	22
17		*Gelidiella acerosa*	33
18		*Gelidiopsis variabilis*	30
19		*Gracilaria corticata*	21
20		*Grateloupia lithophila*	21
21		*Hypnea valentiae*	33
22		*Laurencia papillosa*	23

Seaweed Endemic to Islands

About 25 seaweed species were found to be growing exclusively on the islands. Among them, 12 species of red algae, 9 species of green algae and 4 species of brown algae are endemic to the islands (Table 1.7)

Table 1.7 Seaweed species exclusively recorded from the islands.

Group	Seaweed species
Green algae (Chlorophyceae)	*Anadyomene stellata, Avrainvillea erecta, Avrainvillea amadelpha, Codium indicum, Codium isthmocladum, Codium iyengarii , Codium tenue, Struvea anastomosans,* and *Udotea flabellum,*
Brown algae (Phaeophyceae)	*Chnoospora implexa, Dictyopteris australis, Dictyopteris delicatula,, Sargassum tenerrimum*
Red algae (Rhodophyceae)	*Asparagopsis taxiformis, Claudea elegans, Dasya flagellifera, Gracilaria arcuata, Gracilaria canaliculata, Gracilaria debilis, Haloplegma duperreyi, Haloplegma preissii, Heterosiphonia muelleri, Laurencia microcladia, Lopocladia lallemandii* and *Tolypocladia glomerulata*

Threatened and Endangered Seaweed Species

The seaweeds of the SE coastline are harvested for extraction of agar and algin. More than 2000 people are involved in harvesting the seaweed biomass. *Gracilaria edulis, Gracilaria verrucosa, Hypnea musciformis and Hypnea valentiae* are harvested for agar extraction. These 4 species have become endangered as a result. Likewise, *Gelidiella, Gelidium, Sargassum and Turbinaria* have been continuously harvested for the last several decades. These resources are under serious threat (Table 1.8).

Table 1.8 Threatened and endangered seaweed species of the southeast Indian coast

Endangered species	*Gracilaria edulis, Gracilaria verrucosa Hypnea musciformis, Hypnea valentiae*
Threatened species	*Gelidiella acerosa, Gelidium pusillum, Sargassum wightii, Sargassum tenerrimum, Turbinaria conoides, Turbinaria ornata Turbinaria decurrens*

DIVISIONS OF SEAWEEDS

CHLOROPHYTA

Acrosiphonia J. Agardh

***Acrosiphonia* J. Agardh**

Filaments profusely branched, often matted at least in lower part of thallus by short curved branches or adventitious rhizoids. Cells cylindrical, usually very much longer than broad, but varying considerably. Growth by division of apical cells. Cells uninucleate or multinucleate.

Key for the species

Filaments interwoven but not by hook like branches ……..*A.orientalis*

Acrosiphonia J. Agardh

REFERENCES: J. Agardh (1846a: 103), J. Agardh (1848b: 2, pl. XVII).

A

B

Acrosiphonia J. Agardh

Acrosiphonia orientalis **(J. Agardh) P. Silva**

***Description*:** Plants caespitose, upto 6.5 cm high, attached by a mat of rhizoidal branches arising from the lower parts of filaments which form rope like stalks to the lobes of the thallus, filaments repeatedly branched, branches subdichotomous, pseudodichotomous or tridichotomous; lower cells of branches elongated, 180- 300 µm broad the top and gradually tapering towards apex. Cells with numerous pyrenoids and filled with abundant starch grains.

Uncommon: attached to rocks in subtidal region.

Distribution: Manapad, Kuduthalai, Vattakottai.

A. Habit of the plant______

B. Broad filaments in basal part

C (Thin filaments in the upper part)

Anadyomene Lamouroux

Anadyomene Lamoux

Plant flabellate with short stipe, which is formed from several interwoven elongate, parallelly arranged rhizoids; flabellate branches produced in one plane adhering laterally, distinct at margins. Cells club shaped, pale, arranged closely to form a continuous expanse. Asexual reproduction by quadriflagellate zoospores; sexual reproduction by biflagellate gametes.

Key for the species

Blades with 1-2 layers of cells. – *Anadyomene stellata*

Anadyomene Lamouroux

REFERENCES: V. Krishnamurthy & H. Joshi (1970: 8), Untawale, Dhargalkar & Agadi (1983: 14), Jagtap (1985a: 46, 50), Kaliaperumal *et al.* (1992: 12), Kalimuthu, Kaliaperumal & Ramalingam (1992: 122), Silva *et al.* (1996: 755).

A

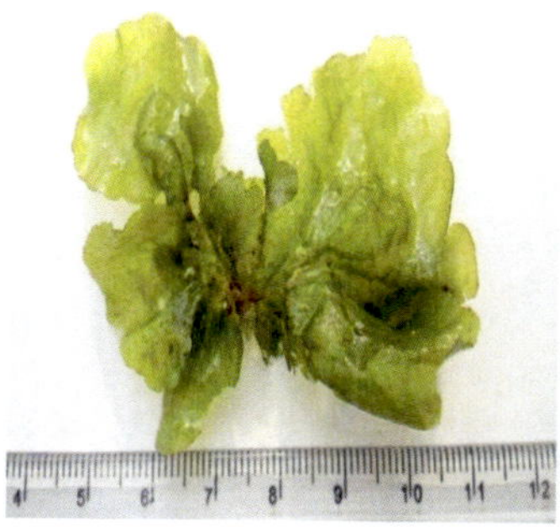

B

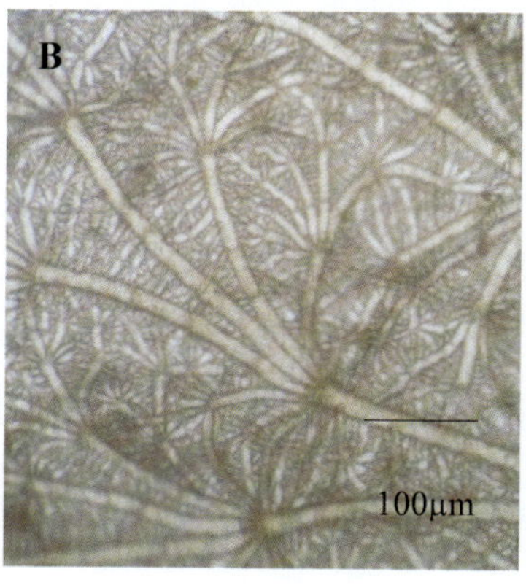

***Anadyomene* Lamouroux**

***Anadyomene stellata* (Wulfen) C. Agardh**

Description: Plants 3-4 cm in height, broader, up to 5 cm consisting of a number of flabellate lobes with undulate margins; attached to coral stones by short stipe; up to 2 mm long rhizoids. Each lobe consists of 7 branches, closely adhering to form a continuous blade; each branch consists of a long clavate cell producing numerous small unicellular branches all around, different branches adhering laterally with no inter cellular space.

Rare: attached to coral stones in sub littoral region.

Distribution: Pullivasal island, Hare island and Koswari island.

A. Habit of the plant

B. Main blade showing parallel cells between main veins.

Avrainvillea Decaisne

Avrainvillea Decaisne

A non-calcified thallus formed by a distinct and well developed bulbous portion buried in soft sediment and an erect, dull green and usually fan shaped portion. The thallus is composed of a system of siphonous, bifurcated tubes with cell wall thickening at the bifurcations, but without cross walls and lime deposition; filaments are not clearly organized into medulla and cortex; chloroplasts are discoid and contain one pyrenoid each.

Key for the species:-

1. Stalk of thallus short, erect branches irregularly lobed - *A amadelpha*

1. Stalk of thallus long, erect branches usually not lobed but occasionally showing deep lobes - *A. erecta*

Avrainvillea Decaisne

REFERENCES: A. Gepp & E. Gepp (1911: 42 – 44, 139, pl. XIV: figs. 112, 113), Borgesen (1940: 54), Borgesen (1953: 10), S. Dixit (1968: 15), Untawale, Reddy & Deshmukhe (1989: table 1), Silva *et al.* (1996: 875).

Avrainvillea
(Montagne) *A*
Gepp
Description: I
dark green in
from an exte
stipe cylindric
part, compres
blade. Stipe w
upto the fan t
blades, 2-4 cn
soft and spon
rounded margi
Uncommon: In
low water mark
Distribution:
Valai island,
and Upputhann
A. Habit sho
stipe and blade
B. Microsco
siphonous tube

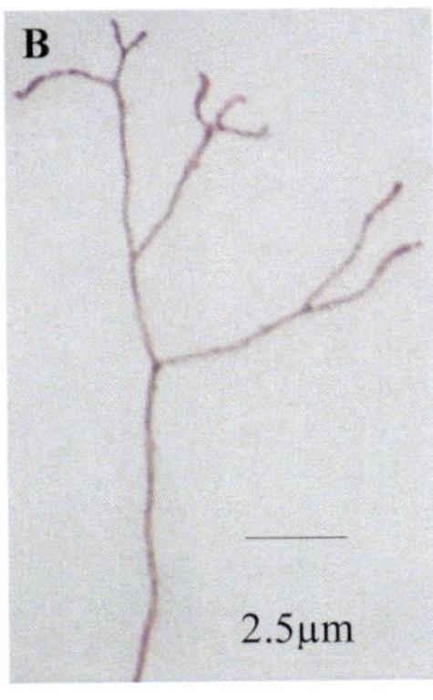

***Avrainvillea* Decaisne**

***Avrainvillea amadelpha* (Montagne) A. Gepp & E. Gepp**
Description: Plant gregarious, dark green in colour, arising from an extensive holdfast; stipe cylindrical in the basal part, compressed towards the blade. Stipe with 2 cm wide upto the fan to wedge shaped blades, 2-4 cm long and wide, soft and spongy with smooth rounded margin; intertwined
***Uncommon*:** In rock crevices in low water mark.
Distribution: Krusadai island, Valai island, Anaipar island and Upputhanni island
A. Habit showing cylindrical stipe and blades
B. Microscopic view of siphonous tube.

Avrainvillea Decaisne

REFERENCES: Awashthi (1989: 78), D. Rao *et al*. (1991: table 1, 2), Jagtap (1992: 57), Kalimuthu, Kaliaperumal & Ramalingam (1992: 122), Menon *et al*. (1993: 59), Silva *et al*. (1996: 875).

***Avrainvillea* Decaisne**

***Avrainvillea erecta* (Berkeley)**
A. Gepp & E. Gepp

***Description*:** Plant dark green in colour, upto 8 cm tall, attached by elongated, interwoven basal mass of rhizoids with apical fan shaped sessile and filamentous foliar portion; foliar portion composed of branched filaments about 4 cm long and 6 cm broad, spongy or hairy.

Common: attached on rocks and muddy substratum.

Distribution: Manali island, Valai island, Valimunai island and Upputhanni island

A. Habit with stipe and fan shaped lamina

B. Microscopic view of siphonous tube

Boergesenia J. Feldmann

***Boergesenia* J. Feldmann**

Thallus formed by tufts of unbranched cylindrical erect axes, very turgid; attached to the substratum by filamentous, segmented rhizoids. Structurally it is formed by club-shaped vesicles with a large central vacuole filled with water and solutes, and a thin peripheral layer of cytoplasm containing many nuclei and chloroplast.

Key for the species -

Light green, erect, Club- shaped branches and thallus consisting cluster of clavate vesicles - *B.forbesii*

Boergesenia J. Feldmann

REFERENCES: Untawale, Reddy & Ambiye (1989: table 1), Jayasankar, Ramalingam & Kaliaperumal (1990: table 1), Jagtap (1992: 57), Kalimuthu, Kaliaperumal & Ramalingam (1992: 122), Jagtap (1993: table 1), Silva *et al.* (1996: 788).

Boergesenia **J. Feldmann**

A

Boergesenia J. Feldmann

Boergesenia forbesii **(Harvey) J. Feldmann**
Description: Thallus yellowish green in colour; vesicular narrowing downwards into a stalk with annular constrictions; attached to substratum by rhizoids; vesicles club shaped expanding upwardly all around the primary vesicle, resulting in a cluster of vesicle.
Common: found attached to coral stones.
Distribution: Munaikadu, Villoonditheertham, Thalathoppu, Manapad, Kanyakumari, Kovalam and all islands of Gulf of Mannar.
A. Habit of the plant.

Boodlea G. Murray & De Toni

Boodlea G. Murray and De Toni

Thallus developed from a primary vesicle. Subsequent divisions of primary vesicles give rise to secondary vesicles; primary vesicle continued as main axis and each segment produces branches. Branches spatially separated and attached together by haptera. Branching irregular more than one plane resulting in sponge like net. Asexual reproduction by quadriflagellate, zoospores produced in an unmodified cell of thallus.

Key for the species

Branches from each cell of secondary filaments one to many, irregularly produced

…..B. composita

Boodlea G. Murray & De Toni

REFERENCES: Jagtap (1987: table. 1), Chaugule, Quadri & Goswamy (1989: 113, figs. 54, 55), V. Krishnamurthy *et al*. (1990: fig. 4), Jagtap (1992: 57), Jagtap (1993: table I), Silva *et al.* (1996: 789).

Boodlea G. Murray & De Toni

***Boodlea composita* (Harvey) Brand**

***Description*:** Plant soft, sponge like masses consisting of intertwined filaments; primary filaments with few ramifications. Secondary filaments richly branched, each cell producing 1to 2 or more branches, irregularly produced and extending in all directions; opposite; ultimate branches single, on lower side. Primary filaments are 160-300 µm in diameter; ramuli with 70-100 µm in diameter, branch ends connected to adjacent filaments by tenacula.

Common: attached to rocks and small corals in lower intertidal region.

Distribution: Munaikadu, Rameswaram, Seeniappadharga, Nochiurani, Pudumadam, Thalathoppu, Kanjirankudi, Valinokkam, Manapad, Leepuram, Kovalam, Kanyakumari, Shingle island, Krusadai island, Valai island, Appa island, Anaipar island and Upputhanni island.

A. Habit of the plant

B. Branchlets connecting to adjacent filament by tenacula

C. Primary and secondary branches.

***Bryopsis* Lamouroux**

Genus Bryopsis Lamouroux

The plants are composed of a series of siphon like filaments without cross walls; only the gametangia is separated by cross walls. The creeping filaments produced erect, pinnate (or) feather- like branches, sometimes with spirally disposed branchlets. Each filament consists of a large central vacuole surrounded by cytoplasm containing many nuclei, chloroplast, and a cell wall. The lens shaped chloroplast contains one pyrenoid each and can accumulate starch. Species identification is based on the morphology of the gametophytic phase, since the alternate sporophytic phase of some species remains unknown.

Key for the Species:-

1. Branch lets inconspicuous, sparse; not forming distinct organized fronds. – *B. ramulosa*
1. Branchlets conspicuous, numerous; forming distinct organized fronds.
2. Branchlets 1 to 1.5cm length, shortening towards apex; main axes less than 200μm diameter. – *B. plumosa*
2. Branchlets 2.5mm of, uniform length, main axes greater than 200 μm diameter.
3. Branchlets in opposite pair. – *B. pennata*
3. Branchlets appearing unilateral. – *B. pennata var.secunda*

Bryopsis Lamouroux

REFERENCES: Nair *et al*. (1986: 735), G. Rao & Umamaheswara Rao (1986: 103), Chaugule, Quadri & Goswamy (1989: 11), Nair *et al*. (1990: table 2), Jagtap (1992: 32), Silva *et al.* (1996: 807).

Bryopsis Lamouroux

Bryopsis pennata **Lamouroux**
Description: Plant dark ~~green~~ in colour; 4-9 cm in height; fronds erect, feather like, consisting of cluster of simple or branched axis measures 300 μm broad at the base, 450 μm at the middle and 240μm towards the apex. Rhizoids elongate, lower part of axis naked, ramuli small, cylindrical distichously or rarely unilaterally disposed on the frond, apex obtuse, constricted at the base.
Uncommon: attached on rocks in low water mark.
Distribution: Nochiurani, Pudumadam, Thalathoppu, Kanyakumari, Kovalam and Hare island.
A. Habit of the plant
B & C. Lateral branches with pinnate branching (in two opposite rows.)

Bryopsis Lamouroux

REFERENCES: Harvey (1858: 31, pl, XLV. A: figs. 1 – 3).Collins & Harvey (1917: 62), S. Dixit (1940a: 87 – 88), S. Dixit (1968: 14), Jagtap (1993: table 1), Silva et al. (1996: 807).

Bryopsis Lamouroux

Bryopsis pennata **Lamouroux var. *secunda* (Harvey) Collins & Harvey**

***Description*:** Thallus filamentous, bushy; 10 cm in height; dark green in colour; frond 2-8 mm wide, sparingly branched, main axis 240 µm diameter, lateral branches 75 µm in diameter of uniform length, constricted at base, appearing unilateral (from one side only) in upper half of branches, lower half often naked.

Rare: on rocks in lower intertidal region.

Distribution: Thalathoppu

A. Habit of the plant

B. Frond showing unilateral branching.

Bryopsis Lamouroux

REFERENCES: Kesava Rao & Indusekhar (1989: 38 table 1), Untawale, Reddy & Deshmukhe (1989: table 1), Jayasankar, Ramalingam &Kaliaperumal (1990: table 1), Jayasankar & Ramalingam (1993: 42), Silva *et al.* (1996:808), Oza (1997: 39).

A

Bryopsis plumosa (Hudson) C. Agardh

Description: Plants light to olive green in colour, growing as patches; 6-9 cm in height; erect fronds naked below, regularly plumose above with basal rhizomatous portions; branching pinnate with long branches below and short ones above giving arrow head appearance.

Common: attached on rocks in upper to lower intertidal region.

Distribution: Villoonditheertham, Ariyankundu, Panjal, Leepuram, Kovalam, Chinnamuttam Kanyakumari, Shingle island, Valai island and Appa island.

A. Habit of the plant

Bryopsis Lamouroux

REFERENCES: Montagne (1842: 16, pl. III: fig. 2), M. Dave & Parekh (1975: 42), R. Parekh, Maru & Dave (1977: 360), M. Dave & E. Lewis (1979: 172 – 175, tables 1, 2), Silva *et al*. (1996: 808).

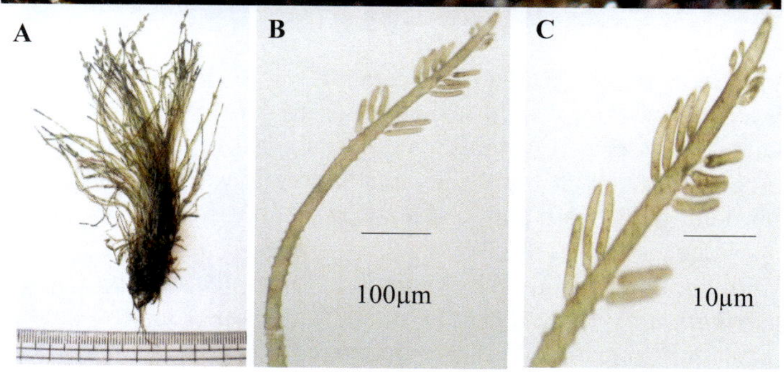

Bryopsis Lamouroux

Bryopsis ramulosa **Montagne**

Description: Thallus forming filamentous tufts, 2-4 cm high; dark green in colour; primary axis often without secondary branchlets, frond long, fine, 8 mm long; branchlets 65- 120 µm diameter, short, scattered to 2 mm long, constricted at base, unilateral or two in opposite rows, apices rounded, rhizoids fibrous.

Common: on rocks in intertidal to 1 meter deep.

Distribution: Pamban, Seeniappadharga, Ervadi, Panjal, Chinnamuttam, Vattakottai, Kanyakumari, Kovalam, and Valimunai island

A. Habit of the plant.

B. & C. Frond showing scattered branching.

***Caulerpa* Lamouroux**

Thallus characterized by a macroscopic coenocytic thallus without cell division. A horizontal, creeping main branch produces bunches of rhizoids on the underside. Production of pillar like extension (expansion or extension?) of the cell wall that crosses the lumen of the cell in large numbers. No lime deposition on the thallus. The erect branches morphology of this genus is very plastic and may vary significantly within the same species, depending on the environmental conditions. The variability in the morphology of the erect branches has led some authors to describe different forms and varieties of species.

Key for the species:

1 Assimilator cylindrical, simple or radially arranged ... 2
1 Assimilator flattened, simple or branched ... 11
2 Ramuli distichously arranged ... 3
2 Ramuli radially arranged ... 5
3 Assimilator axis not jointed, ramuli opposite or sub opposite ..... *C.corynephora*
3 Assimilator axis jointed ... 4
4 Joint bearing two, more (or) less lenticular, flat pinnules .....C. *lentellifera*
4 Joint bearing two rounded pinnules with short stalk .....*C. fergusonii*
5 Ramuli cylindrical, dichotomously forked arranged in verticils .....*C. verticillata*
5 Ramuli simple, not in verticils ...6
6 Ramuli cylindrical with swollen tip ... *C.laetevirens*
6 Ramuli spherical or clavate ...7
7 Ramuli spherical or clavate with swollen tip ... *C.racemosa*
7 Ramuli abruptly spherical 8
8 Branches abruptly spherical, evenly spaced On long axis ...*C.racemosa var. occidental*
8 Ramuli constricted at stalk 9
9 Ramuli constricted with sphere interface ...*C.microphysa*
9 Ramuli flattened at top or peltate 10
10 Ramuli distinctly peltate ..... *C. peltata*
10 Ramuli well-spaced ..... *C.chemnitzia*
10a. Ramuli peltate, crowded ....*C.imbricata*

Caulerpa Lamouroux

11 Assimilator flattened, single on dorsal side of rhizome With two series of lobes on the margin ...*C..scalpelliformis*
11 Pinnules leafy, marginal in two series 12
12 Pinnules narrow, linearly curved and closely set ...*C.sertularioides*
12 Pinnules sickle shaped and curved 13
13 Pinnules sickle shaped with constricted base Overlapping ..*C. taxifolia*
13 Pinnules not sickle shaped 14
14 Pinnules broad, closely set and overlapping incurved with acuminate tip ...*C. mexicana*
14 Pinnules spinous in 2 or 3 series 15
15 Pinnules teeth like, assimilator axis straight ....*C. cupressoides*
15 Pinnules spine like, assimilator axis twisted ....*C. serrulata*

Bryopsidales-Caulerpaceae

Caulerpa Lamouroux

Caulerpa Lamouroux

Caulerpa corynephora **Montagne**
Description: Plant dark green in colour, coenocytic, occurring in patches; 8-10 cm tall with prostrate rhizomes and erect assimilators; assimilators with radially arranged clavate and compressed ramuli.
Uncommon: attached on rocks in the intertidal and lower intertidal region.
Distribution: Nalupanai, Nochiurani, Valinokkam, Kuduthalai, Idinthakarai, Chinnamuttam, Pullivasal island, Krusadai island and Valimunai island
A. Habit of the plant
B. Upright branches with compressed ramuli.

Caulerpa Lamouroux

REFERENCES: Jayasankar, Ramalingam & Kaliaperumal (1990: table 1), D. Subramanian (1991: 70), Kalimuthu, Kaliaperumal & Ramalingam (1992: 121), Jayasankar & Ramalingam (1993: 42), D. Subramanian (1993: 207), Silva *et al*. (1996: 816).

Caulerpa Lamouroux

***Caulerpa cupressoides* (Vahl) C. Agardh**

***Description**:* Frond erect, covered with 3-4 longitudinal parallel columns of short branchlets upto 25 cm in height; grass green in colour; branching dichotomous to irregular. Branchlets stiff, conical, 1-2 mm long, upcurved, apices pointed. Stolons creeping; 1-1.5 mm in diameter. rhizoids numerous, stalked

Common: on rocks and sandy bottoms up to 3 meter depth.

Distribution: Munaikadu, Nalupanai, Ariyankundu, Pamban, Thonithurai, Keelakarai light house, Keelakarai, Ervadi, Sadaimuniyan valasai, Valinokkam, Pullivasal island, Valai island, Valimunai island and Koswari island

A. Habit of the plant with 3-4 parallel branches

B & C: Ramuli with clavate margin.

Caulerpa Lamouroux

REFERENCES: Untawale, Dhargalkar & Agadi (1983: 11), Jayasankar, Ramalingam & Kaliaperumal (1990: table 1), Thomas & Subbaramaiah (1991: table 2), Kalimuthu, Kaliaperumal & Ramalingam (1992: 121), Rama Rao *et al.* (1993: 106), Silva *et al.* (1996: 820).

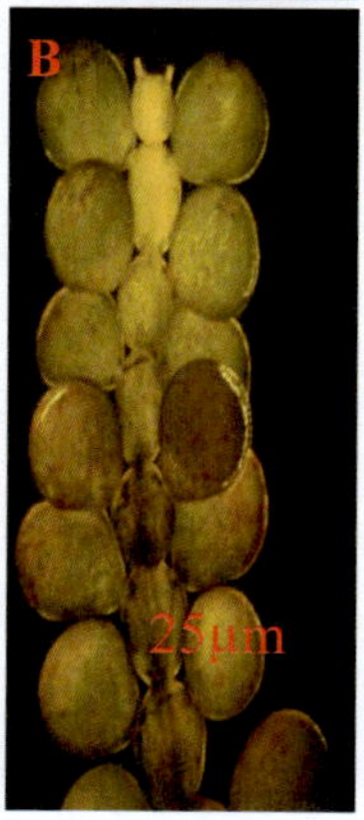

***Caulerpa* Lamouroux**

***Caulerpa fergusonii* G. Murray**

Description: Frond has stolons 1.5-2mm in diameter, densely branched in the central parts; attached by numerous tufts of rhizoids; upright branch composed of slightly compressed, unbranched rachis; 3-4 cm in height; most of the branchlets widening towards the apical part.

Common: on coral stones in subtidal region.

Distribution: Keelakarai. Ervadi, Sadaimuniyan valasai, Valinokkam, Manapad, Kuduthalai, Idinthakarai, Pullivasal island, Manali island, Hare island, Valai island, Appa island, Valimunai island, Anaipar island and Koswari island

A. Habit of the plant

B. Upright branch with cylindrical ramuli.

Caulerpa Lamouroux

Caulerpa imbricata G. Murray

Description: Plants growing in dense very blush green clumps; erect branches densely set by radially placed side branchlets of similar morphology over the whole length of rachis; side branchlets trumpet shaped, shortly stalked.

Uncommon: on coral stones in low intertidal and shallow subtidal region.

Distribution: Keelakarai, Pullivasal island, Krusadai island and Valimunai island.

A: Habit of the plant

Bryopsidales-Caulerpaceae

Caulerpa Lamouroux

A

B

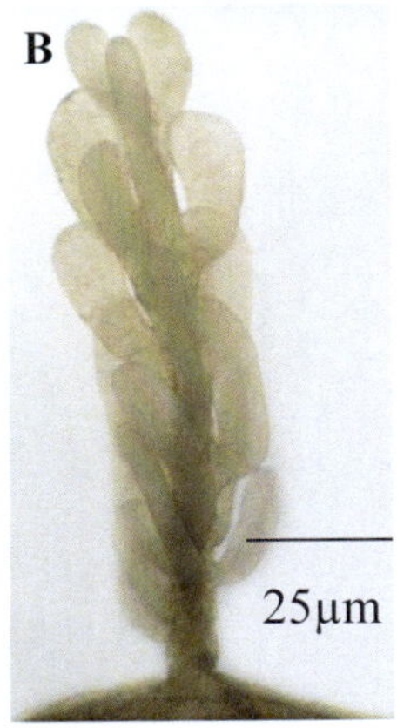

Caulerpa Lamouroux

Caulerpa laetevirens **Montagne**
***Description**:* Plant dark green in colour, coenocytic, occurring in patches; 5- 8 cm tall with prostrate rhizomes and erect assimilators; assimilators elongated club shaped, ramuli occurring in rosette type.
Rare: occurring on rocks in the intertidal to lower intertidal region.
Distribution: Distribution: Nalupanai, Valinokkam and Kuduthalai.
A. Habit of the plant
B. Individual branch with club shaped ramuli.

Caulerpa Lamouroux

REFERENCES: J. Agardh (1835: 173), Hieronymus (1895: 24), Weber-van Bosse (1898: 380 – 381, pl. XXXIV: fig. 1), Boregesen (1953: 9 – 10), Untawale, Reddy & Agadi (1983: 16), Silva *et al.* (1996: 824).

Caulerpa Lamouroux

***Caulerpa lentellifera* J. Agardh**

Description: Plant dark bluish green in colour; stolons terete; erect fronds densely set, upto 3 cm long and unbranched even upto 11 cm in length; cylindrical, densely covered with spherical or subspherical ramuli of 1-2 mm diameter; more imbricately placed in 5-8 rows

Common: on coral stones and pebbles in low water mark.

Distribution: Pamban light house, Villoonditheertham, Thonithurai, Pudumadam, Bharathi nagar, Pullivasal island, Krusadai island, Hare island, Valai island, Appa island, Valimunai island, Anaipar island, Salli island, Upputhanni island, Van island and Koswari island.

A. Habit of the plant

B. Erect branch with spherical ramuli.

Caulerpa Lamouroux

REFERENCES: Umamaheswara Rao (1972: 221), Untawale, Reddy & Agadi (1983: 12), Chaugule, Quadri & Goswamy (1989: 109, figs. 11, 12), D. Subramaniyan (1991: 70), Subramaniyan (1993: 207), Silva *et al*. (1996: 825).

***Caulerpa* Lamouroux**

***Caulerpa mexicana* Sonder ex Kützing**

Description: Plants small, 3 cm in height and 4- 10mm wide in wave exposed areas, elongated form 15-20 cm high in calm protected habitat, grass green in colour; stolon 500 μm broad; central axis (midrib) upto 1 mm broad; branchlets opposite, flattened, upcurved, apices tapering, pointed. Stolons creeping, rhizoid delicate.

Common: attached to small coral fragments or pebbles upto 15 meter deep.

Distribution: Nochiurani, Keelakarai, Ervadi, Hare island, Valai island, Valimunai island, Anaipar island, Salli island, Upputhanni island and Koswari island.

A. Habit showing upright branches with thick midrib.

B. Apical portion of branch with up curved, opposite branchlets.

Caulerpa Lamouroux

REFERENCES: Weber-van Bosse (1898: 362 – 381, pl. XXXIII: fig. 5), Silva *et al*. (1996: 826), Mohammed Gulshad *et al*. (1998: 15).

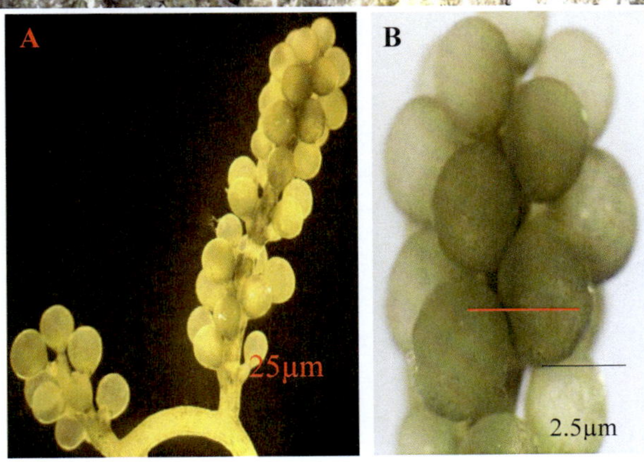

***Caulerpa* Lamouroux**

***Caulerpa microphysa* (Weber-van Bosse) J. Feldmann**

Description: Plant bright to dark green in colour, upto 5 cm tall with slender creeping rhizomes and cylindrical assimilators; ramuli much crowded on the assimilators; attached to the axis, sessile or stalked with constriction

Common: attached to rocks, coral fragments or pebbles upto 10 meter depth.

Distribution: Nalupanai Pudumadam, Bharathi nagar, Kanyakumari, Kovalam, Manali island, Hare island, Mully island, Valai island, Appa island, Valimunai island, Salli island, Van island and Koswari island

A. Upright branches with sessile ramuli

B. Upper part of branch with crowded ramuli.

Caulerpa Lamouroux

REFERENCES: Thomas & Subbaramaiah (1991: 205, table 2), Kalimuthu, Kaliaperumal & Ramalingam (1992: 121), Kamat *et al.* (1992: table 1), Jayasankar & Ramalingam (1993: 42), D. Subramanian (1993: 30), Silva *et al.* (1996: 828).

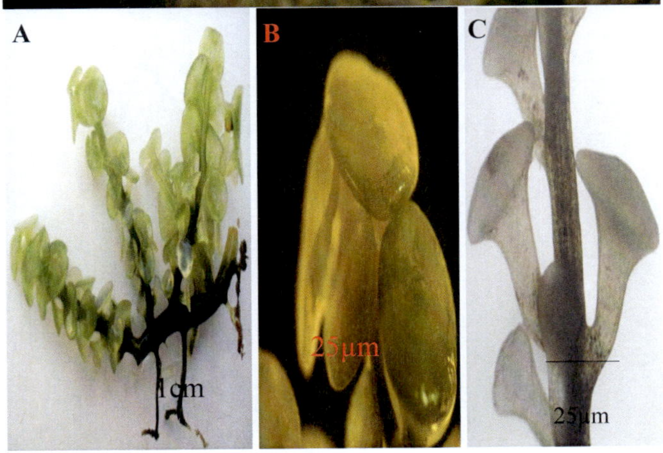

Caulerpa Lamouroux

Caulerpa peltata Lamouroux

Description: Plants growing in patches; with distinct prostrate rhizome and short, erect assimilators, ramuli short and terminating in a plate like disc; disc upto 2-5 mm in diameter and 3- 8 mm thick.
Common: in rocks and coral stones in sheltered places.
Distribution: Munaikadu, Nalupanai, Villoonditheertham, Rameswaram, Pamban, Thonithurai Nochiurani, Pudumadam, Sallimunthal, Bharathi nagar, Ervadi, Sadaimuniyan valasai, Valinokkam Kanyakumari, Kovalam and all 20 islands of Gulf of Mannar.

A. Habit of plant
B. Apical part of branch with plate like ramuli
C. Ramuli with stalk.

Caulerpa Lamouroux

REFERENCES: Jayasankar & Ramalingam (1993: 42), V. Kumar (1993a: 139, 140), V. Kumar (1993b: 47), Sivanand & D. Subramanian (1993: 30, fig. 23), D. Subramanian (1993: 207), Silva *et al*. (1996: 832).

Caulerpa Lamouroux

Caulerpa racemosa (Försskål) J. Agardh

Description: Plant pale green in colour, growing as patches; prostrate rhizomes and erect assimilators; often much crowded; beads like branchlets 1-5 cm in height, spherical, not constricted at base, stalk 1-2 mm in diameter; 2-5 mm long. Stolons creeping, 2-3mm in diameter; often branched; numerous rhizoids.

Common: attached on rocks

Distribution: Munaikadu, Pamban light house, Nalupanai, Ariyankundu, Villoonditheertham, Rameswaram, Pamban, Thonithurai, Velankanni nagar, Seeniappadharga, Nochiurani, Pudumadam, Sallimunthal, Keelakarai light house, Keelakarai, Bharathi nagar, Ervadi, Sadaimuniyan valasai, Valinokkam, Kovalam Kanyakumari, and all 20 islands of Gulf of Mannar.

A. Habit of the plant

B. Plant with prostrate rhizome and erect assimilators

C. Branch with beads like ramuli

Caulerpa Lamouroux

REFERENCES: Weber-Van Bosse (1898: 370 – 373, pl. XXXI: figs. 5 – 7), V. Krishnamurthy & H. Joshi (1970: 6), K. Srinivasan (1973: v42), Nair, Sobha & Arunachalam (1982: 267 table 1), Untawale, Dhargalkar & Agadi (1993: fig. 11), Sobha & Nair (1985: 21), Chagule, Quadri & Goswamy (1989: 109, fig. 14).

Caulerpa Lamouroux

***Caulerpa racemosa* (Försskål) J. Agardh var. *chemnitzia* (Esper) Weber-van Bosse**

Description: Plants with assimilators upto 12 cm high, erect, irregularly branched with radially arranged ramuli which are pestle shaped; ramuli 2 mm broad and 5 mm long at apices.

Rare: attached on rocks.

Distribution: Rameswaram, Velankanni nagar and Kanjirankudi.

A. Habit of the plant.

B. Branch with pestle shaped ramuli.

Caulerpa Lamouroux

Caulerpa racemosa (Försskål) J. Agardh var. _occidentalis_ (J. Agardh) Børgesen

Description: Plant yellowish green in colour, growing as patches, stolons upto 2 mm in diameter with long erect assimilators; 2-10 cm tall or more with radially arranged spherical ramuli, occasionally mushroom shaped, 2-4 mm diam. to 1 cm long; ramuli with long cylindrical stalks, ovate or globular apices with convex ends.

Uncommon: attached on rocks and other hard substratum in the intertidal region

Distribution: Most of the islands of Gulf of Mannar.

A. Habit of the plant.

Caulerpa Lamouroux

REFERENCES: Jayasankar & Ramalingam (1993: 42), V. Kumar (1993a: 139, 140), V. Kumar (1993b: 47), D. Subramanian (1993: 207), Dave & Parekh (1994: 59), Silva *et al*. (1996: 838).

Bryopsidales-Caulerpaceae

***Caulerpa* Lamouroux**

***Caulerpa scalpelliformis* (R. Brown ex Turner) C. Agardh**

Description: Plants stoloniferous; stolen cylindrical, giving branched rhizoids below and erect stalked assimilator fronds upto 15 cm in length, ramuli opposite, flat, incurved upwards upto 2 mm broad and 8 mm long, dentate near the apex, extending up to mid part of ramuli.

Common: attached to the rock and coral stones.

Distribution: Thonithurai, Velankanni nagar, Seeniappadharga, Nochiurani, Kuduthalai,Pudumadam, Ervadi, Valinokkam, Manapad, Idinthakarai, Panjal, Chinnamuttam, Vattakottai, Kanyakumari, Kovalam, Valimunai island and Upputhanni island.

A. Habit of the plant

B. Branch with flat ramuli.

C. Upper part of branch with opposite, up curved ramuli.

Caulerpa Lamouroux

REFERENCES: Pant & Untawale (1989: table4, 6), Untawale, Reddy & Deshmukhe (1989: 78, table 1), V. Krishnamurthy *et al.* (1990: figs. 4, 6, 7), Thoams & Subbaramaiah (1991: 202, table 2), Jagtap (1992: 57), Silva *et al.* (1996: 844).

A

Caulerpa Lamouroux

Caulerpa sertularioides **(S. Gmelin) Howe**
Description: Plants with stolon of 1.5 mm diameter, much branched rhizoids and erect assimilatory filaments feathery, 5- 20 cm in height; bright to dark green in colour; rarely branched, branching generally at the base of assimilators; branchlets opposite, incurved above, needle shaped with pointed apices.
Common: on rocks and coral stones.
Distribution: Arasaneripattinam, Kodikarai, Nalupanai, Villoonditheertham, Ariyankundu, Pamban, Thonithurai, Seeniappadharga, Nochiurani, Thalathoppu, Keelakarai light house, Keelakarai, Bharathi nagar Sadaimuniyan valasai, Valinokkam, Shingle island, Krusadai island, Manali island, Hare island, Mully island, Appa island, Anaipar island, Salli island Upputhanni island and Van island
*A.*Habit of the plant with feathery assimilators.

Caulerpa Lamouroux

REFERENCES: Jagtap (1987: table 1), Untawale, & Jagtap (1989: table 1), Jagtap (1992: 57), Kalimuthu, Kaliaperumal & Ramalingam (1992: 121), Silva *et al*. (1996: 842), P. S. N. Rao & Marcel Tigga (2000: 14).

Caulerpa Lamouroux

Caulerpa serrulata **(Försskål) J. Agardh**

Description: Thallus grass green in colour, small, flat, often twisted or spiraling to 4 cm high; branching widely dichotomous; fronds 2-4 mm wide, 2-3.5 cm long 1.5 -2.5 mm thick; un branched or forked; margins toothed (serrulated); teeth wedge shaped, 1.3mm wide 1 mm long, apices pointed. Stalk usually inconspicuous but sometimes growing upto 1.5 mm in diameter. Stolons creeping; rhizoids stalked.

Common: on dead coral stones and sand covered rock surfaces upto 5 meter deep.

Distribution: Nalupanai, Rameswaram, Thonithurai, Mully island, Upputhanni island, Van island and Koswari island.

A. Habit of the plant

B. Branch with serrulated margin.

Caulerpa Lamouroux

REFERENCES: Kamat *et al*. (1991: 54), Vidyavathi & Sridhar (1991: fig. 2), Kalimuthu, Kaliaperumal & Ramalingam (1992: 121), Jagtap (1993: table 1), Jayasankar & Ramalingam (1993: 18), Silva *et al*. (1996: 845).

***Caulerpa* Lamouroux**

***Caulerpa taxifolia* (Vahl) C. Agardh**

Description: Plant with stolon upto 1.5 mm diameter bearing branched rhizoids below and erect stalked assimilator fronds above; plant yellowish to dark green in colour; upto 20 cm in height; 8-13 cm broad; pinnately divided, pinnules opposite, compressed, sickle shape with constricted base, overlapping along edges with pointed apices.

Common: attached to rocks in lower intertidal and subtidal region.

Distribution: Munaikadu .Nalupanai, Villoonditheertham, Rameswaram, Pamban, Thonithurai, Velankanni nagar, Seeniyappadharga, Pudumadam, Kanjirankudi, Keelakarai light house, Bharathi nagar, Ervadi, Sadaimuniyan valasai, Kooduthalai, Idinthakarai, Vattakottai, Kanyakumari, Kovalam and all 20 islands of Gulf of Mannar.

A. Habit of the plant.

B. Branch showing sickle shape pinnule.

Caulerpa Lamouroux

REFERENCES: Untawale, Dhargalkar & Agadi (1983: 12, fig. 15), Chennubhotla, Kaliaperumal & Kalimuthu (1987: 100 – 101), Chagule, Quadri & Goswamy (1989: 107, figs. 2 – 4), D. Subbramanian (1991: 70), D. Subbramanian (1993:207), Silva *et al.* (1996: 849).

Caulerpa Lamouroux

***Caulerpa verticillata* J. Agardh**

Description: Plants stoloniferous; stolons 500 μm in diameter bearing rhizoids below and erect frond upto 2-6 cm in length above, abundantly branched with branches and branchlets; dark green in colour; determinate branchlets dichotomously branched; ramuli cylindrical, 0.75- 1.25 mm in length, tapering at apices; apices mucronate.

Uncommon: on coral stones covered with fine sand.

Distribution: Nalupanai, Villoonditheertham, Bharathi nagar, Kanyakumari, Kovalam, Shingle island, Manali island, Hare island, Valai island, Appa island, Anaipar island Van island and Koswari island.

A. Habit of the plant.

B. Erect branches with cylindrical ramuli.

Caulerpa Lamouroux

A

***Caulerpa* Sp.**

Description: Plant green in colour, occurring in patches, small, 2-5 cm in height, assimilators erect and crowded; ramuli narrow, elongated and with rounded apex.

Rare: Attached on coral rocks in upper intertidal region.

Distribution: Seeniappa dharga

A. Habit of the plant

Chaetomorpha Kützing

***Chaetomorpha* Kutzing**

Thallus filamentous, unbranched either as a loose mat of entangled filaments or bunches; attached by rhizoids produced from a basal cell, usually distinct from other cells and much larger; the cells are large with a thick cell wall and can be seen with naked eye or a hand lens; chloroplasts reticulate or free

Key for the species-

1. Filaments attached …2
1. Filaments free floating …5
2. Filaments thin, never exceeding 100 µm in diameter …3
2. Filaments 450 – 700 µm in diameter …4
3. Cells not swollen or constricted at septa . ...*C. indica*
4. Cell wall thick, 45- 100 µm in diameter … *C. antennina*
4. Cell wall thin, less than 45µm in diameter … *C. aerea*
5. Cells thick, more than 300 µm in diameter …*C. crassa*
5. Cells thinner, less than 300 µm in diameter … 6
6. Filaments 100-250 µm in diameter, cells slightly swollen in middle …. *C. linum*

Chaetomorpha Kützing

REFERENCES: Gajaria & Patel (1988: 3-4, figs. 1-3, 29), Jayasankar, Ramalingam & Kaliaperumal (1990: table 1), S. Balakrishnan, Ravichandran & Kaliaperumal (1992: 65), Kaliaperumal, Rajagopalan & Chennubhotla (1992: 106), Kalimuthu, Kaliaperumal & Ramalingam (1992: 121), Silva *et al.* (1996: 759).

A

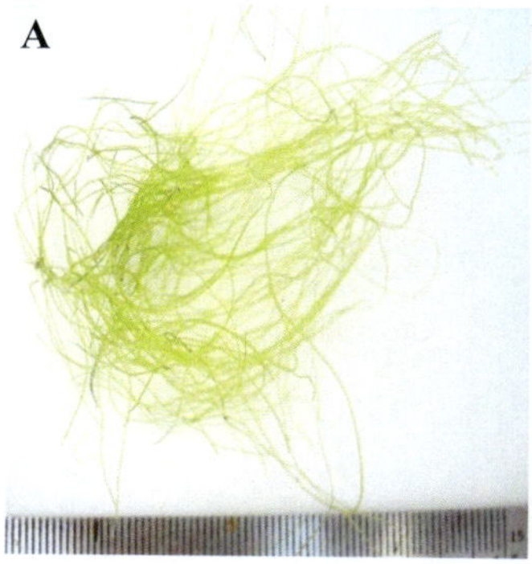

B

Chaetomorpha Kützing

Chaetomorpha aerea (Dillwyn) Kützing

***Description**:* Thallus grass green to yellowish green in colour, unbrached, attached to sand covered rocks, upto 10 cm in height; cells cylindrical, about 450 to 660 µm in width and 2 times as long as wide. Cells producing swarmers become barrel shaped or sub-globose.

Uncommon: attached on rocks and hard substratum in intertidal region.

***Distribution**:* Ennore, Muttukadu, Kovalam, Mahabalipuram, Pondicherry and Pudumadam.

A. Habit of the plant with unbranched filaments.

B. Typical filament cell.

Chaetomorpha Kützing

REFERENCES: Jagtap (1992: 57), Kalimuthu, Kaliaperumal & Ramalingam (1992: 121), Maya & Nair (1992: 26), Jagtap (1993: table 1), Jayasankar & Ramalingam (1993: 42), Suresh *et al*. (1993: 113), Silva *et al*. (1996: 760), Sampathkumar *et al*. (1999: 24).

***Chaetomorpha antennina* (Bory de Saint-Vincent) Kützing**

***Description*:**Plants dark green in colour; 1- 10 cm high, erect, isolated, characteristic brush like tufts, composed of rigid, unbranched septate filaments, dark to bright green in colour. Attached to the substratum by rhizoidal sprout; rhizoidal system is profusely branched, stoloniferous. Chloroplast reticulate or free, but close together; basal cells thick with annular constrictions. Filaments of reproductive specimen with white tips that erode easily, resulting in diminishing the size of the plant.

Common: attached on rocks at high water tides.

Distribution: Kasimedu, Velankanni nagar, Seeniappadharga, Nochiurani, Pudumadam, Valinokkam, Panjal, Leepuram, Vattakottai and Shingle island.

A. Habit of the plant showing brush like filaments

Chaetomorpha Kützing

REFERENCES: C. Agardh (1824: 99), Sonder, (1854: 1), Kutzing (1845: 204), Borgesen (1936: 64), S. Dixit (1968: 13), K. Krishnamurthy & Kannan (1982: 34), Silva *et al*. (1996: 762).

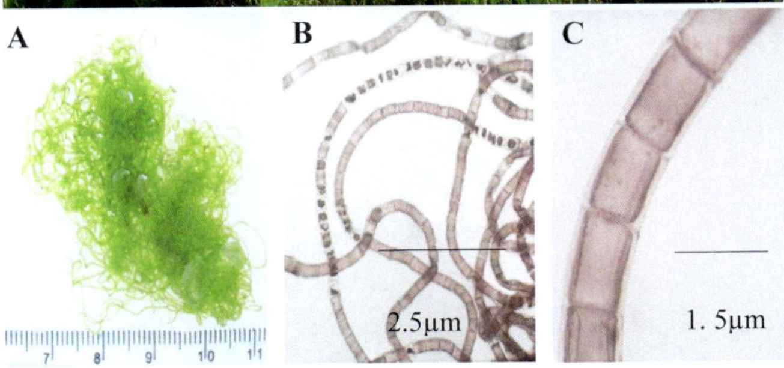

Chaetomorpha Kützing

***Chaetomorpha crassa* (C. Agardh) Kützing**

Description: Plants forming loose lying clumps or entangled with other algae; thick, curly, unbranched, mostly dark green filaments; cells 450- 650 µm in diameter, broad, cylindrical, sometimes slightly swollen.

Common: in intertidal zone, most frequently in low to middle intertidal pools

Distribution: Kottaiapttinam, Ervadi, Panjal, Chinnamuttam, Kanyakumari, Kovalam, Mully island, Anaipar island, Salli island, Upputhanni island, Van island and Koswari island.

A. Habit of the plant

B. Curled filaments

C. Filaments showing individual cells with thick walls.

Chaetomorpha Kützing

REFERENCES: V. Krishnamurthy & H. Joshi (1970: 3), Radhakrishnan (1976: table 2), Sreenivasa Rao & K. Parekh (1981: 578), Untawale, Dhargalkar & Agadi (1983: 9), Bhakuni *et al.* (1992: 514), Silva et *al.* (1996: 764).

Chaetomorpha indica (Kützing) Kützing

Description: Filaments rigid, 90-100µm in thickness; cells equal to or slightly longer than broad; Cells are not swollen or constricted at septa.

Rare: attached on rocks in supra littoral region.

Distribution: Tranquebar

A. Habit of the plant

B. Filaments showing thin walls and slightly longer than broad

Chaetomorpha Kützing

REFERENCES: Gajaria & R. Patel (1988: 4 – 5, figs. 15, 16, 27), Untawale & Jagtap (1989: table 1), Untawale, Reddy & Deshmukhe (1989: table 1), Mohankumar, Rajendran & Kannan (1991: 219), Jagtap (1992: 57), Silva *et al.* (1996: 765).

***Chaetomorpha* Kützing**

***Chaetomorpha linum* (O. F. Müller) Kützing**

Description: Filaments long, free floating or entangled with other algae with cylindrical cells, sometimes little swollen in middle, 100-250 μm in breadth and 1-2 times as long as broad.

Common: free floating or sometimes entangled with other algae.

Distribution: Arasaneripattinam, Nalupanai, Valinokkam, Kooduthalai, Idinthakarai, Leepuram ,Shingle island Pullivasal island, Krusadai island, Salli island and Upputhanni island.

*A.*Habit of the plant with thick filaments

B. Filaments with little swollen in the middle

***Cladophora* Kützing**

Thallus a tuft of uniseriate, copiously branched, green filaments attached to the substratum by a basal cell supported by rhizoidal extensions from the lowermost cells; cells are large and multinucleate, with thick cell walls and numerous densely grouped chloroplast, each containing one pyrenoid, sometimes loosely associated in a net like structure.

Key for the species

1. Plants not in cluster or turf, branches from base di or trichotomous ...*C.prolifera*
2. Branching regular, thallus sponge like ...*C.socialis*
2. Branches of main axis sparse, upper branches erect, straight, fasciculate ...*C.vagabunda*

Cladophora Kützing

REFERENCES: Agadi & Untawale (1978: 63, table 2), Untawale, Dhargalkar & Agadi (1983: 10), Untawale & Jagtap (1989: table 1), Untawale, Reddy & Deshmukhe (1989: table 1), Silva *et al*. (1996: 779), Asolkar *et al*. (1996: 27 – 31).

Cladophora Kützing

Cladophora prolifera **(Roth) Kützing**

Description: Plants usually upto 12 cm in height, occasionally even upto 30 cm, dark green, sometimes brownish in colour; filaments thick, cartilaginous to sub membranous, di or trichotomously branched. Cells 250-320 µm broad, 4 times as long as broad in lower branches.

Rare: attached to rocks and tide pools.

Distribution: Villoonditheertham and Valai island

A. Habit

B. Apex of branchlet

C. Cell arrangement at the base of the branch.

Cladophora Kützing

REFERENCES: Kutzing (1849: 416), Borgesen (1946: 28 – 29), Borgesen (1948: 19 – 20), 781) G. Rao & Umamaheswara Rao (1986: 103), Sri Ramana & Venkata Rao (1990: 259), Silva *et al.* (1996:

Cladophora Kützing

Cldophora socialis **Kützing**
***Descriaption*:** Thallus sponge like with intricately interwoven filaments, frequently and indistinctly dichotomous, branches elongate, cells 55-75 µm in diameter; 3- 6 times as long.
Rare: attached to rocks
Distribution: Kanyakumari and Kovalam.
A. Habit of the plant
B. Arrangement of cells in the main filament and branches.

Cladophora Kützing

REFERENCES: Linnaeus (1753: 1167), Van den Hoek (1963: 144), Jagtap (1993: table 1), Silva *et al.* (1996: 782).

Cladophora Kützing

Cladophora vagabunda **(Linnaeus) van den Hoek**
***Description*:** Plants light to dark green in colour; 30- 40 cm in height, soft, densely branched; main axis sparingly branched with alternate branches; cells of the main axis 4-6 times as long as broad with highly lamellated cell walls; cell walls and main filaments bears a loosely branched fascicle like tufts.
Common: on rocks
Distribution: Kovalam, Karankadu, Ammapattinam, Munaikadu, Villoonditheertham Rameshwaram, Pudumadam, Ervadi, Valinokkam, Kanyakumari and Kovalam.

A. Habit of the plant
B. Cells arrangement
C. Enlarge view of cells

Cladophora Kützing

***Cladophora* Sp.**

Description

Plants light green in colour; 5- 10 cm in height, soft, densely branched; main axis has dense branches. Branches irregular in all directions; cells of the main axis 2-3 times as long as broad

***Rare*:** grown on rocks

Distribution: Nalupanai

A. Habit of the plant

B. Plant showing irregular branches

***Cladophoropsis* Borgesen**

Thallus consists of tubular uniseriate filaments of segments formed by segregate division; attached to the substratum by rhizoidal branches; branching alternate, opposite or sometimes irregular in all directions; segments formed irregularly, longer and shorter segments intermingled in an unorganized manner.

Key for the species

1. Free living filaments, branches not regularly curved, less than 0.1mm in diameter. - *C. sundanensis*

Cladophoropsis Børgesen

REFERENCES: V. Krishnamurthy & H. Joshi (1970: 8), Subbaramaiah (1971: 21), Chauhan & Mairh (1978: 33), Untawale, Dhargalkar & Agadi (1983: 1), Chowdary & Singh (1984: 178 – 180, 182, figs, 6, 6a, 7), Silva *et al.* (1996: 793).

***Cladophoropsis* Børgesen**

***Cladophoropsis sundanensis* Reinbold**

Description : Thallus green in colour, tufted, forming small clumps, filaments soft, reaching a height of 2-3 cm, primary axis not distinct, branching unilateral with irregular septation; cells 2-8 times as long as broad, apices obtuse.

Rare: attached to stones in middle to lower littoral region.

Distribution: Kovalam, Manapadu, Kanyakumari, Kovalam and Koswari island.

A. Habit of the plant.

B. Branches with long cells showing irregular septa

Codium Stackhouse

Codium Stackhouse

The spongy thallus is composed of entangled, siphonous, coenocytic filaments organized in large cylindrical to flattened and usually bifurcated branches. The dissected thallus shows narrow, tube-like branched filaments with constriction and thickened cell walls which lack cross walls. The cylindrical filaments of variable diameter are entangled in the center of thallus; at the periphery they expand into club-shaped tubes forming "utricles". Starch is deposited within the chloroplast and there is no lime deposition.

Key for the species-

1. Thallus prostrate with cushion forming …….2
1. Thallus erect. …….3
2. Vesicles clavate with rounded apex …….*C.repens*
2. Vesicles clavate with thickened apical wall …….*C.arabicum*

Codium Stackhouse

REFERENCES: P. Silva (1952: 382 – 87, figs. 11 – 13, pl. 34b), P. Silva (1962: 207 – 208. Fig. 5), S. Dixit (1968: 16), Jagtap (1992: 57), Jagtap (1993: table 1), Silva *et al.* (1996: 850).

Codium Stackhouse

***Codium arabicum* Kützing**

***Description*:** Plant dark green in colour; thallus expands to form encrusted layers, closely attached to the substratum, lobed at margins, firm; 2.5-10 mm thick, 10 cm in diameter, utricles in large clusters, clavate or cylindrical, 44-175 μm in diameter, 300-600 μm long; round apex, gradually tapering at base; gametangium elipsoid to cylindrical in shape.

Common: attached to the coral stone upto 10 meter deep.

Distribution: Bharathi nagar, Krusadai island, Mully island, Anaipar island.

A. Habit of the plant

B. Utricles

Codium Stackhouse

REFERENCES: M. Dave, Parekh & Chauhan (1990: 43, 44, tables 1, 2), Tewari, Thampan & H. Joshi (1990: 33 – 37, figs. 4 – 6), Silva *et al*. (1996: 854), Kaliaperumal *et al*. (1998: 141 – 151), Shanmugam *et al*. (2000: 35), Shanmugam *et al*. (2001: 49 - 52).

Codium Stackhouse

Codium dwarkense **Børgesen**

Description: Plants light green in colour, attach to the substratum by a disc and bears many erect cylindrical axes; these axes are 16 cm or more in height, cylindrical, 2-4 mm broad, many times regularly dichotomously divided at every 2-3 cm or even more between the divisions; angle between the branches narrow, straight and upwardly directed. Utricles clavate.

Common: attached to coral stones upto 10 meter deep.

Distribution: Ervadi, Kuduthalai, Idinthakarai, Chinnamuttam, Kanyakumari, Kovalam and all islands of Gulf of Mannar.

A. Habit of the plant

B. Branches with narrow angle

C. Utricles.

Codium Stackhouse

REFERENCES: S. Dixit (1968:16), V. Krishnamurthy & H. Joshi (1970: 7), Untawale, Dhargalkar & Agadi (1983: 13, fig. 16), Chaugule, Quadri & Goswamy (1989: 111, figs. 31 – 36), Untawale, Reddy & Ambiye (1989: table 1), Silva *et al*. (1996: 857).

Codium Stackhouse

Codium indicum **S. Dixit**

Description: Thallus erect, dark green in colour, upto 10 cm in height; fronds mostly regularly dichotomous, sometimes irregularly dichotomous. Fronds wide upto 1 cm in diameter, flat at the point of dichotomy and slightly swollen at the end. Utricles mostly cylindrical to club shaped.

Uncommon: growing upto 10 meter deep.

Distribution: Hare island, Mully island, Appa island and Salli island.

A. Habit of the plant

B. Utricle

Codium Stackhouse

REFERENCES: Sundararajan & Annakkili (2000: 73).

Codium Stackhouse

Codium isthmocladum **Sundararajan & Annakkili**
Description: Thallus light green in colour, as tightly compact hemispherical clumps, upto 20 cm in height, branch dark green, surface smooth, soft, covered by delicate hairs; branching dichotomous near tips, irregular below. Branches soft, cylindrical, 2-6 mm diam. Occasionally flattened. Utricles cylindrical, club shaped; occasionally constricted below.
Rare: on coral and other hard surfaces upto 10 m deep.
Distribution: Mulli island, Valai island
A. Habit of the plant
B. Upper branches
C. Utricle

Codium Stackhouse

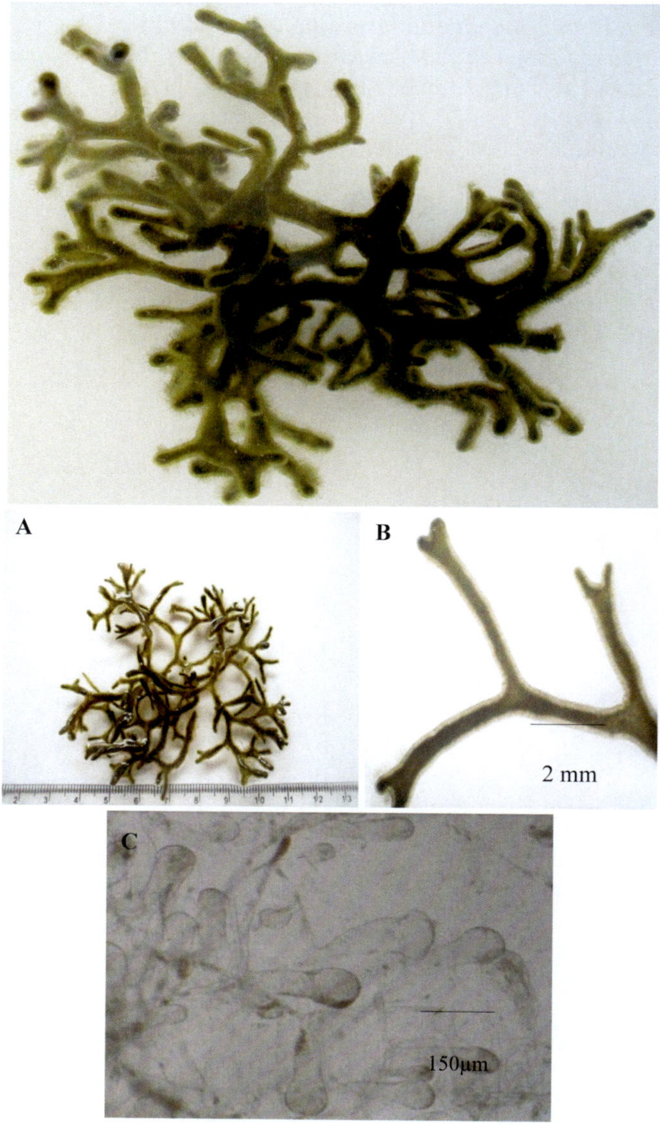

Codium Stackhouse

Codium iyengarii **Vickers**

***Description*:** Plant light green in colour. Thallus erect, up to 10 cm in height. Frond cylindrical, narrow about 1 mm in diameter, irregularly dichotomous and slightly swollen at the place of dichotomous. Branches smooth, covered by delicate hairs throughout. Utricles club shaped.

Uncommon: growing upto 10 meter deep.

Distribution: Mully island, Valai island and Salli island

A. Habit of the plant

B. Branches

C. Utricle

Codium Stackhouse

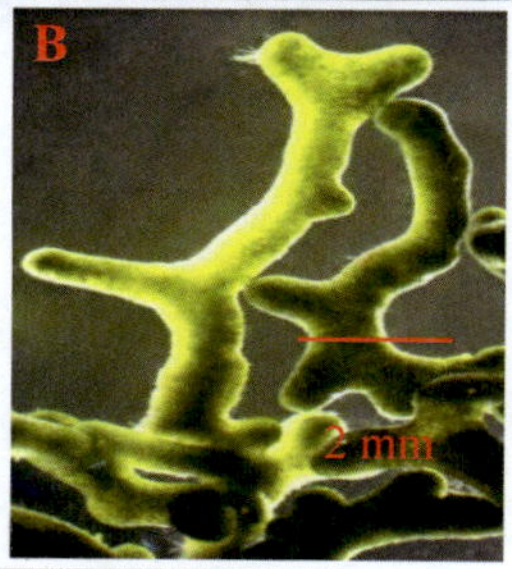

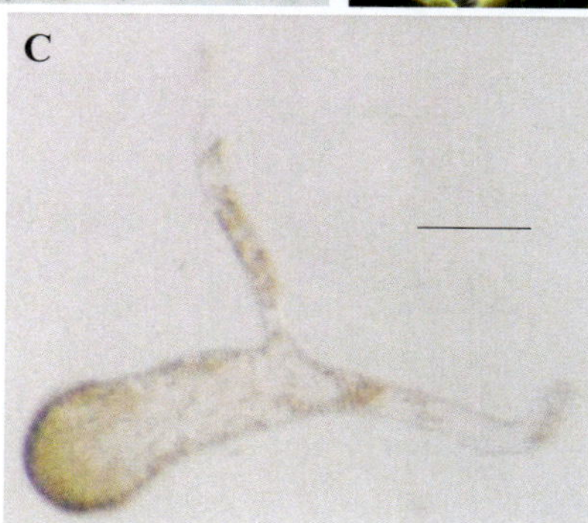

Codium Stackhouse

Codium repens **Vickers**

Description**:** Thallus creeping, gregarious, often forming mats or patches upto 40 cm diameter, pale dull to dark green in colour, branching widely dichotomous to irregular. Branches tough, cylindrical to slightly flattened, 1-3 mm in diameter, often fusing together, apices blunt, rounded. Utricles variable, cylindrical to oval, club shaped.

Uncommon: growing or clinging over the rocks.

Distribution: Panjal, Kanyakumari, Kovalam, Hare island, Appa island, Anaipar island, Salli island, Upputhanni island, Van island and Koswari island.

A. Habit of the plant
B. Branches
C. Utricle

Codium Stackhouse

REFERENCES: Kutzing (1849: 501), P. Silva *et al*. (1959: 142), Silva *et al*. (1996: 863).

Codium Stackhouse

Codium tenue **(Kützing) Kützing**
***Description*:** Plant light to dark green in colour, thallus erect, long, upto 14 cm in height. Frond cylindrical, 1 to 2 mm in diameter, regularly dichotomous. Branches smooth covered by delicate hairs throughout. Utricles, long and cylindrical in shape, slightly constricted at the base.
Rare: grown upto 10 meter deep.
Distribution: Salli island and Upputhanni island.
A. Habit of the plant
B. Branches
C. Utricles

***Dictyosphaeria* Decaisne**

The thallus is hollow or solid, globose or flattened, formed by vesicular segments resembling large, polygonal, honeycomb-like multinucleate cells; attached by or tentacular cells and fixed to the substratum by rhizoids produced from basal vesicles.

Key for the species

1.Thallus sac like, hollow or irregularly lobed …*D. cavernosa*

1. Thallus spherical when young, flattened when mature ….. *D. versluysii*

Dictyosphaeria Decaisne

REFERENCES: Jagtap (1987; table. 1), Untawale & Jagtap (1989: table 1), Jayasankar, Ramalingam & Kaliaperumal (1990: table 1), Kalimuthu, Kaliaperumal & Ramalingam (1992: 122), Jagtap (1993: table 1), Silva *et al*. (1996: 794).

Dictyosphaeria Decaisne

***Dictyosphaeria cavernosa* (Försskål) Børgesen**
***Description*:** Plants attached by means of rhizoids, dark green in colour; hemispherical in shape often compressed or lobed, 1 to 1.5 cm in diameter. Thallus consists of single layer of large polygonal cells. Vesicles 1-3 mm broad, 5-10 mm long; branching from the sides or more usually from the end of the cells.
Common; attached on rocks and coral stones in the intertidal zone.
Distribution: Thonithurai,Shingle island, Pullivasal island, Krusadai island, Hare island, Mully island, Valai island, Valimunai island, Van island and Koswari island.
A. Habit
B. Transverse section of the thallus.

***Dictyosphaeria* Decaisne**

A

B

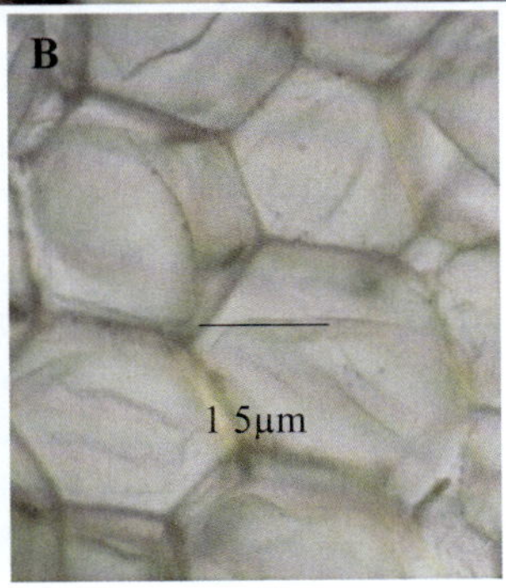

Dictyosphaeria Decaisne

Dictyosphaeria versluysii **(Försskål) Weber-van Bosse**
Description: Thallus spherical; solidly composed of macroscopic cells, 3-5 cm in diameter; light green in colour. Primary cells angular or polyhedral in surface view, appearing honey comb like, adhering to each other by minute hapteroid cells. Hapteroid cells 30 – 60 µm in diameter. Rhizoids generally short, unbranched.
Common: attached on rocks and coral stones.
Distribution: Sadaimuniyan valasai, Shingle island, Pullivasal island, Krusadai island, Hare island, Valai island, Valimunai island, Van island and Koswari island.
A. Habit
B. Transverse section of the thallus

Enteromorpha Link

***Enteromorpha* link**

Plants usually in groups, forming light green carpets; thallus generally tubular, though in some species the cavity is only visible in cross section, with only one layer in cross section; cells contain one chloroplast and 1-7 pyrenoids. Plants attached to the substratum through a bunch of rhizoidal proliferations produced as an elongation of basal cells, sometimes forming cushion like. Reproduction through biflagellate gametes or quadriflagellate zoospores, released through a small opening. Usually found on rocks, partially buried in sand or silt in the upper intertidal. Most species are very tolerant to environmental variations and this genus therefore tends to be the dominant alga in polluted harbours, estuaries and brackish water.

Key for the species -

1. Frond not proliferous, tubular, compressed above, branching from gradually tapering stalk like portion. …*E. compressa*
1. Frond with long stipe, 50µm thick; cells in section taller than broad … *E. linza*

Enteromorpha Link

REFERENCES: Ragothaman & Haranatha Babu (1992: 91), Jagtap (1993: table 1), Jayasankar & Ramalingam (1993: 42), Silva *et al.* (1996: 729), Rattan Kumar & Naidu (1997: 43), Kaliaperumal *et al* (1998: 141 – 151), Ruma Pal (2000: 51 – 60).

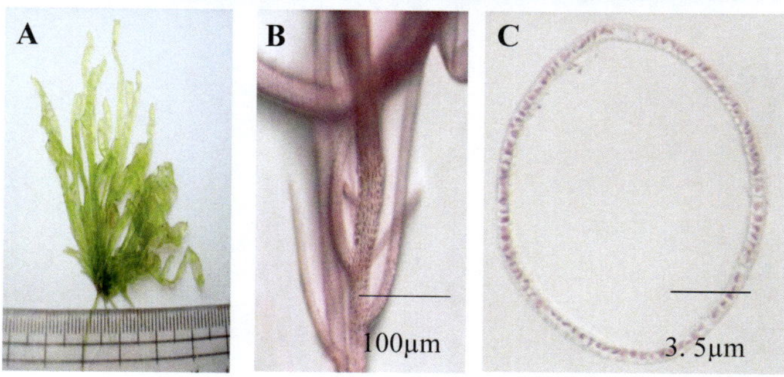

Enteromorpha Link

Enteromorpha compressa **(Linnaeus) Nees**

Description: Plants upto 11 cm in height; yellowish green in colour, profusely branched from stalk like basal portion, narrow at base and gradually expanding into rounded apices; tubular at base; compressed above. Cells in surface view elongate. Cells uninucleate, chloroplast platelike, with one pyrenoid.

Common: attached to rocks in mid littoral region.

Distribution: Ennore, Muttukadu, Kovalam, Mahabalipuram, Ammapattinam, Munaikadu Ariyankundu, Villoonditheertham , Thonithurai, Velankanni nagar, Seeniappadharga, Nochiurani, Pudumadam, Thalathoppu, Sallimunthal, Kanchirankudi, Keelakarai , Bharathinagar, Ervadi, Sadaimuniyan valasai, Manapad, Panjal, Valinokkam, Chinnamuttam, Vattakottai, Kanyakumari and Kovalam.

A. Habit of the plant.

B. Cell arrangement of branch and branchlets.

C. Transverse section of thallus.

Enteromorpha Link

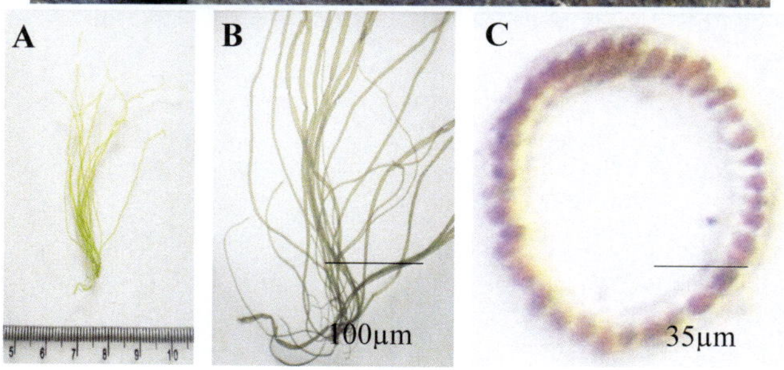

Enteromorpha Link

Enteromorpha linza **(Linnaeus) J. Agardh**

Description**:** Plants simple upto 45 mm in height; 6 mm broad, tapering to 2 mm at the base, attached by a disc, fronds several arising from common base, each frond cylindrical with tubular basal portion gradually expanding above and becoming flat. Cells in surface view linear.

Uncommon: attached to rocks in intertidal zone.

Distribution: Muttukadu, Kovalam, Pondicherry and Vattakottai

A. Habit of the plant

B. Arrangement of branches.

C. Transverse section of thallus.

Halimeda Lamouroux

Halimeda Lamouroux

The thallus looks like an articulated series of branches made up of green disk like segments which are rigid and tough due to the deposition of lime; the basal segment ends in a dense tuft of rhizoids. Under microscope one can see that the segments are composed of a system of entangled, siphon like filaments devoid of transverse cell walls. The thallus is coenocytic, with a bunch of siphons running longitudinally from one segment to the next. The central filaments produce lateral filaments that when reaching the surface of the thallus, expand into utricles; cell wall thickenings occur at the joint of filament bifurcations. Species recognition is based on thallus structure, the shape of the segments and the hold fast; the degree of calcification can also be used as comparison between different species.

Key for species:-

1. Nodal medullary filaments remaining separate (or) joint in to, three, four; plant attached to rocks and stones - 2
1. Nodal medullary filaments uniting as a single group, the filaments communicating by pores; plants growing in sand (or) mud. - 6
2. Branching generally in more than one plane – *H. opuntia*
2. Branches essentially in one plane ……3
3. Thallus flaccid, decumbent; segments small,less than 9mm long, 15mm broad, sub – cylindrical (or) flat – *H. gracilis*
3. Thallus erect, not flaccid, segments upto 29 mmlong and 42 mm broad ….. 4
4. A mall cushion segment interposed between all the segments of the plant – *H. cuneata*
4. Small cushion segments absent …… 5
5. Secondary utricles not inflated, segments predominantly remiform ……… *H. tuna*
6. Segments large, broad and flat; upto 20 mm long and 40 mm broad ….*H. macroloba*
6. Segments smaller, flat or cylindrical upto 12 mm Long and 16 mm broad ……7
7. Segments predominantly remiform with outer margin entire. Plants upto 10 cm tall …. *H. simulans*

Halimeda Lamouroux

REFERENCES: Krauss (1846: 214), Kutzing (1849: 505), Biswas (1949b: 38), Silva *et al*. (1996: 866).

***Halimeda* Lamouroux**

***Halimeda cuneata* Hearing**

Description: Thallus light green in colour; grows upto 15 cm high, arising from a distinct holdfast 0.5–1 cm long, calcification light, smooth surface; branching dichotomous to occasionally polychotomous. The basal segment sub cylindrical, upper segments cuneate to sub-cuneate or discoid, to 16 mm long, 18 mm broad and 0.5–1 mm thick; occasionally with a small cushion segment at the node. Cortex 2–4 layers of utricles, outermost remaining firmly attached after decalcification, Nodal medullary filaments united in groups of 2 or 3, the fused units entangled.

Rare: grows on rocky substratum in lower intertidal region

Distribution: Manapad, Kuduthalai.

A. Habit of the plant

B. Apical segment

C. Utricles

D. Transverse section showing cortex and medullary filaments.

Halimeda Lamouroux

REFERENCES: Bhakuni *et al.* (1992: 515), Kalimuthu, Kaliaperumal & Ramalingam (1992: table 1), Jayasankar & Ramalingam (1993: 42), Kerkar (1994: table 1), Silva *et al.* (1996: 867), Haneefa Koya *et al.* (1998: 10).

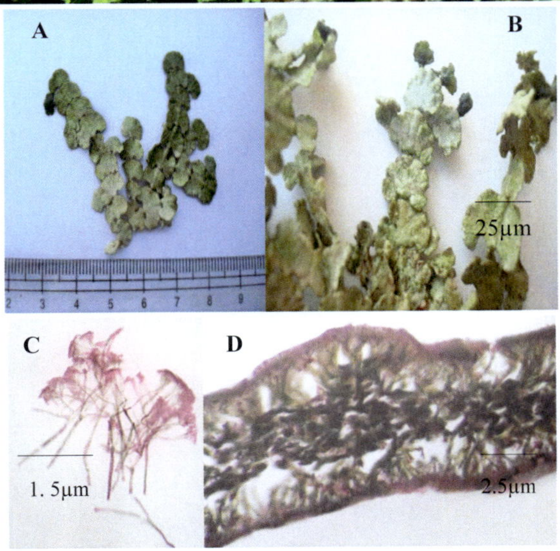

Halimeda Lamouroux

Halimeda gracilis **Harvey ex J. Agardh**

***Description*:** Plant whitish to bluish green in colour; mostly large, population form thick cushion on the substratum. Thallus 20-25 cm long, basal part stiff, upper part supple and moving along the waves, branching sparse but some segments supporting 4-5 branches; attachment by rhizoids at several places; segments strongly calcified and brittle, in most population small, cuneate, flabellate, diamond to round shaped, smooth, 2-3 mm long and 2-3 mm wide.

Common: on rocky substratum or in sheltered lagoons.

Distribution: Munaikadu, Pamban light house, Nalupanai, Ariyankundu Villoonditheertham, Thonithurai, Velankanni nagar, Pudumadam, Ervadi, and all 20 islands of Gulf of Mannar

A. Habit of the plant

B. Arrangement of branches

C. Utricle arrangement

D. Transverse section of blade with cortex and medullary filaments

Halimeda Lamouroux

REFERENCES: Kaliaperumal *et al.* (1992: 12), Kalimuthu, Kaliaperumal & Ramalingam (1992: 122), Jayasankar & Ramalingam (1993: 42), Rama Rao *et al.* (1993: 106), Kaliaperumal *et al.* (1995: 183), Silva *et al.* (1996: 869).

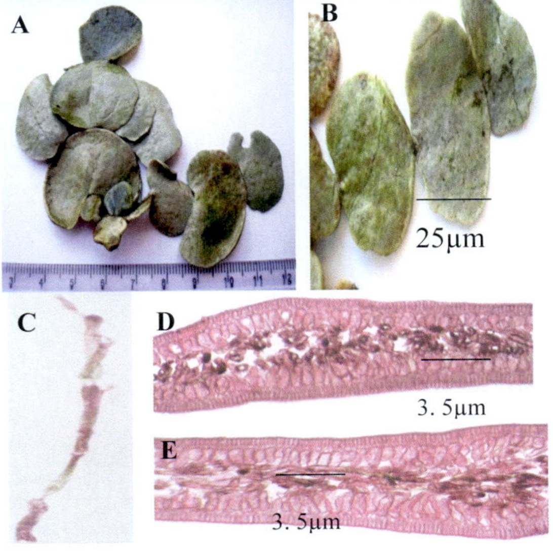

Halimeda Lamouroux

Halimeda macroloba **Decaisne**

Description Thallus fan shaped, attached to the substratum by a cylindrical holdfast of 3 cm in height; grows upto 11 cm in height; lower most segments are compressed; upper segments mostly remiform. Segments measure 0.5-1.5 cm in height and 1.5 to 2.5 cm in breadth; margin smooth, sometimes slightly undulate; node between segments narrow; medullary filaments trichotomously branched. Cortex is made of three layers of utricles.

Common: growing submerged in water on mud covered coral stones.

Distribution:Pamban, Thonithurai, Velankanni nagar, Pudumadam Valinokkam, Manapad,, Seeniappadharga, Bharathinagar, Sadaimuniyan valasai, and all islands of Gulf of Mannar.

A. Habit of the plant
B. Apical segment
C. Utricle
D. Transverse section
E. Longitudinal section

Halimeda Lamouroux

REFERENCES: Untawale & Jagtap (1989: table 1), Bhakuni et al. (1992: 515), Jagtap (1992: 57), Jagtap (1993: table 1), Kerkar (1994: table 1), Silva *et al*. (1996: 871).

Halimeda Lamouroux

***Halimeda opuntia* (Linnaeus) Lamouroux**

***Description*:** The plant grows upto 30 cm in height; branches sparsely, dichotomous and trichotomous; segments smaller and highly variable; length of segments ranges from 3-5 mm and 2-6 mm broad and 1 mm thick; margin entire. The medullary filaments are trichotomously branched; cortical cells have 3-4 layers of utricle.

Common: attached on rocks in the lower intertidal regions

Distribution: Thonithurai, Manapad, Leepuram, Krusadai island, Manali island, Van island and Koswari island.

A. Habit of the plant

B. Apical segment

C. Utricles

D. Transverse section of the segment showing cortex and medullary filament

Halimeda Lamouroux

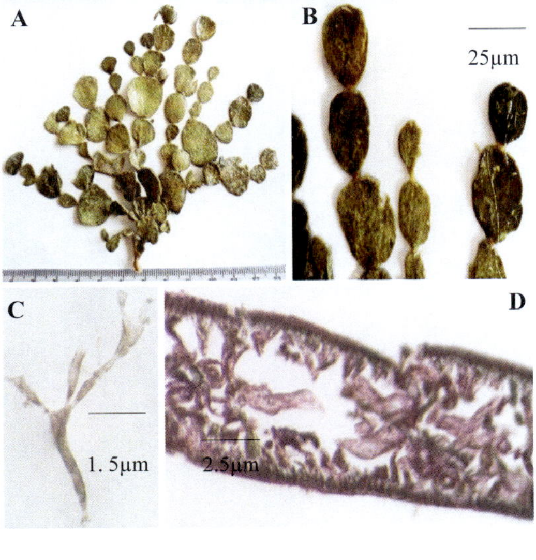

Halimeda Lamouroux

Halimeda simulans **M. Howe**

Description:Plants upto 10 cm in height; attached to substratum by flattened holdfast; thallus moderately calcified; calcification higher in lower segments; braches in a single plane, trichotomous; segments do not have much variations, segments 2 mm long, 2-5 mm in breadth; margin entire. In cross section, medulla is surrounded by cortex, utricles elongated and does not show structural variations.

Rare: attached on rocks in lower intertidal region.

Distribution: Thonithurai

A. Habit of the plant

B. Apical segment

C. Utricles

D. Transverse section of the segment showing cortex and medullary filament

Halimeda Lamouroux

REFERENCES: Jagtap (1993: table 1), V. Kumar (1993a: 139, 140), V. Kumar (1993b: 47), M. Dave & R. Parekh (1995: 16), Silva *et al.* (1996: 873), Sundararajan, Selvarahu & V. Krishnamurthy (1996: 21 – 33)

Halimeda Lamouroux

***Halimeda tuna* (Ellis & Solander) Lamouroux**

***Description*:** Plant upto 8 cm in height; attached to the substratum by compressed holdfast; branches di or trichotomous in a single plane; occasionally four or more branches originate from single segment. Segments near the holdfast is cuneate; others are discoidal; margin smooth. Segments 6-13 mm in height, 5-23 mm in breadth and 1-2 mm in thickness. Calcification is very mild and thallus looks green. Anatomy shows central medulllary filaments di or trichotomously branched. Utricles are many in cortical cells.

Common: attached to rocks in intertidal region.

Distribution: Thonithurai, Kanyakumari, Kovalam, Krusadai island and Koswari island

A. Habit of the plant

B. Apical segment

C. Utricles

D. Transverse section of the segment showing cortex and medullary filament

***Microdictyon* Decaisne**

Plants form small flabellate expanses formed by abundantly branching filaments, which form a minute network; main filament looks like prominent veins; lower segments producing many number of thick walled rhizoids. Cells cylindrical; ultimate branches grow vertically towards others and connected by thickened cellulose rings. Asexual reproduction by quadriflagellate zoospores; sexual reproduction by biflagellate gametes.

Key for the species:

Plants net-like, with a main axis and branches of cylindrical cells, end of branches attached together by cellulose ring (or) tenaculum. – *M. tenuius*

Microdictyon Decaisne

REFERENCES: V. Krishnamurthy & H. Joshi (1970: 8), Untawale, Dhargalkar & Agadi (1983: 5), Chennubhotla, Kaliaperumal & Kalimuthu (1987: 102, 103), Untawale, Reddy & Ambiye (1989: table 1), Kalimuthu, Kaliaperumal & Ramalingam (1992: 122), Silva *et al*. (1996: 757).

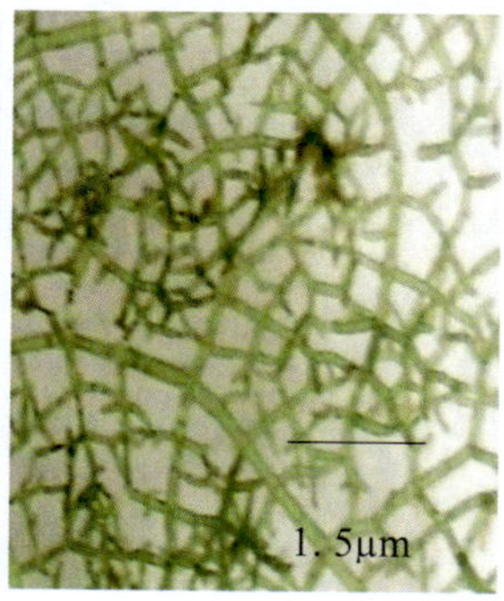

***Microdictyon* Decaisne**

***Microdictyon tenuius* J. Gray**

Description: Thallus is 5-10 cm in height; 6-8 cm broad; a foliose expanse; formed by repeatedly branched filaments; attachment by long slender rhizoids that forms a short stipe; main filaments consists of larger cells; ultimate branches unicellular, often connected by tenacula which are not separated by septa. Cells cylindrical.

Rare: attached to rocks and coral stones upto 20 meter depth

Distribution: Arasaneripattinam, Kodikkarai, Krusadai island, Hare island and Koswari island.

A. Habit of the plant

B. Arrangement of cells in mesh like network of blade.

***Neomeris* Lamouroux**

Thallus cylindrical to club shaped, 1.5cm long; the erect axis produces a delicate tuft of filaments. With the exception of the tip, the thallus is whitish due to lime deposition and has whorls of densely disposed laterals, with the terminal portion expanded and laterally cemented by lime.

Key for the species:

Gametangia fused in set of 5-12 and the structure may be ovate to oblong. - *N. annulata*

Neomeris Lamouroux

REFERENCES: Untawale, Dhargalkar & Agadi (1983: 13), Chowdary & Singh (1984: 171 – 171, figs. A – F, 1 – 10a), Untawale, Reddy & Ambiye (1989: table 1), Kalimuthu, Kaliaperumal & Ramalingam (1992: 121), Menon *et al.* (1993: 59), Silva *et al.* (1996: 889).

Neomeris ann
Description:
cylindrical to
1.5cm long;
produces a d
filaments. Wit
of the tip,
whitish du
deposition anc
densely disp
with the te
expanded a
cemented by li
Common: att
and dead coral
Distribution:
Arasaneripatti
Karankadu, A
Kodikkarai,
Villoonditheert
Thonithurai,
islands of Gulf

A. Habit of the
B. Individual
lime depositior

Neomeris Lamouroux

Neomeris annulata **Dickie**

Description: Thallus cylindrical to club shaped, 1.5cm long; the erect axis produces a delicate tuft of filaments. With the exception of the tip, the thallus is whitish due to lime deposition and has whorls of densely disposed laterals, with the terminal portion expanded and laterally cemented by lime.

Common: attached to rocks and dead coral stones.

Distribution: Arasaneripattinam, Karankadu, Ammapattinam, Kodikkarai, Villoonditheertham, Thonithurai, and all 20 islands of Gulf of Mannar.

A. Habit of the plant

B. Individual branch with lime deposition

Penicillus Lamouroux

REFERENCES: Umamaheswara Rao (1969: 40), Umamaheswara Rao (1970: 409, figs. 2, 5, 8), Umamaheswara Rao (1974a: 302), S. Dixit (1980: 62), Untawale, Dhargalkar & Agadi (1983: 14), V. Krishnamurthy *et al.* (1990: figs. 4, 7).

Penicillus Lamouroux

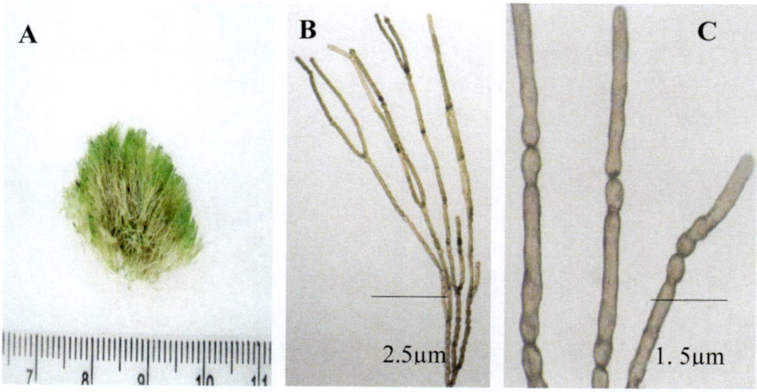

Penicillus sibogae V. Krishnamurthy & Thomas

Description: The plant is light green in colour. The plants have a long stalk, up to 5 cm, and an oblong to spherical 'brush' of filaments of 3 cm. The plants are substantial calcified. The filaments are slender, but because of the calcification rather tough. Slight magnification reveals their repeated dichotomous branching. The stalks appear spongy under magnification. They are slightly constricted at their base, where the bulbous mass of rhizoids arise. The plant is coenocytic.

Rare: Attached to rocks in intertidal and lower intertidal region.

Distribution: Seeniappa dharga

A. Habit of the plant

B. Filaments showing repeated dichotomous branches

C. Close view of filaments with bulbous structure.

***Rhizoclonium* Kützing**

Thallus a slender mass of entangled, unbranched filaments, rhizoids may sometimes resemble branches; cells cylindrical, 1-6 diameter long, apical cells often swollen.

Key for the species -

1 Cell wall constricted at septa. ……*R. hieroglyphicum*

1 Cell wall not constricted at septa …... *R. riparium*

Rhizoclonium Kützing

A

B

C

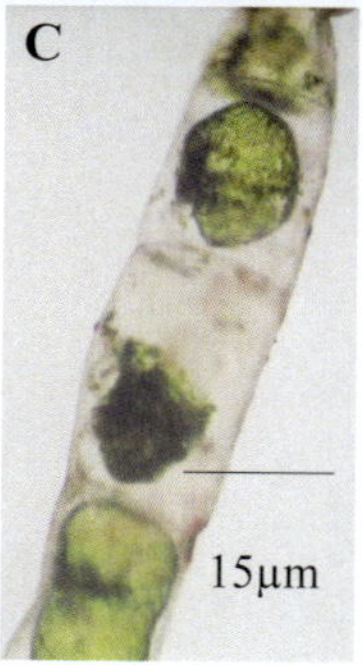

Rhizoclonium hieroglyphicum **(C. Agardh) Kützing**

Description: Bright or pale green in colour; coespitose; filaments with frequent rhizoidal branches, short and produced laterally; cells 12-35µm broad, 2 to 3 times long as broad; constricted at septa.

Rare: attached on rocks.

Distribution: Valinokkam

A. Habit of the plant

B. & C Unbranched filament.

Rhizoclonium Kützing

REFERENCES: Roth (1806: 216 – 217), Krauss (1846: 215), Harvey (1949: [1846 – 1851]: pl. CCXXXVIII), Biswas (1949a: 75, pl. V: fig. 48), S. Dixit (1970: 105), Silva *et al.* (1996: 786).

Rhizoclonium riparium **(Roth) Harvey**

Description: Plants yellowish green in colour; crisp, caespitose; interwoven into loose lying or dense mats; cells 15-28 µm in diameter, 2-3 times long as broad in younger cells. Rhizoid like branches numerous, short.

Rare: on moist soils and rocks in high intertidal zone.

Distribution: Manapad

A. Habit of the plant

B & C. Unbranched filament.

***Struvea* Sonder**

Thallus is net like with long stipe, attached by rhizoids which are in branches. Stipe with annular constrictions continued as main axis bearing series of vesicles, longer at base, progressively shortening above; each vesicle producing opposite branches, which in some species have long basal cells and repeated constriction of primary axis; branches usually in one plane; interconnected by tenacula produced on end cells.

Key for the Species:

Frond with primary axis and secondary axis, each with paired branches,

Secondary axis anastomosing by tenacula …..*S. anastomosans*

Struvea Sonder

REFERENCES: Nair *et al.* (1986:736), Jagtap (1987: table 1), Chaugule, Quadri & Goswamy (1989: 113, fig. 56), Untawale, Reddy & Ambiye (1989: table 1), Srivastava (1996: 161 – 162), Silva *et al.* (1996: 798).

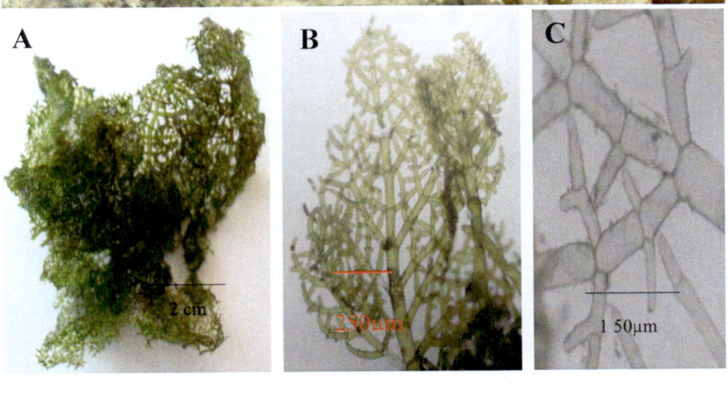

Struvea Sonder

Struvea anastomosans* (Harvey) Piccone & Grunow ex Piccone

Description: Frond 4-5 cm in height, consists of primary axis with long basal cell, narrowing at base, with annular constrictions, attached by rhizoids; secondary axis with long basal cells and branching as primary axis. Branches in opposite pairs in every distal cell of both primary and secondary axis - curved upwards; axis and ramuli interconnected forming net like frond. Lower branches much separated while upper branches crowded and intertwined.

Rare: attached to coral stone in deep water.

Distribution: Pullivasal island, Appa island.

A. Habit of the plant

B. Separated lower branch and crowded upper branch

C. Branches and branchlets interconnected.

***Trichosolen* Montagne**

Plant transparent green in colour; shining. Plant usually 5-10 cm tall; erect, with cylindrical axis, densely covered with pinnae from all sides; pinnae mostly unbranched; 2-3 mm long, 160- 250µm diameter, loosely arranged on axis; gametangia oblong. Fronds are completely coenocytic without septa.

Key for identification

Axis cylindrical and pinnae unbranched - *T.mucronatus*

Trichosolen Montagne

REFRENCES: Gopalakrishnan (1971: table II), Untawale, Dhargalkar & Agadi (1983: 11), Kaliaperumal & Kalimuthu (1987: 100 – 101), Chaugule, Quadri & Goswamy (1989: 111, figs. 20 – 30), Kamat *et al.* (1992: table 1), Silva *et al.* (1996: 811).

Trichosolen Montagne

Trichosolen mucronatus (Børgesen) W. R. Taylor

Description: Plant yellowish green in colour, 5-10 cm tall; erect, branched thallus, axis cylindrical, densely covered with pinnae from all sides; pinnae unbranched; 2-3 mm long, 160- 250μm diameter, loosely arranged on axis; gametangia oblong. Fronds are transparent green, shining and completely coenocytic without septa.

Rare: attached to dead coral stones.

Distribution: Manapad

A. Habit of the plant.

B & C. Branches of the plant.

D. Branch covered with pinnae.

Udotea Lamouroux

Udotea Lamouroux

Thallus fan like or funnel shaped, with a stalk and bunch of rhizoids that penetrate the sediment; substantial deposition of lime. Thallus structure consists of siphonous, forked filaments with constrictions at the base of the forks.

Key for the species:

1. Thallus heavily calcified, blade 7cm long, thick and lateral appendages repeatedly branched with stubby, finger-like blunt apices - *U. caribaea*
1. Thallus moderately calcified, blade 30cm long leathery and lateral appendages irregularly spaced with short, rounded apices - *U. flabellum*

Udotea Lamouroux

REFERENCES: Vijaya Parthsarthy & V. Krishnamurthy (2000: 15 – 17).

Udotea Lamouroux

Udotea flabellum **(Ellis & Solander) Howe**

Description: The plant is dark green in colour; stipe cylindrical and extends into a fan shaped flabellate that grows more than 12 cm high. Cluster of rhizoids hold the plant upright and securely in the substratum. The rhizoids extend several inches below the substratum. Many siphons intertwined together and form the thallus.

***Uncommon*:** attached to the mud in deep water region.

Distribution: Manapad, Hare island, Valai island, Upputhanni island Van island and Koswari island.

A. Habit showing stipe, siphonous filament

B. Cross section of filaments

C. Longitudinal section of filaments with cortex and medulla

Udotea Lamouroux

REFERENCES: K. Srinivasan (1973: 35, pl. XXXV), S. Dixit (1980: 62), V. Krishnamurthy (1980: 52), Untawale, Dhargalkar & Agadi (1983: 14), V. Krishnamurthy & Balasundaram (1990: 8), Silva *et al*. (1996: 884).

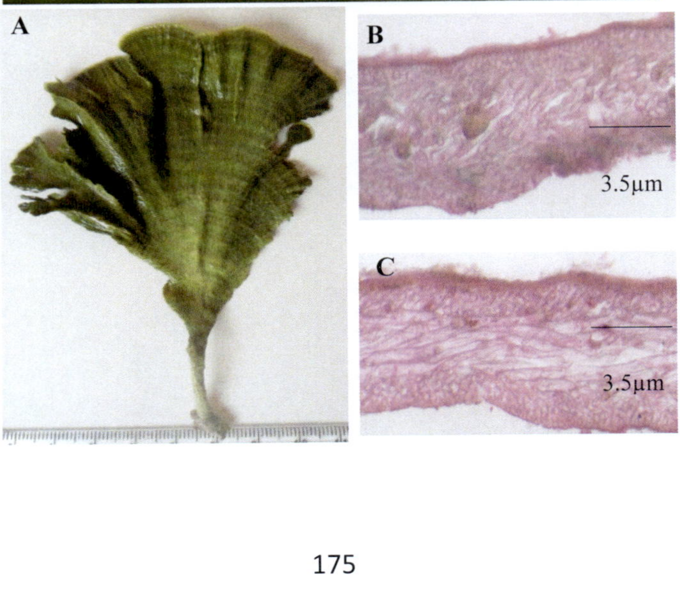

Udotea Lamouroux

Udotea caribaea **Litter & Litter**

Description: The frond has cylindrical stipe, gradually expanding into flabellate blade measuring up to 4 cm in height. One or two lobes appear on the blade above the stipe and these expand parallel to the primary lobe and overlaps it. The basal part of stipe ends in a mass of fibrous rhizoids. The thallus is wholly made up of many intertwined dichotomously branched siphons which are constricted at the region of branching.

Rare: attached to the mud covered coral stones in deep water.

Distribution: Mully island.

A. Thallus showing stipe and siphonous filaments

B. Cross section of siphonous filament with loosely arranged medulla (M)

Ulva Linnaeus

***Ulva* Linnaeus**

Thallus formed by either a broad leaf - entire or with perforations, or by a tuft of narrow strips; structurally the thallus is composed of two layers of cells. The cells contain one plastid and one or more pyrenoids each; attached to the substratum by holdfast formed by rhizoidal proliferation of the basal cells. Reproduction as in Enteromorpha. Some species tolerate wide environmental variations and pollution, while others are restricted to the more constant environment of deep waters. In lagoon or very productive areas, free floating populations originating from a torn portion of a thallus may occur. A cosmopolitan genus also known as sea lettuce.

Key for the species:-

1. Frond reticulate with spinose projections into meshes ….. *U. reticulata*
1. Frond not reticulate ….. 2
2. Frond deeply divided into several long linear lobes …. 3
2. Frond not deeply divided into several long linear lobes …. 5
3. Frond attached by rhizoidal disc …..*U. fasciata*
3. Frond attached by rhizoidal stipe …. 4
4. Frond in section shows 2 layers of cells arranged in 8µm apart ……*U. covelongensis*
5. Frond attached by solid rhizoidal stipe ……. *U.rigida*
5. Frond attached by rhizoidal disc ....*U. lactuca*

Ulva Linnaeus

REFERENCES: V. Krishnamurthy & H. Joshi (1969: 126, 128 figs. 4, 10, 16 pl. 1, table 1), S. Dixit (1970; 103 – 104), V. Krishnamurthy & H. Joshi (1970: 3), Untawale, Dhargalkar & Agadi (1983: 7), Silva *et al.* (1996: 743).

Ulva Linnaeus

Ulva covelongensis Krishnamurthy & Joshi

Description: Plants attached to rocks with conspicuous stipe; plants usually long, 12 cm or more; deeply divided into several linear ribbon like blades with wavy margin, obtuse in apex; gradually tapering towards the base. Cells in surface view vary in size and shape; mostly polygonal.

Rare: attached to rocks in upper intertidal region.

Distribution: Kovalam and Pondicherry.

A. Habit of the plant

B. Transverse section of blade

C. Longitudinal section of blade

Ulva Linnaeus

REFERENCES: Silva *et al*. (1996: 743), Ganesan *et al*. (1997: 17), Oza & Mairh (1997: 39), Rattan Kumar & Naidu (1997: 43), Kaliaperumal *et al*. (1998: 141 - 151), Anbazhagan *et al*. (2001: 23).

Ulva Linnaeus

Ulva fascisata Delile

Description: Thallus bright green in colour; composed of tough blades, irregularly divided. Fronds upto 30 cm long; 1-3 cm broad, gradually tapering towards their apices; undulated especially in the margins and sometimes longitudinally contorted. Cells in cross section showed 2 adhering layers; isodiametric or broader than long. Pyrenoid is single per cell; occasionally two.

Common: attached to rocks in high intertidal zone, as well in pools with continuous wave swept. Frequently hanging down from vertical walls.

Distribution: Muttukadu Kovalam, Kuduthalai, Manapad, Idinthakarai, Panjal, Leepuram, Chinnamuttam, Vattakottai, Kanyakumari and Kovalam.

A. Habit of plant

B. Transverse section of plant showing 2 layers of cells

C. Longitudinal section of thallus

Ulva Linnaeus

REFERENCES: Sindhu & Panikkar (1995c: 123, figs. 1/6, 1/8, 1/9), Sekar, Thangaraju & Rengasamy (1995: 49 – 52), Silva *et al.* (1996: 745), Thangaraju & Rengasamy (1997: 41), Kaliaperumal *et al.* (1998: 141 – 151), Ruma Pal (2000: 51 – 60).

Ulva Linnaeus

Ulva lactuca **Linnaeus**

Description**:** Fronds much broader than long, upto 20 cm long, markedly undulated all over, bright green in colour; thallus attached by basal disc, 2-4 mm in diameter; cells in surface view angular; containing single or 2 pyrenoids. Transverse section of blade is 60-60 μm thick in middle part, upto 80-100 μm in the basal part. Cells isodiametric or slightly elongated.

Common: Attached to rocks in lower intertidal region.

Distribution: Nalupanai, Kottaipattinam, Munaikadu, Thalathoppu, Keelakarai, Manapad, Panjal, Chinnamuttam and in all 20 islands of Gulf of Mannar

A. Habit of the plant

B. Plant attached on rock

C. Transverse section of the blade

D. Longitudinal section of the blade

Ulva Linnaeus

REFERENCES: Jagtap (1992: 57), Kalimuthu, Kaliaperumal & Ramalingam (1992: 121), Jagtap (1993: table 1), Menon *et al.* (1993: 59), Sindhu & Panikkar (1995: 123 - 125, figs.2/1, 2/4), Silva *et al.* (1996: 749)

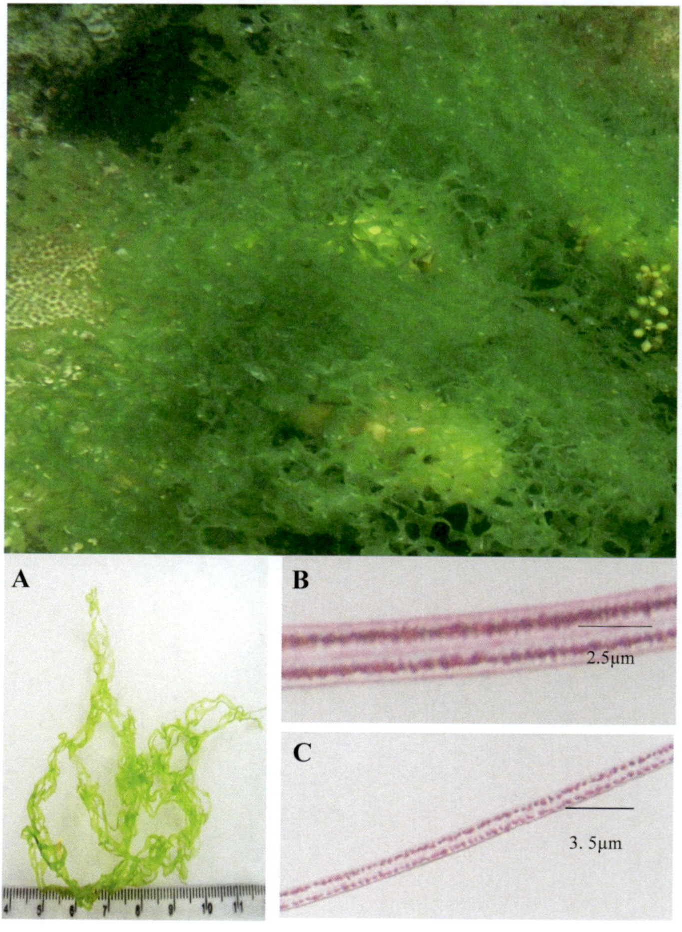

Ulva Linnaeus

Ulva reticulata **Försskål**

Description: Thallus bright green in colour with net like structure; without conspicuous holdfast; oval to oblong or rectangular in shape; lamina perforated with larger and smaller holes side by side; lacinae 6-15 mm in width; margin of lacinae with microscopic teeth; cells in surface view are polygonal, vertically elongate, 26-36 μm long and 12-16 μm broad, arranged in two layers. Chloroplast plate like situated near the outer wall of cell

Common: mostly as epiphytic on other algae in sub littoral to lower littoral.

Distribution: Kottaipattinam, Rameswaram, Leepuram, and in all 20 islands of Gulf of Mannar.

A. Habit of the plant

B. Transverse section of blade

C. Longitudinal section of blade

Ulva Linnaeus

REFERENCES: Parekh & Chauhan (1987: 31), Umamaheswara Rao (1987: 21), Untawale, Reddy & Deshmukhe (1989: table 1), Jagtap (1992: 57), Silva *et al*. (1996: 750), Ruma Pal (2000: 51 – 60).

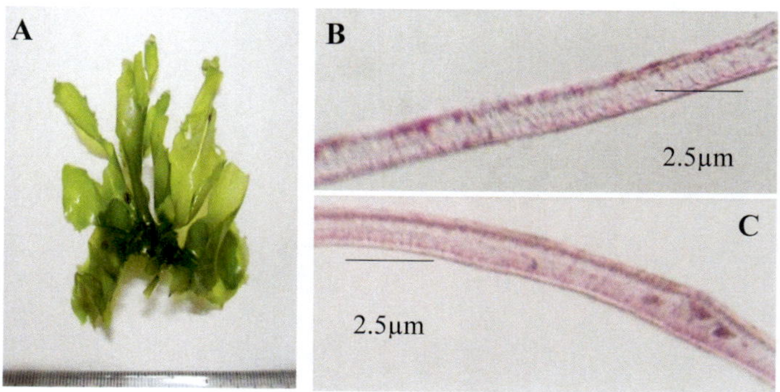

Ulva Linnaeus

Ulva rigida C. Agardh

Description: Plants forming small rosettes of relatively stiff blades, 2-3 cm in diameter at air exposed sites and 10 cm diameter in more sheltered pools; stipe is distinct but solid; blades undulated, lobed with smooth surface and small, marginal teeth; dark green in colour. Cells arranged in two layers in cross sectional view. Cells uninucleate; chloroplast cup shaped; pyrenoid single or occasionally two.

Uncommon**:** attached to rocks in lower intertidal region.

Distribution: Panchal, Kanyakumari.

A. Habit of the plant

B. Transverse section of blade

C. Longitudinal section of blade.

***Valonia* C. Agardh**

All species are vesicular, consisting of macroscopic, multinucleate segments, each with a large central vacuole; vesicles are very firm and turgid when the algae are alive; if ruptured, they are reduced to just cell walls and the attached thin layer of cytoplasm and chloroplasts.

Key for the species-

Thallus prostrate, creeping, cells elongated, cylindrical, often arched (or) curved, uncrowded. – *V. utricularis*

Valonia C. Agardh

REFERENCES: Umamaheswara Rao (1969: 38), V. Krishnamurthy H. Joshi (1970: 9), S. Singh, Jose & Chowdary (1981: 65 – 68, figs. 1, 3), Untawale, Dhargalkar & Agadi (1983: 15), Jagtap (1992: 57), Silva *et al*. (1996: 802).

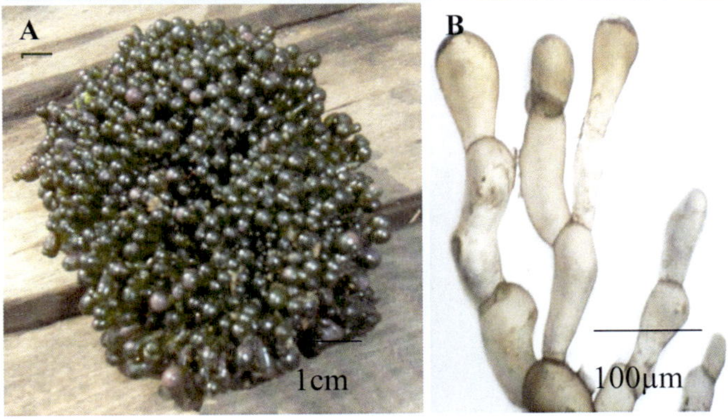

Valonia C. Agardh

***Valonia utricularis* (Roth) C. Agardh**

Description: Plant dark green in colour; cushion like tuft of cylindrical to clavate vesicles, growing to a height of about 5 cm; vesicles profusely branched and interconnected by hapteroid growth of lenticular cells; vesicles narrowing towards base; lower part of plant decumbent.

Uncommon: Attached to rocks and coral stones in intertidal region.

Distribution: Villoonditheertham, Thonithurai, Kanyakumari, Kovalam, Shingle island, Hare island and Mulli island.

A. Habit of the plant

B. Microscopic view of branches.

Valoniopsis Borgesen

***Valoniapsis* Borgesen**
Thallus filamentous, cushion like, prostrate and entangled; with large, multinucleate cells.

Key for the species-

Vesicles cylindrical, forming a branched filament, branches
Cluttered -*V. pachynema*

Valoniopsis Borgesen

REFERENCES: G. Martens (1968: 24, 62 – 63, pl. IV: fig. 2), G. Martens (1869: 235), G. Martens (1871: 465), K. Srinivasan (1961: 229), V. Krishnamurthy & H. Joshi (1970: 5), Untawale, Dhargalkar & Agadi (1983: 11)

Valoniopsis Borgesen

Valoniopsis pachynema (G. Martens) Børgesen

Description: Thallus forming low dense tufts, attached to substrata by rhizoid like branches; branches cylindrical, arcuate, irregularly ramified; filaments repeatedly branched, attached by septate in primary and secondary branches; younger branches erect, older branches spreading. Branches upto 1 mm thick. Chloroplast numerous with a large pyrenoid.

Common: attached to rocks in lower intertidal and subtidal regions.

Distribution:

Pamban, Thonithurai, Nochiurani, Sadaimuniyanvalasai, Valinokkam, Manapad, Kuduthalai, Idinthakarai, Panjal, Leepuram , Vattakottai and all islands of Gulf of Mannar.

A. Habit of the plant

B. Branching pattern

C. Individual branch

PHAEOPHYTA

Chnoospora J. Agardh

Thallus macroscopic and well branched, cylindrical to compressed, attaches firmly on rocks by a basal disk; structurally composed of a dense medulla of vertically elongated cells without chloroplast and a cortex of smaller pigmented cells. Fertile plants produce elongated plurilocular organs among a dense tuft of filaments.

Key for the species

1. Thallus compressed, grooved on both surfaces …*C. bicanaliculata*
1. Thallus grooved on one surface only ….2
2. Thallus intricately branched , branches cylindrical to compressed, attenuated at the apex …..*C. implexa*
2. Thallus regularly dichotomously branched, branches cylindrical, flattened at the forking, not attenuated …. *C. minima*

Chnoospora J. Agardh

REFERENCES: V. Krishnamurthy & Thomas (1977: 44, 46, 103: figs. 3 – 7, 27), Untawale, Dhargalkar & Agadi (1983: 20), Nair *et al*. (1986: 736), Maya & Nair (1992: 26), Silva *et al*. (1996: 625).

Chnoospo
V.
&Thomas
Descriptio
cm in heig
cylindrical
flattering
flattened,
well defin
two flat s
10-15 µr
35-40 µ
gametangi
in the groo
single lir
loculi.
Uncommo
rocks in in
Distributio
Kovalam
A. Habit o
B. Branche
C. Trans
thallus wit
D. Long
showing
cells.

Chnoospora J. Agardh

Chnoospora bicanaliculata **V. Krishnamurthy & Thomas**

Description**:** Plants upto 15 cm in height, stipe long, sub-cylindrical, gradually flattering above, branches flattened, clustered with two well defined grooves on the two flat sides; cortical cells 10-15 µm, medullary cells 35-40 µm, plurilocular gametangia in extensive sori in the grooves, consisting of a single linear series of 8 loculi.

Uncommon: Attached on rocks in intertidal zone.

Distribution: Kanyakumari, Kovalam

A. Habit of the plant

B. Branches with flat surface.

C. Transverse section of thallus with hairs on edges

D. Longitudinal section showing clear medullary cells.

Chnoospora J. Agardh

REFERENCES: Nair *et al*. (1981: table 1), Untawale, Dhargalkar & Agadi (1983: 20), Pant & Untawale (1989: table 6), Kalimuthu, Kaliaperumal & Ramalingam (1990: table 2), Jagtap (1993: table I), Silva *et al*. (1996: 625).

Chnoospora
Agardh
Description:
intricately br
cylindrical
branches attє
axils betv
rounded or
Thallus parє
superficial
rectangular
progressivel
Hairs arisi
depression.
Uncommon
rocks in
habitats
Distribution.
Krusadai isla
A. Habit of tl
B. Cylindric
attenuated ti
C. Trans
showing
cells, larger i
D. Longituc

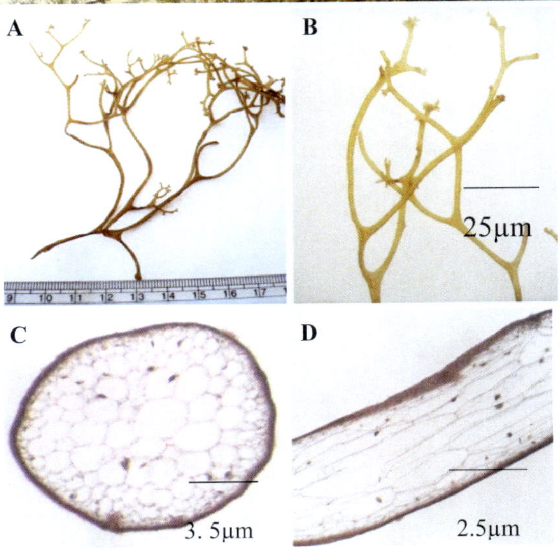

Chnoospora J. Agardh

Chnoospora implexa J. Agardh
Description: Plants intricately branched, branches cylindrical to compressed; branches attenuated at apices; axils between branches rounded or obtusely angled. Thallus parenchymatous and superficial layer is small, rectangular and cells progressively larger inwards. Hairs arising from deep depression.
Uncommon: attached to rocks in wave exposed habitats
Distribution: Single island, Krusadai island, Hare island.
A. Habit of the plant
B. Cylindrical branches with attenuated tips
C. Transverse section showing parenchymatous cells, larger inwards
D. Longitudinal section.

Chnoospora J. Agardh

REFERENCES: Subba Rao *et al.* (1985: 32), Nair *et al.* (1986: 736), Chennubhotla, Kaliaperumal, Kalimuthu (1987: 104 – 105, fig. 5B), Umamaheswara Rao (1987: 22), Chennubhotla, Mathew & Joseph (1990: 1), Silva *et al.* (1996: 626).

Chnoospor
(Hearing) I
Description
dichotomou
height. Bra
slightly fla
forking v
apices. Ha
superficial
prominent.
of small
rectangular,
elongated
gametangia
uniseriate.
Common:
in exposed
Distributio
Pamban
Villoondith
Ariyankund
Kuduthalai,
Krusadai isl
Mully islan
Valimunai
island.

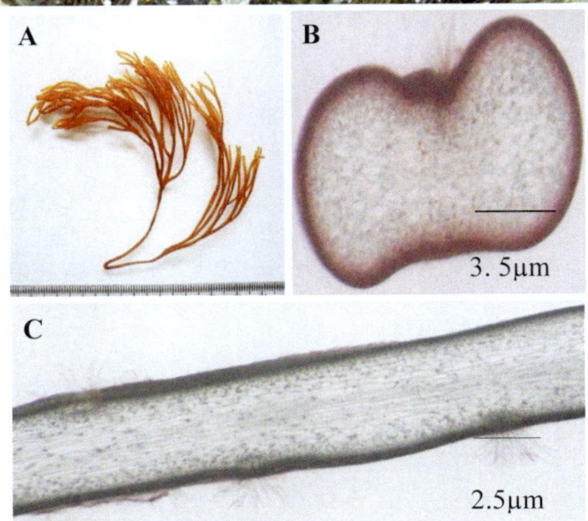

A. Ha
dichotomou
B. Tran
showing hai
C. Longi
showing el
the middle

Chnoospora J. Agardh

Chnoospora minima (Hearing) Papenfuss

***Description*:** Plants regularly dichotomous; 6-10 cm in height. Branches cylindrical, slightly flattened near the forking with attenuated apices. Hairs arising from superficial depressions, prominent. Thallus composed of small superficial cells, rectangular, enclosing larger elongated cells. Plurilocular gametangia clavate, uniseriate.

Common: attached to rocks in exposed places

Distribution: Munaikadu, Pamban light house, Villoonditheertham, Ariyankundu, Valinokkam Kuduthalai, Idinthakarai, Krusadai island, Hare island, Mully island, Valai island, Valimunai island and Salli island.

A. Habit showing dichotomous branches

B. Transverse section showing hairs.

C. Longitudinal section showing elongated cells in the middle

***Colpomenia* (Endlicher) Derbes & Solier**

Light brown plants with a vesicular, sometime compressed, hollow thallus; structurally composed of 2-3 layers of cells; the inner one with larger cells without chloroplast and the outer one made of smaller cells with chloroplast; hairs with basal growth are common in the cortex. Plurilocular organs arranged in groups on the surface, associated with sterile filaments.

Key for the species

Thallus irregularly spherical, sinuous, without war like projections ….. *C.sinuosa*

Colpomenia (Endlicher) Derbes & Solier

REFERENCES: Pant & Untawale (1989: tables 4, 5), Untawale & Jagtap (1989: table I), Kalimuthu, Kaliaperumal & Ramalingam (1992: 122), Jagtap (1993: table I), Jayasankar & Ramalingam (1993: 42), Silva et al (1996: 627).

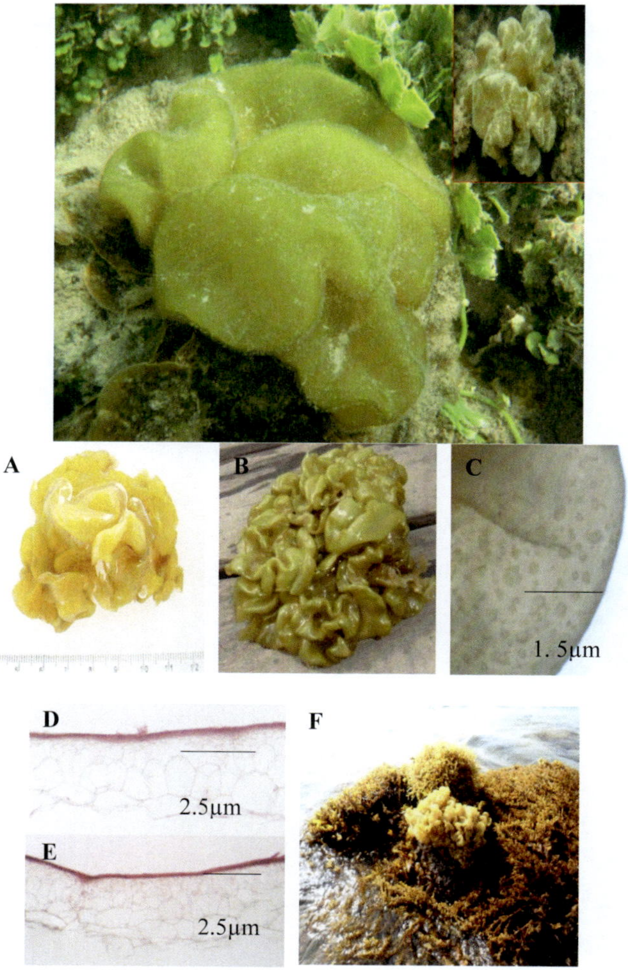

Scytosiphonales-Scytosiphonaceae

Colpomenia (Endlicher) Derbes & Solier

Colpomenia sinuosa **(Mertens ex Rotyh) Derbes & Solier**
Description: Plants yellow brown in colour. Thallus thin, membranous, extending upto 10 cm in diameter, roughly spherical to irregular in shape; peripheral cells single layered, polygonal in shape, compactly arranged with plastids and assimilatory function; inner cells large, rounded and irregular in shape colourless; plurilocular gametangia uniseriate.
Common: attached to rocks and hard substratum upto 10 meter deep.
Distribution: Nalupanai, Sallimunthal, Kanchirankudi, Keelakarai light house, Keelakarai, Bharathinagar, Ervadi, Manapad, Kuduthalai, Idinthakarai, Leepuram , Vattakottai and all 20 islands of Gulf of Mannar.

A & B. Habit of the plant
C. Surface view
D. Transverse section showing surface hairs
E. Longitudinal section
F As epiphytic alga

***Cystoseira* C. Agardh**

Large brownish plants usually several decimeters long; attached by discoid base from which a few well branched main axes are produced; these are cylindrical or flattened, covered by a short and spiny lateral branches inflated with vesicles, single or in successive series. Thallus structure parenchymatous with several layers of cells; medullary cells are smaller, with thick wall cells and devoid of chloroplasts; the medulla is separated from the epidermis layer by few layers of larger, subcortical cells; the single layer of epidermis cells is distinctive with radially elongated cells and many chloroplasts; hairs are present on the thallus surface. The reproductive structures, oogonia and anthredia are arranged in slightly modified branches.

Key for the species

1. Plants without spiny appendages on receptacles,
   Vesicles in chains of more than two ……*C. trinodis*

Cystoseira C. Agardh

REFERENCES: Kalimuthu, Kaliaperumal & Ramalingam (1988: 17 – 20, fig. 1), D. Rao *et al*. (1991: 122), Kalimuthu, Kaliaperumal & Ramalingam (1992: 122), Kaliaperumal & Kalimuthu (1993: 9, 11), Silva *et al*. (1996: 649), Babu & Rengasamy (2001: 30).

Cystoseir
(Forsskål
Descripti
cm in hei
more. Pla
colour;
branches
dentate, v
shape, e
middle of
Vesicles u
than two.
Uncomm
rocks an
upto 5 me
Distributi
island and
A. Habit o
B. Branc
two vesicl
C. Transve
D. Longitu

Cystoseira C. Agardh

***Cystoseira trinodis* (Forsskål) C. Agardh *Description*:** Plants upto 30 cm in height, sometimes even more. Plants dark brown in colour; sparsely branched, branches articulated; margin dentate, vesicles oblong in shape, embedded in the middle of the swollen wings. Vesicles usually two or more than two.

Uncommon: attached to the rocks and coral substratum upto 5 meter deep.

Distribution: Krusadai island and Hare island.

A. Habit of the plant

B. Branches having one or two vesicles.

C. Transverse section.

D. Longitudinal section.

***Dictyopteris* Lamouroux**

Narrow, ribbon shaped branched thallus with a conspicuous midrib; lateral veins in some species; growth from a group of small apical cells; with the exception of the midrib, the wing is generally composed of two layers of cells though this may vary with species. Reproduction through gametangia and sporangia on the thallus surface.

Key for the species

1. Thallus more than 10 cm in height 2
1. Thallus not more than 10 cm in height ...D.delicatula
2. Margin sinuate and dentate 3
3. Veins originating from the mid rib all through the thallus; sporangia in sori ... D.australis
3. Veins absent, sporangia occurring as narrow belt, on either side of the midrib. ... D.woodwardia.

Dictyopteris Lamouroux

REFERENCES: Dhargalkar (1985: 80, fig.1), (1992: 133), Jagtap (1987: table 1), Pant & Untawale (1989: table 4, 7), Untawale, Reddy & Deshmukhe (1989: table 1), Kamat *et al*. (1991: 54), Silva *et al*. (1996: 580).

Dictyopteris Lamouroux

Dictyopteris australis (Sonder) Askenasy

***Description*:** Thallus light brown in colour, up to 25 cm tall, 1-2cm in width; attached by a small disc; broad ribbon shaped thallus, repeatedly either dichotomous or trichotomous. Mid rib very prominent in main axis as well as in branches, present on both sides of the thallus. Margins irregularly dentate, apices narrow. Thallus multilayered in mid rib region and two layers in the wing. Reproductive bodies in recurved arched groups.

Rare: attached on coral and hard substratum in deep waters.

Distribution: Koswari island.

A. Habit of the plant

B. Thallus with prominent midrib and reproductive bodies

C. Transverse section

D. Longitudinal section

Dictyopteris Lamouroux

REFERENCES: Nair *et al.* (1986: 736), Chennubhotla, Kaliaperumal, Kalimuthu (1987: 102 – 103), Jagtap (1987: table 1), Untawale, Reddy & Ambiye (1989: table 1), Kalimuthu, Kaliaperumal & Ramalingam (1992: 122), Silva *et al.* (1996: 581).

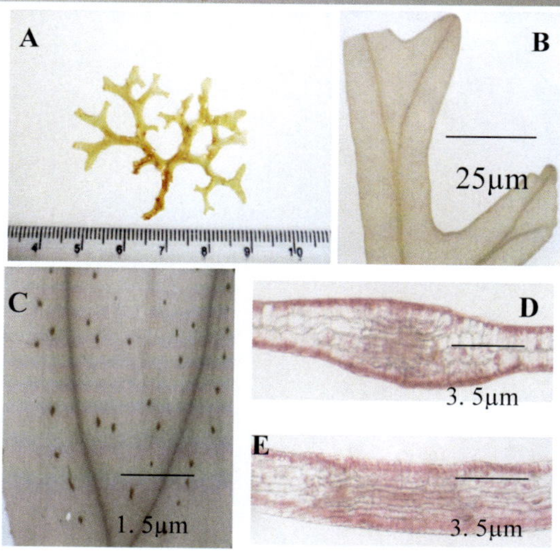

Dictyopteris Lamouroux

Dictyopteris delicatula **Lamouroux**

Description**:** Fronds small, upto 5 cm in height, light brown in colour, attached to the substratum by very small disc. Thallus dichotomously or irregularly branched; branches upto 3 mm broad; margin entire, more or less wavy; apices rounded; mid rib more prominent in lower portion of thallus. Thallus having 4-6 layers of cells in midrib region and two layered in the wings; reproductive bodies irregularly scattered, occurring on both surface of the thallus.

Rare: Attached on coral and other hard substratum in deeper waters.

Distribution: Mully island.

A. Habit of the plant.

B. Thallus with prominent midrib

C. Thallus with reproductive bodies.

D. Transverse section

E. Longitudinal section

Dictyopteris Lamouroux

REFERENCES: Misra (1967: 233), K. Srinivasan (1969: 35, pl. XXX), Goplalakrishnan (1970: 63), V. Krishnamurthy & H. Joshi (1970: 10), Untawale, Dhargalkar & Agadi (1983: 20), Chennubhotla, Kaliaperumal & Kalimuthu (1987: 102 – 103), Silva *et al*. (1996: 580).

Dictyopteris Lamouroux

***Dictyopteris woodwardia* (R. Brown ex Turner) C. Agardh**
***Description*:** Fronds yellowish brown in colour, 15 to 30 cm in height; attached to the substratum by a small disc; thallus irregularly dichotomously branched. Branches 0.7 to 1 cm wide; margin dentate; tapering apices; midrib conspicuous; lamina often torn. Thallus many layered in the midrib region and 2-4 layered in the wings; cells in midrib region is thick. Hairs occurring on both sides, little away from the midrib. Reproductive bodies on both surfaces; tetrasporangia sunken and upper portion protruding out.
Uncommon: growing in deeper waters.
Distribution: Kanyakumari, Kovalam and Van island
A. Habit of the plant
B. Thallus with dentate margin and reproductive bodies
C. Thallus with midrib.
D. Transverse section with sunken tetrasporangia
E. Longitudinal section.

Dictyota Lamouroux

Ribbon like branched thalli; very similar in its appearance to the genus Dictyopteris from which it can be distinguished by the absence of a midrib and the presence of a single conspicuous apical cell at the tip of each strip; whenever two such cells are present, a new bifurcation is taking place. Structurally, the thallus is composed of a cortical layer of small cells with chloroplasts on each side of the blade and one, rarely more, layer of larger medullary cells with few or no chloroplasts. The thallus is attached by rhizoids produced extensively at the basal portion of the main strip. Groups of gametangia and sporangia are found on the thallus surface as well as filamentous, uniseriate hairs.

Key for species identification

1. Margin of thallus entire …..2
1. Margin of thallus dentate, ciliolate 7
2. Plant upto 5 cm in height and 0.5-1.5mm in breadth …..3
2 Plant more than 5 cm height …. 4
3 *Plant* stupose ……*D.linearis*
3. Plant not stupose, not creeping …….*D. ceylanica*
4. Plant 5-10 cm in height, lower branch wide upper branches tapering abruptly ……..*D. divaricata*
4 Plant upto 12 cm in height upper branches tapering gradually ……*D. dichotoma*
Ultimate dichotomy branch equal ……6
5 One part of ultimate dichotomy simple and another divided …… *D.cervicornis*
5. Segments uniformly broad ……*D. pinnatifida*
6 Segments uniform, ended with or without spine …..8
6. Plants 12-15 cm tall, margin serrate with minute teeth Apex broadly rounded …… *D.maxima*
7 Plants 15-20 cm tall, margin dentate which are usually projecting, apex acute ……*D.ciliolata*
7. End of the branches with spine ……*D. spinulosa*
8 End of the branch without spine ……*D. crispate*

Dictyota Lamouroux

Dictyota c
Descriptio
in dense
because
intertwine
brown in c
height; bra
dichotomo
one comp
divided
undivided;
thallus h
layer of t
chromatop
of larger
Reproduct
scattered
the branch
Uncommo
rocks in
deep water
Distributi
house,
Nochiuran
Kanchiran
light hou
Bharathina
Sadaimuni
Shingle isl
A. Habit
B. Dichoto
C. Thallus
bodies
D. Transve
E. Longitu

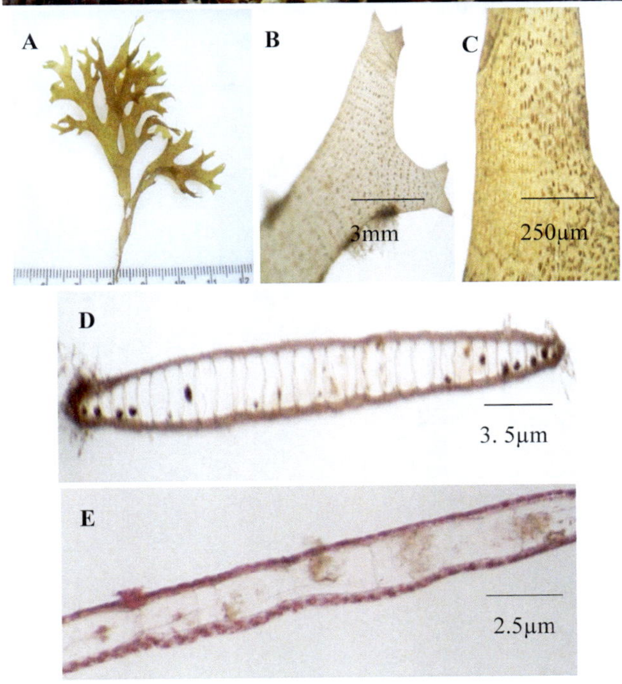

E. LS

Dictyota Lamouroux

Dictyota cervicornis **Kützing**
Description: Thallus growing in dense tufts, entangled because of twisted and intertwined branches, light brown in colour; 10-25 cm in height; branches more or less dichotomous, rather upwards; one component of branches divided while other undivided; margin entire, thallus having superficial layer of thin walled, richly chromatophores; middle layer of larger cells rectangular. Reproductive bodies scattered on both surface of the branches.
Uncommon: Attached to rocks in lower intertidal to deep water
Distribution: Pamban light house, Thonithurai, Nochiurani, Thalathoppu, Kanchirankudi, Keelakarai light house, Keelakarai, Bharathinagar, Ervadi, Sadaimuniyan valasai, Shingle island.
A. Habit
B. Dichotomous branching
C. Thallus with reproductive bodies
D. Transverse section
E. Longitudinal section

Dictyota Lamouroux

REFERENCES: Kutzing (1859: 11, pl. 25: fig.1), G. Martens (1868: 80 – 81), Borgesen (1936: 78, fig. 7), Misra (1966b: 139 – 140, fig. 72), Misra (1967: 233), S. Dixit (1968: 16), Silva *et al.* (1996: 588).

Dictyota cey
Description
as tufts, oliv
basal portior
disc, upper
less dichoto
uniform, ab
margin
rounded. T
of superficia
cells enclo
layered me
cells thin;
large, vari
Reproductiv
the mid
branches.
Common: a
and hard su
intertidal reg
Distribution
house, Vi
Ariyankundu
Nochiurani,
Kanyakumar
Krusadai i
island,
Upputhanni
island and K
A. Habit of t
B. Transvers
C. Longitudi

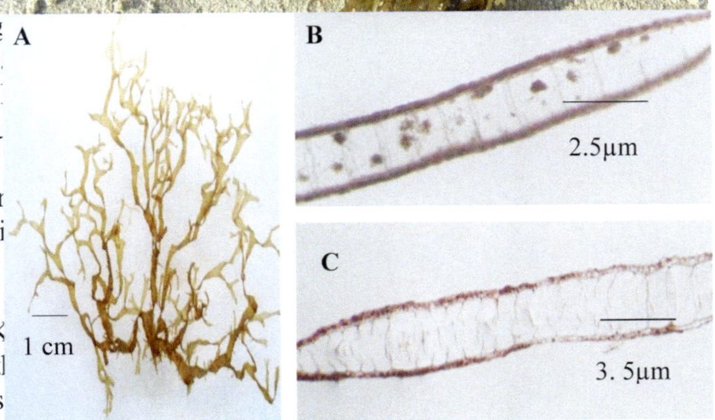

Dictyota Lamouroux

Dictyota ceylanica Kützing

Description Plants growing as tufts, olive green in colour; basal portion having a minute disc, upper portion more or less dichotomous. Branches uniform, about 1 mm broad, margin entire, apices rounded. Thallus composed of superficial layer of small cells enclosing a single layered medulla. Superficial cells thin; medullary cells large, variable in shape. Reproductive branches along the mid region of the branches.

Common: attached to rocks and hard substratum in lower intertidal region.

Distribution: Pamban light house, Villoonditheertham, Ariyankundu, Thonithurai, Nochiurani, Panjal, Kanyakumari, Kovalam, Krusadai island, Anaipar island, Salli island Upputhanni island Valai island and Koswari island.

A. Habit of the plant.

B. Transverse section

C. Longitudinal section

Dictyota Lamouroux

REFERENCES: Kutzing (1859: pl. 27: fig. I), Untawale, Dhargalkar & Agadi (1983: 25), Silva *et al.* (1996: 588).

Dictyota ciliol
Description: F
tufts, 15-20 cn
like, dark b
attached by a
portion more
dichotomous, l
mm broad, ar
branches acu
margin has c
small ascen
Reproductive
on both surfa
Tetrasporangia
groups.
Common: gr
lower mid litto
Distribution: P
Villooditheert
Ariyankundu,
Thonithurai,
Sullimunthal,
Keelakarai,
Ervadi, Sa
Pullivasal islar
Hare island, U

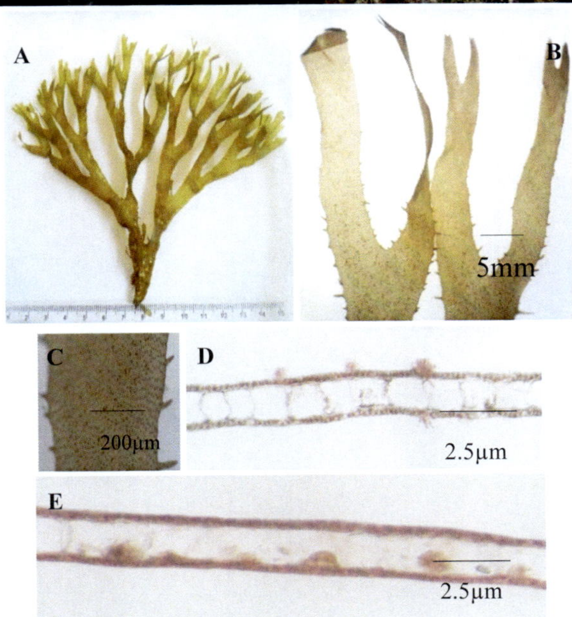

A. Habit of the
B. Dichotomou
C. Thallus sho
margin
D. Transverse
tetrasporangia
E. Longitudina

Dictyota Lamouroux

Dictyota ciliolata **Kützing**

Description: Fronds growing in tufts, 15-20 cm in height, ribbon like, dark brown in colour, attached by a small disc, upper portion more or less regularly dichotomous, branches flat, 7-12 mm broad, angles between the branches acute or rounded, margin has dentate possessing small ascending projections. Reproductive bodies distributed on both surfaces of the thalli. Tetrasporangia in scattered groups.

Common: grown attached in lower mid littoral region

Distribution: Pamban light house, Villoonditheertham, Ariyankundu, Pamban, Thonithurai, Pudumadam, Sullimunthal, Kanchirankudi, Keelakarai, Bharathinagar, Ervadi, Sadaimuniyanvalasai, Pullivasal island, Manali island, Hare island, Upputhanni island.

A. Habit of the plant
B. Dichotomous branching thallus
C. Thallus showing projections in margin
D.Transverse section with tetrasporangia
E. Longitudinal section

Dictyota Lamouroux

Dictyota ***c.***
Lamouroux
Description:
bushy, upto 2
brown in c
mostly altern
wide. Medu
apex pointe
medullary c
rectangular,
distinct tr
Surface cells
Hold fast
Sporangia s
scattered.
Uncommon:
and muddy
moderately
areas.
Distribution:
house,
Kanjirankudi,
Pullivasal isl
Anaipar islan
A. Habit
B. Arrangem
C. Branch wi
D. Transvers
E. Longitudir

Dictyota Lamouroux

***Dictyota crispata* J. V. Lamouroux**

Description: Thallus large, bushy, upto 20cm high, yellow brown in colour, branching mostly alternate, blades 3 mm wide. Medullary cells wide, apex pointed when mature; medullary cells one layered, rectangular, arranged in distinct transverse rows. Surface cells in regular row. Hold fast inconspicuous. Sporangia spherical, densely scattered.

Uncommon: attached on rocks and muddy substratum in moderately wave exposed areas.

Distribution: Pamban light house, Thonithurai, Kanjirankudi, Shingle island, Pullivasal island, Valai island, Anaipar island and Salli island.

A. Habit

B. Arrangement of branches

C. Branch with pointed tip.

D. Transverse section.

E. Longitudinal section.

Dictyotales-Dictyotaceae

Dictyota Lamouroux

REFERENCES: Kaliaperumal *et al*. (1992: 12), ~~V. Kumar~~ (1993a: 139, 140), V. Kumar (1993b: 47), Rama Rao *et al*. (1993: 106), Silva *et al*. (1996: 589), Kaliaperumal *et al*. (1996: 73 – 77).

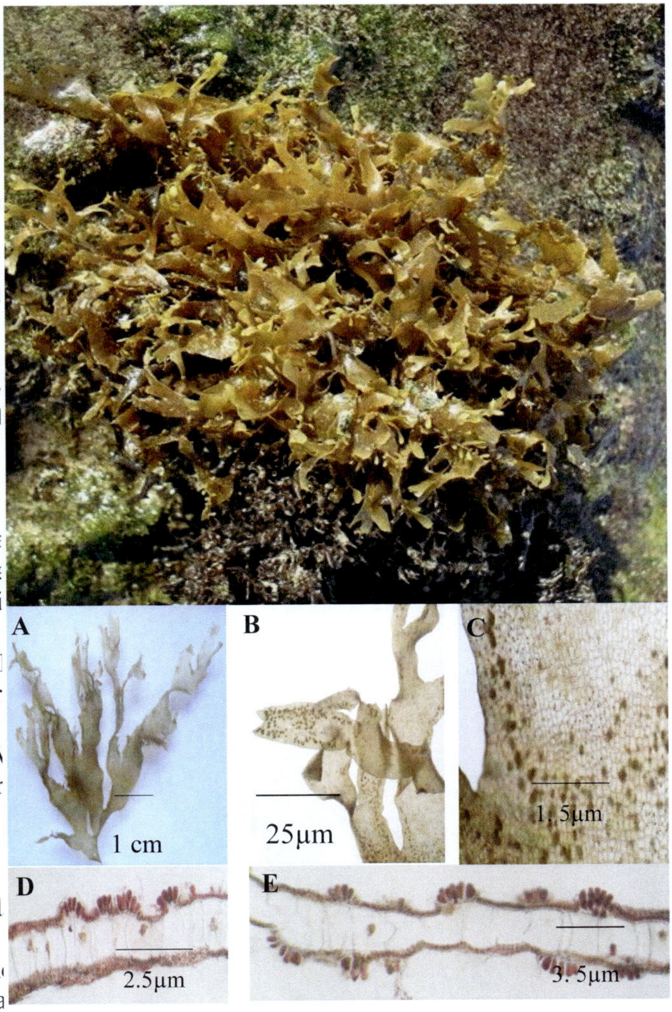

Dictyota dich
Lamouroux
Description: l
yellowish brov
cm in height;
small cuneat
portion is
branches flat,
slightly broad
entire, apices
composed of s
small thin w
many chromat
of large
rectangular
Reproductive
on separate tha
Common: Att
Rocks in midli
Distribution:
Pamban light
Villoonditheer
Rameswaram,
Thonithurai,
Seeniappadhar
Sullimunthal
Keelakarai
Keelakarai,
Ervadi, Sad
Valinokkam,
Leepuram an
Gulf of Manna
A. Habit of the
B. Thallus wit
C. Reproducti
D. Transvers
reproductive b

Dictyota Lamouroux

***Dictyota dichotoma* (Hudson) Lamouroux**

Description: Plant ribbon like, yellowish brown in colour; 6-12 cm in height; basal portion has small cuneate disc, upper portion is dichotomous, branches flat, uniform length; slightly broader at base; margin entire, apices rounded. Thallus composed of superficial layer of small thin walled cells with many chromatophores, medulla of large thick walled, rectangular in shape. Reproductive bodies occurring on separate thalli.

Common: Attached to Rocks in midlittoral region.

***Distribution*:** Munaikadu, Pamban light house, Nalupanai, Villoonditheertham, Rameswaram, Pamban, Thonithurai, Velankanni nagar, Seeniappadharga, Pudumadam, Sullimunthal Kanchirankudi, Keelakarai light house, Keelakarai, Bharathi nagar, Ervadi, Sadaimuniyanvalasai, Valinokkam, Kuduthalai, Leepuram and all islands of Gulf of Mannar.

A. Habit of the plant
B. Thallus with branches
C. Reproductive bodies
D. Transverse section with reproductive bodies
E. Longitudinal section

Dictyota Lamouroux

REFERENCES: V. Krishnamurthy & H. Joshi (1970: 11), Untawale, NDhargalkar & Agadi (1983: 21), M. Dave & Chauhan (1985: table 1), Bheemasankar Rao *et al*. (1991: 1971), Silva *et al*. (1996: 591).

Dictyota divari
Description: F
height, formin
light brown in
irregularly dic
portion of the l
mm in breadth
narrow. Angl
branch wide;
branch entire.
of superficial
quadrate cells.
large, rectang
bodies present
the thallus.
Uncommon: A
mid littoral reg
Distribution:
Ariyankundu,
Pudumadam,
Kanchirangudi
A. Habit of the
B. Thallus sho
surface
C. Surface of tl
D. Transverse s
E. Longitudina
rectangular me

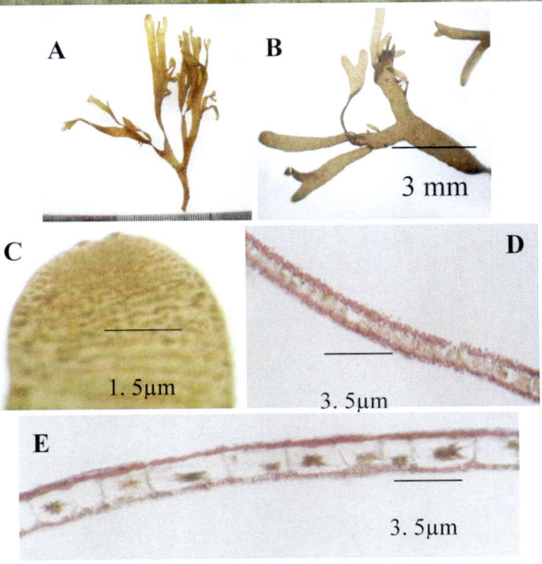

Dictyota Lamouroux

Dictyota divaricata **Lamouroux**

Description**:** Plants 5-10 cm in height, forming entangled tuffs, light brown in colour; branches irregularly dichotomous; lower portion of the branch broad 3-3.5 mm in breadth; upper portion is narrow. Angles between the branch wide; margin of the branch entire. Thallus composed of superficial layer of small, quadrate cells. Cells of medulla large, rectangular Reproductive bodies present on both surface of the thallus.

Uncommon: Attached to rocks in mid littoral region.

Distribution: Pamban, Ariyankundu, Thonithurai, Pudumadam, Sullimunthal and Kanchirangudi

A. Habit of the plant

B. Thallus showing broad lower surface

C. Surface of the thallus

D. Transverse section

E. Longitudinal section showing rectangular medullary cells.

Dictyota Lamouroux

Dictyota
Agardh) Gr
***Description*:**
membranous
dichotomous
sometimes
sometimes
linear with
ultimate seg
divaricate, a
rectangular,
Reproductiv
along the m
thallus.
Rare: Grov
sublittoral re
Distribution
house, Thon
island and K
A. Habit
B. Narrow
irregular dicl
C. Transvers
D. Longitudi

Dictyota Lamouroux

Dictyota linearis **(C. Agardh) Greville**

***Description*:** Fronds membranous, irregularly dichotomous; segments sometimes narrow, sometimes little broader, linear with smooth margin, ultimate segments sub linear, divaricate, apex obtuse, cells rectangular, elongated. Reproductive cells placed along the median part of the thallus.

Rare: Grown on rocks in sublittoral region

Distribution: Pamban light house, Thonithurai, Anaipar island and Koswari island.

A. Habit

B. Narrow branches with irregular dichotomy

C. Transverse section

D. Longitudinal section

Dictyota Lamouroux

REFERENCES: Untawale, Reddy & Ambiye (1989: table 1), Kaliaperumal *et al.* (1992: 12), Maya & Nair (1992: 26), Subba Rao *et al.* (1992: 181), Kaliaperumal *et al.* (1995: 183 – 193), Silva *et al.* (1996: 595).

Dictyota maxin
Description Th
height, olive gr
colour with sn
upper branc
sometimes
branched; br
about 1.5 cm
more or less
provided with
thallus comp
superficial lay
middle laye
superficial lay
or less quadrat
cells are large
Tetrasporangia
or group of 2-3.
Rare: attachec
intertidal region
Distribution:
house, Thonitl
island, Hare is
island

A. Habit of the
B. Thalus show
bodies
C. Transverse s
D Longitudinal

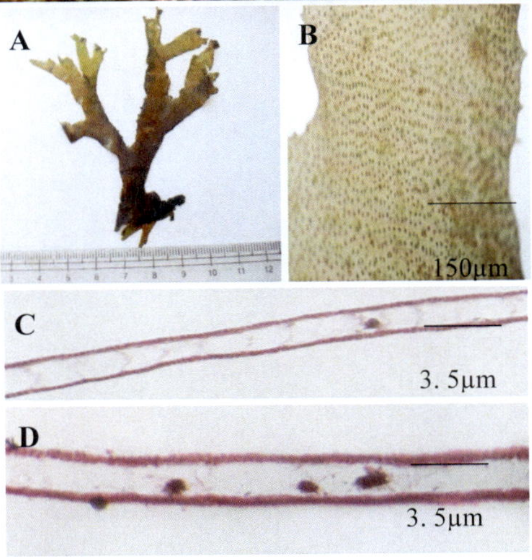

Dictyota Lamouroux

Dictyota maxima **Zanardini**

Description Thalli 12-15 cm in height, olive green to brown in colour with small basal disc; upper branch irregularly, sometimes dichotomously branched; branches broad, about 1.5 cm broad, margin more or less sinuate and provided with minute teeth; thallus composed of two superficial layers enclosing a middle layer of cells; superficial layers small, more or less quadrate; middle layer cells are large, thick walled. Tetrasporangia spherical, single or group of 2-3.

Rare: attached on rocks in intertidal region.

Distribution: Pamban light house, Thonithurai, Shingle island, Hare island and Valai island

A. Habit of the plant

B. Thalus showing reproductive bodies

C. Transverse section of thallus

D. Longitudinal section.

Dictyota Lamouroux

REFERENCES: S. Dixit (1966: 8), Misra & Nair (1966: 140), Misra (1967: 233), V. Krishnamurthy & H. Joshi (1970: 11), Untawale, Dhargalkar & Agadi (1983: 21), Silva *et al.* (1996: 596).

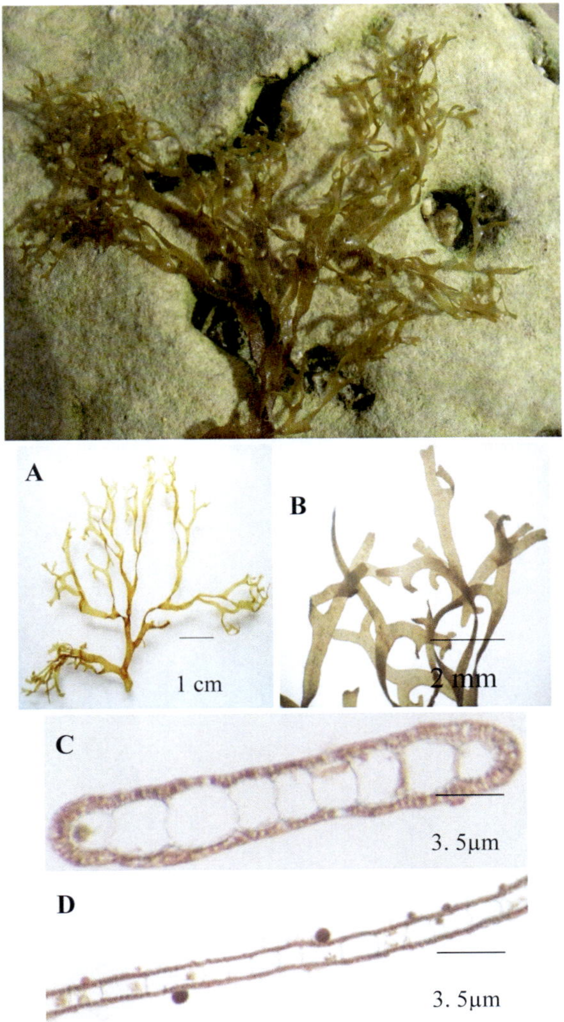

Dictyota pin
Description:
18 cm in
bushy, light
with basal p
cuneate
irregularly
branches fla
mm in
separated,
apices roun
bodies occu
surface of th
Uncommon
pools in t
region.
Distribution
house,
Nochiurani,
Krusadai
island, Hare
island, Val
island, Ana
island,
Koswari isla
A. Habit of t
B. Thallus s
dichotomy
C. Transve
elongated m
D. Longitud
reproductive
sides.

Dictyota Lamouroux

Dictyota pinnatifida **Kützing**

***Description*:** Plants large, 10-18 cm in height, sparsely bushy, light brown in colour with basal portion forming a cuneate disc. Branches irregularly dichotomous; branches flat, uniformly 2-3 mm in width, distantly separated, margin entire, apices round. Reproductive bodies occur all over the surface of the thallus.

Uncommon: grown on rock pools in the mid littoral region.

Distribution: Pamban light house, Thonithurai, Nochiurani, Valinokkam, Krusadai island, Manali island, Hare island, Mully island, Valai island, Appa island, Anaipar island, Salli island, Van island and Koswari island.

A. Habit of the plant

B. Thallus showing irregular dichotomy

C. Transverse section with elongated medulla

D. Longitudinal section with reproductive bodies on both sides.

Dictyota Lamouroux

Dictyota spin
Description:
tufts, bushy ι
height. Thallı
colour. Frond
wide. Branch
alternately to
divided, havi
their surfaces
and have acu
Uncommon:
gravels in lov
subtidal zone
moderate wa
Distribution:
Arasaneripatt
Ponnagaram,
house,
Thonithurai
Ervadi,
Vattakottai aı
A. Habit
B. Dichotom
with spines a
C. Surface of
reproductive
D. Transvers
thallus.

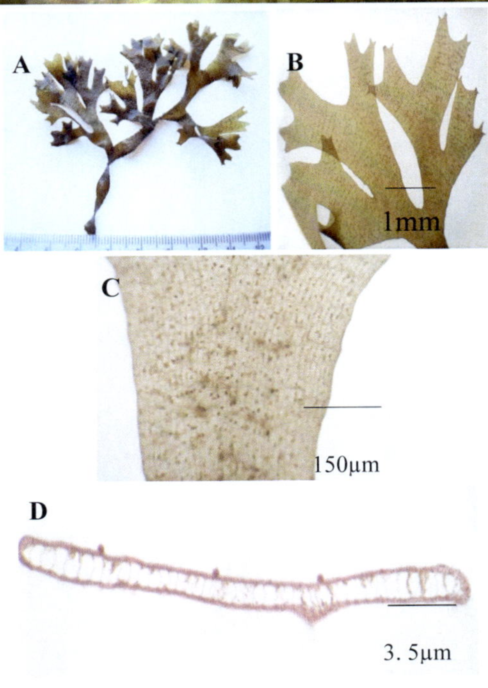

Dictyota Lamouroux

Dictyota spinulosa **harvey**

Description: Plants erect, tufts, bushy upto 15 cm in height. Thallus dark brown in colour. Fronds broad, 1.5 cm wide. Branches are strap like, alternately to dichotomously divided, having spines on their surfaces and margins and have aculeate tips.

Uncommon: on rocks and gravels in lower intertidal to subtidal zone exposed to moderate wave action.

Distribution: Arasaneripattinam, Ponnagaram, Pamban light house, Nalupanai, Thonithurai Bharathinagar, Ervadi, Chinnamuttam, Vattakottai and Salli island.

A. Habit
B. Dichotomous branching with spines at tip
C. Surface of thallus with reproductive bodies.
D. Transverse section of thallus.

***Hormophysa* Kützing**

Thallus, tough, bearing simple laterals; thallus jointed, segments triquetrous; some includes vesicles

Key to the species-

Thallus joint, segments triquetrous, some including vesicle
...*H. cuneiformis*

Hormophysa Kützing

REFERENCES: Forsskal (1775: 191), J. Agardh (1848: 216), G. Murray (1887: 34), Zanardini (1858: 243), Bornet (1888: 18), Lawson (1980: 24), J. Gmelin (1792: 1389).

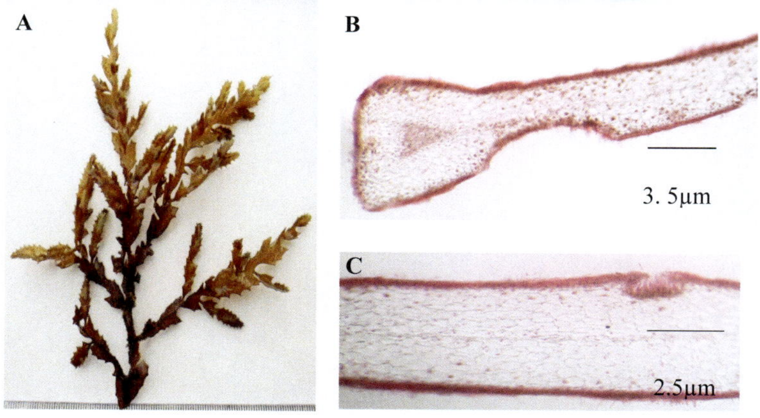

Hormophysa Kützing

Hormophysa cuneiformis **(J. Gmelin) P. Silva**

Description: Plant light to dark brown in colour, 20 -30 cm in height, even more, sparse to densely branched; branches articulated, triangular, articulation narrow at the base, 3-5 mm broad, margin dentate, vesicles ellipsoidal, embedded in the middle of swollen wings.

Uncommon: grown on rocks and coral substratum in subtidal to 10 meter deep.

Distribution: Manapad, Kuduthalai, Idinthakarai, Panjal, Vattakottai, Krusadai island, Hare island and Koswari island.

A. Habit of the plant

B. Transverse section

C. Longitudinal section

***Hydroclathrus* Bory de Saint-Vincent**

The light brown thallus formed by irregular hollow lobes is typically perforated with holes of variable diameter, giving this alga the appearance of a three dimensional net. The thallus consists of 2-3 layers of cells, the inner ones larger and almost devoid of chloroplasts and the outer ones smaller with chloroplast; hairs occur in thallus depressions. Plurilocular gametangia are formed in large spots on the thallus surface.

Key for species-

Thallus in net like ....*H.clathratus*

Hydroclathrus Bory de Saint-Vincent

REFERENCES: Kamat *et al.* (1991: 54), Kalimuthu, Kaliaperumal & Ramalingam (1992: 122), Kamat *et al.* (1992: table 1), Jagtap (1993: table I), Jayasankar & Ramalingam (1993: 42), Silva *et al.* (1996: 630).

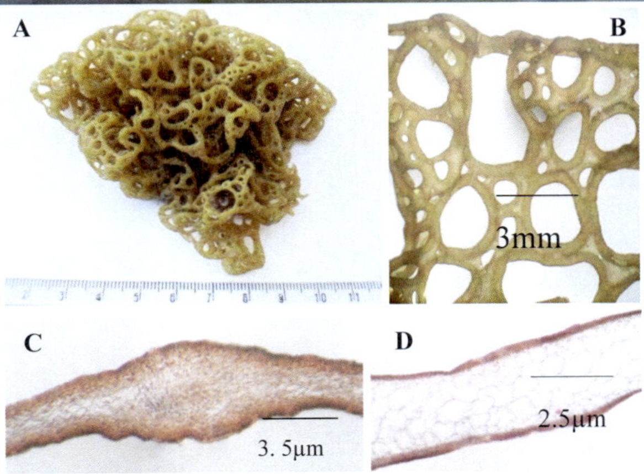

Hydroclathrus Bory de Saint-Vincent

***Hydroclathrus clathratus* (C. Agardh) Howe**

***Description*:** Thallus spherical to irregular in shape, grows into a net like appearance. Plants yellowish to dark brown in colour, grown upto 12 cm long and 4-8 cm broad. Fronds attached by a broad base, sometimes several bases. Thallus composed of cells differentiated into two regions, a superficial layer of small cells and more than one inner layer of large rounded cells.

Common: attached to rocks and coral stones in infra littoral fringe, lower littoral and wave exposed zones

Distribution: Munaikadu, Pamban light house, Nalupanai, Villoonditheertham, Ariyankundu, Kanchirankudi, Pullivasal island, Krusadai island, Hare island, Mulli island Valai island, Valimunai island, Salli island, Van island and Koswari island.

A. Habit of the plant

B. Surface view of perforated thallus

C. Transverse section

D. Longitudinal section showing large rounded cells.

***Iyengaria* Boergesen**

When young, thallus is vesicular and becoming tuberculate when becoming old, hemispherical, thallus surface has branched tuberculate projections; superficial cells small, packed together, inner cells larger, elongated, forming 5-7 layers; plurilocular sporangia cylindrical, mostly uniseriate, occurring in sori.

Key for species identification

Plants sac like structure with even surface …..*I. stellata*

Iyengaria **Børgesen**

REFERENCES: Dave & Chauhan (1985: 46, table 1), Chennubhotla, Kaliaperumal, Kalimuthu (1987: 104 – 105), Tewari & H. Joshi (1988: 394), Mairh, Ramavat *et al.* (1989: 430, table I), Kamat *et al.* (1991: 55), Silva *et al.* (1996: 631).

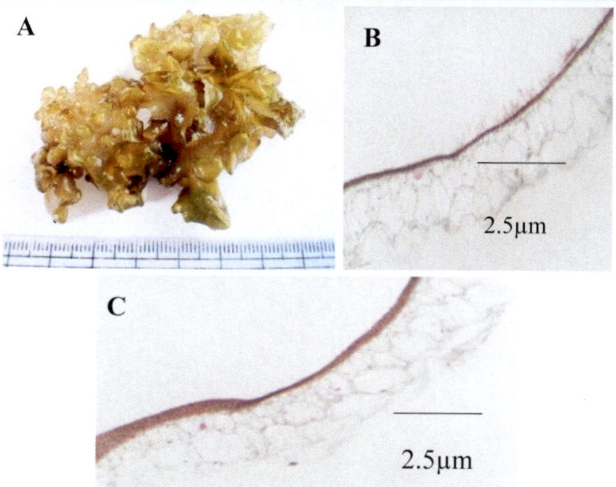

***Iyengaria* Børgesen**

***Iyengaria stellata* (Børgesen) Børgesen**

Description: Plant yellowish brown in colour, when young, it has smooth surfaces, when mature, it became large, hollow, superficial protuberances; attached by rhizoids arising from lower epidermal cells. Fronds composed of superficial palisade like cells enclosing several layers of larger cells.

Rare: attached to rocks in mid intertidal region

Distribution: Anaipar island, Upputhanni island and Van island.

A. Habit of the plant

B. Transverse section showing larger cells in inner region

C. Longitudinal section.

***Lobophora* J. Agardh**

Brown, fan shaped blade with a firm texture; usually attached to the substratum by a bundle of rhizoids produced at the basal portion, but in some populations the thallus may be more prostrate and attached to the substratum by rhizoids produced from its ventral surface. Longitudinal growth from a margin of apical cells. Structurally, the thallus is composed of a central layer of larger cells, 3-10 layers of smaller medullary cells and a layer of cortical cells full of chloroplasts on each side of the blade. Reproductive cells in small sori on the upper face of the thallus are interspersed with narrow sterile cells.

Key for the species-

Thallus with 9-11 layers of cells .....*L.variegata*

Lobophora J. Agardh

REFERENCES: Chacko, Mahadevan & R. Ganesan (1955: 11), S. Dixit (1966: 8), V. Krishnamurthy & H. Joshi (1970: 11), Untawale, Dhargalkar & Agadi (1983: 22), Chennubhotla, Kaliaperumal, Kalimuthu (1987: 102 – 103), Subba Rao *et al*. (1992: 182).

Lobophor
(Lamouro
Oliveria
Descriptio
brown in
circular to
young, 3-:
thalli erect
15 cm in l
in breadth;
by rhizoid
basal par
Thallus
superficial
epidermis,
intermedia
wall isod
medullary
large thin
cells. R
distributed
the lobe.
Common:
infra littor
Distributi
Kanchiran
light hou
Bharathi
Sadaimuni
Idinthakar
Vattakotta
Kovalam a
Gulf of M
A&B. Hab
C. Thallus
zone
D. Transv
E .Longit

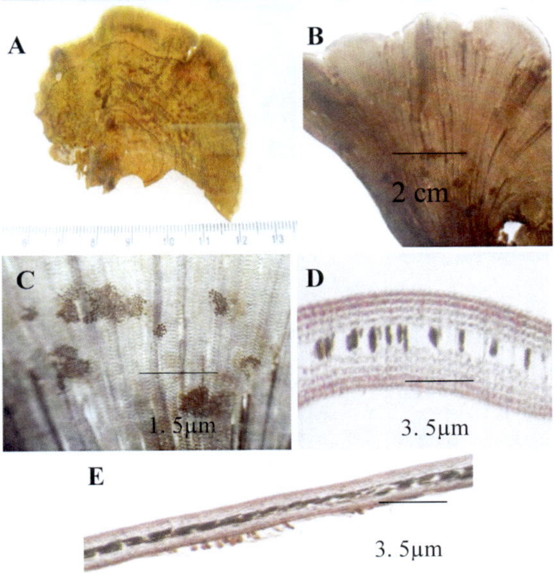

Lobophora J. Agardh

***Lobophora variegata* (Lamouroux) Womersley ex Oliveria**

Description Plants dark brown in colour, prostrate, circular to fan shaped when young, 3-5 cm across; older thalli erect and fan shaped, 5-15 cm in height and 4-13 cm in breadth; prostrate, attached by rhizoids arising from the basal part of the fronds. Thallus composed of a superficial layer of thin wall epidermis, three to four intermediary layers of thick wall isodiametric cells, a medullary layer consisting of large thin wall rectangular cells. Reproductive layer distributed on both sides of the lobe.

Common: growing in the infra littoral region.

Distribution: Sullimunthal, Kanchirankudi, Keelakarai light house, Keelakarai, Bharathi nagar, Ervadi, Sadaimuniyanvalasai, Idinthakarai, Leepuram, Vattakottai, Kanyakumari, Kovalam and all 20 islands of Gulf of Mannar

A & B. Habit of the plant

C. Thallus with reproductive zone

D. Transverse section

E. Longitudinal section with oogonial sori.

Padina Adanson

Thallus brown, sometimes partially white due to a thin layer of lime deposition; the fan shaped blades occasionally split into segments, arranged in bundles; attached by a mat of rhizoids produced from the narrow, stipe like basal portion; the blades appear clearly zonated due to regular rows of deciduous, uniseriate hairs; growth from a marginal row of apical cells, typically in-rolled. Structurally the blades consist of two or more layers of cells. Reproductive organs appear as darker dots in successive, parallel layers associated or intercalated with the rows of hairs.

Key for the species

1. Frond with rows of hairs on one or both surfaces, calcified 2
1. Frond with rows of hairs but surfaces not calcified ....
2. Calcification heavy on lower surface, little on Upper surface 3
2. Calcification slightly on both surface ......*P. boryana*
3. Thallus with only two layers of cells ..... *P. distromatica*
3. Thallus with two to four layers in older portions .....*P. pavonica*
4. Thallus 3 layered in mature portion ..... *P.boergesenii*
4. Thallus 4 – 6 layered in mature portion .....*P. tetrastromatica*

Padina Adanson

REFERENCES: Kalimuthu, Kaliaperumal & Ramalingam (1992: 122), S. Balakrishnan & K. Krishnamurthy (1993: 227 – 233, figs. 1 - 14), Jayasankar & Ramalingam (1993: 42), Silva *et al.* (1996: 601), S. Balakrishnan & K. Krishnamurthy (1998: 115 – 117).

A

B

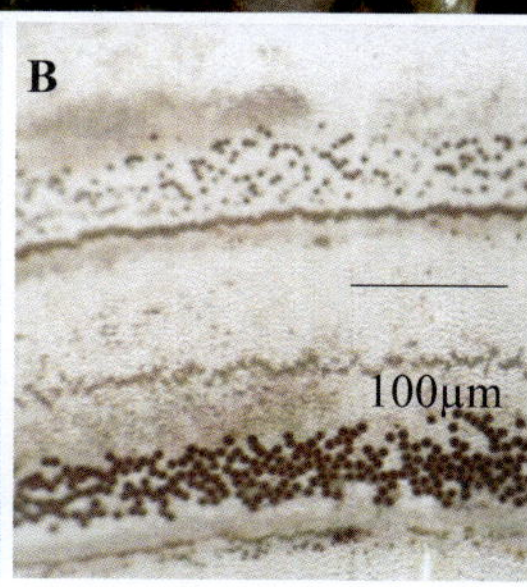

C

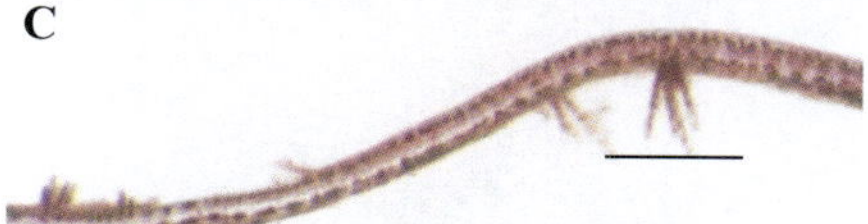

3. 5µm

Padina Adanson

Padina boergesenii **Allender & Kraft**

***Description*:** Plant light to dark brown in colour; 6-15 cm in height, 5-12 cm in breadth; basal portion has small rhizoid; thallus divided into broad lobes; three cells thickness. Thallus composed of upper and lower layer; upper composed of small rectangular, thin layer, lower consisting of large, rectangular, thick walled cells. Hairs developing all over the thallus as concentric rings. Reproductive parts occurring in the central part of the thallus between two hairy zones.

Common: growing on rocks of mid-littoral region

Distribution: Kovalam, Rameswaram, Thonithurai, Velankanni nagar, Thalathoppu, Sallimunthal, Kanyakumari, Kovalam, Shingle island, Krusadai island, Upputhanni island and Van island.

A. Habit of the plant

B. Thallus with reproductive bodies between hairy zones

C. Transverse section with hairs.

Padina Adanson

REFERENCES: W. R. Taylor (1966: 355 – 356, fig. 2), V. Krishnamurthy & H. Joshi (1970: 11), Untawale, Dhargalkar & Agadi (1983: 21), Chennubhotla, Kaliaperumal, Kalimuthu (1987: 102 – 103), Silva *et al*. (1996: 602).

Padina Adanson

Padina boryana **Thivy**

Description: Plant light brown to dark brown in colour; 8-10 cm in height and 2-4 cm in breadth. Basal rhizoidal portion continuous as narrow stipe expanding up into a fan shaped blade. Margin of the blade entire; apex in-rolled; concentric row of hairs and tetrasporangia are present on the surface of the thallus with every row of hairs. Reproductive bodies in rows alternate with rows of hairs. Tetrasporangia occur just above each row of hairs.

Uncommon: growing on rocks of the mid littoral region.

Distribution: Ariyankundu, Thonithurai, Kanchirankudi, Hare island, Mully island, Valimunai island, Salli island and Upputhanni island.

A. Habit of the plant

B. Surface view of thallus with reproductive bodies

C. Longitudinal section with oogonial sori.

Padina Adanson

REFERENCES: Hauck (1887 [1886 – 1889]: 43), De Toni (1895: 244), A. Gepp & E. Gepp (1908: 185), A. Gepp & E. Gepp (1909: 395), E. Lewis (1962: 90, tables I - III), Rengasamy (1991: 185 – 188, figs. 1 - 8), Silva *et al*. (1996: 603).

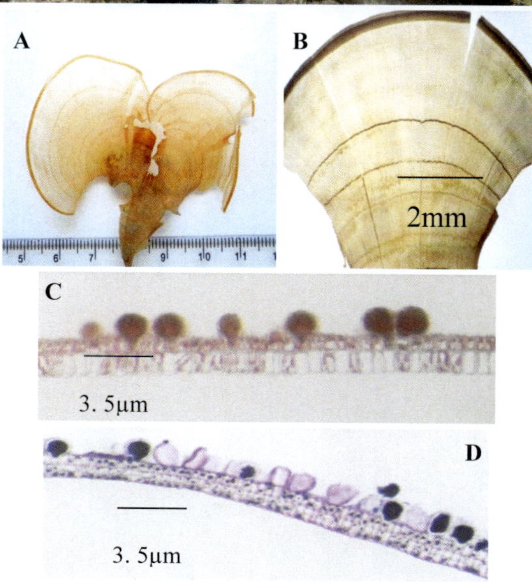

Padina Adanson

Padina distromatica **Hauck**

***Description*:** Thallus arranged in rosettes arising from a rhizoidal cushion, growing to a height of 4-5 cm, stipe 1-1.5 cm; blade split down to the middle, split portion suborbicular, thin, overlapping on the margins, margin entire, heavily calcified on the lower surface. Apex involute. Blades distromatic throughout. Tetrasporangia present only on the upper region of the thallus forming a fertile zone alternate with piliferous zone. The tetrasporangia are not symmetrically disposed.

Common: on rocks and other hard substrate in the mid littoral region.

Distribution: Pamban light house, Nalupanai, Villoonditheertham, Ariyankundu, Thonithurai Pudumadam, Keelakarai, Bharathi nagar, Ervadi, Manapad, Hare island Salli island, Van island and Koswari island

A. Habit of the plant

B. Thallus showing reproductive bodies

C. Transverse section showing distromatic

D. Longitudinal section with tetrasporangial sori

Padina Adanson

REFERENCES: Untawale, Reddy & Ambiye (1989: table 1), Kaliaperumal Kalimuthu, & Ramalingam (1990: table 2), Kannupandi *et al.* (1992: 12), Subba Rao *et al.* (1992: 181), V. Kumar (1993a: 139 – 140), Silva *et al.* (1996: 605).

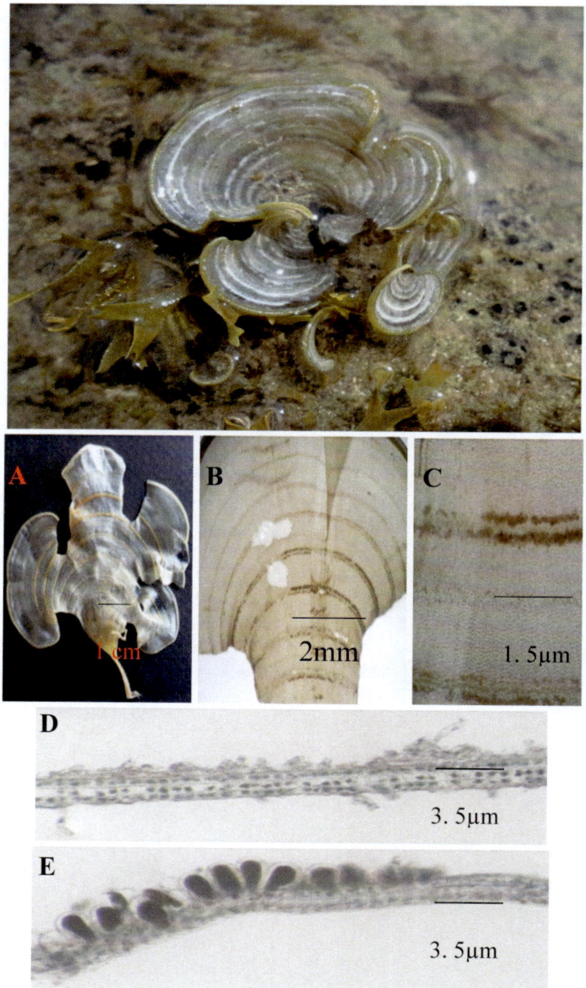

Padina Adanson

***Padina pavonica* (Linnaeus) Thivy**

***Description*:** Thallus brown in colour; 5-8 cm broad, with short stipe; blade simple or dichotomously divided, down to the base into flabellate segments, 2.5- 4 cm wide, strong lime on the lower surface and less on the upper surface, sometime both surfaces encrusted; piliferous layer on both surfaces alternate; Tetraporangial sori usually on the upper surface in older parts, less common on lower surface.

Uncommon: attached to rocks in mid littoral region.

Distribution: Pullivasal island, Hare island, Appa island, Anaipar island, Salli island, Van island and Koswari island.

A. Habit of the plant

B & C. Thallus with tetrasporangial sori

D: Transverse section

E. Longitudinal section showing tetrasporangial sori.

Padina Adanson

REFERENCES: Padmini Sreenivasa Rao (1991: 37, tables 1, 2), Rajendran *et al*. (1991: table2), Maya & Nair (1992: 26), Ramaiah & Chandramohan (1992: 74, 75), Sobha, Surendran & Nair (1992: 77), Silva *et al*. (1996: 607).

Padina Adanson

Padina tetrastromatica **Hauck**

Description**:** Plant reddish brown in colour, 6-15 cm high; 10-25 cm broad, attached by a small disc; thallus flat, irregularly dichotomous to alternately branched; margin entire, angle between branches wide. Tetrasporangial sori occurring on both side of the piliferous zones.

Common: growing on the rock pools of mid littoral region.

Distribution: Munaikadu, Thonithurai, Panjal, Kanyakumari, Kovalam, Pullivasal island and Upputhanni island.

A. Habit of the plant

B & C. Thallus showing tetrasporangial sori on both side of piliferous zone.

D. Transverse section

E. Longitudinal section showing sporangia (S)

***Rosenvingea* Borgesen**

Thallus macroscopic, light brown, formed by cylindrical to flattened hollow branches; branching irregular; branches consist of a few layers of large cells facing the inner cavity and smaller cells with chloroplasts at the external surface. Plurilocular gametangia are produced in dense groups at the thallus surface.

Key to the species

Thallus not constricted, margin not incurved and surface irregularly rugose .... *R.sanctae-crucis*

Rosenvingea Børgesen

REFERENCES: Borgesen (1914: 22 – 25, figs. 14 - 17), Umamaheswara Rao (1969: 42), Umamaheswara Rao (1974c:60, figs, 3 – 5), S. Dixit (1980: 67), Untawale, Dhargalkar & Agadi (1983: 20), Silva *et al*. (1996: 634).

Rosenvi
Børgese
Descript
cylindric
compres
in heigh
rugose,
basal di
arranged
attenuate
apices.
small g
the enti
clavate, s
entire surf
Rare: att
the mid lit
Distributi
A. Habit o
B. Thallu
branches.
C. Transv
D. Longit

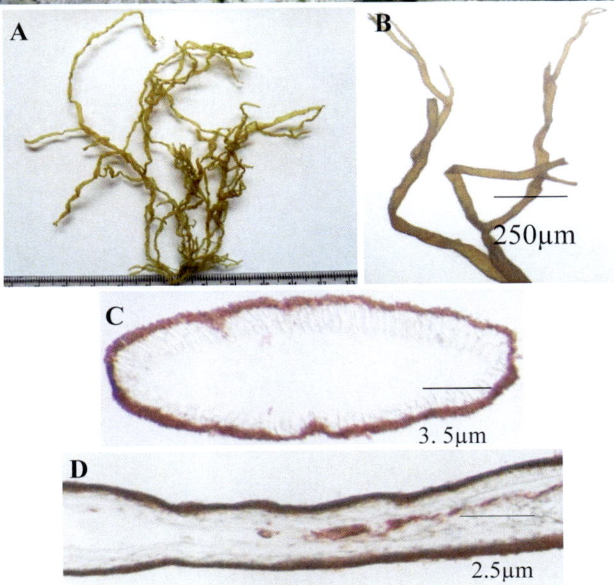

Rosenvingea Børgesen

Rosenvingea sanctae-crucis **Børgesen**

Description: Plants cylindrical, rarely slightly compressed, maximum 20 cm in height, surface irregularly rugose, attached by means of basal disc. Branches sparsely arranged; frequently attenuated at bases and apices. Hairs single or in small group, scattered over the entire surface. Sporangia clavate, scattered over the entire surface of the plant.

Rare: attached on rocks in the mid littoral region.

Distribution: Nalupanai

A. Habit of the plant

B. Thallus with cylindrical branches.

C. Transverse section

D. Longitudinal section

Sargassum C. Agardh

Sargassum C. Agardh

Brownish thallus, usually attached by a discoid, rarely stoloniferous, base from which a short primary branch is produced that gives rise to a few long secondary branches, the secondary branches are covered with leaf like short branches with mid rib; structure parenchymatous with a medulla of elongated cells without chloroplasts and a cortex of a few layers of smaller chloroplast rich cells; rounded or elliptical vesicles are common on most species; hairs are produced in pits scattered on the thallus surface. Reproduction by oogonia and anthredia produced in pits on special small receptacular branchlets.

Key for the species

1 Blades narrowly linear .... *S.acinarium*
1 Blades ovate with serrate margin 2
2 Receptacular branches cylindrical 3
2 Receptacular branches triquertous 4
3 Blades ovate with obtuse apex .... *S. cinereum*
3 Blades linear or lanceolate with acute 5
4 Blades linear with well-defined midrib, irregularly serrate ...... *S.tenerrimum*
4 Blades linear with minutely serrated margin 6
5 Blades linear with receptacular branches not twisted ...... *S.carpophyllum*
6 Receptacle with serrate margin 7
6 Receptacle smooth minutely spinulose 10
7 Stem compressed or flattened 8
7 Stem circular or triangular in cross section 9
8 Branches cylindrical below and flattened above ... *S.swartzii*
8 Branches flattened throughout ....*S.wightii*
9 Branches mucranate throughout, blade lanceolate ....*S. polycystum*
10 Cryptostomata uniseriate on either side of the midrib .... *S.odontocarpum*
10 Cryptostomata densely scattered .... *S.gracile*

Sargassum C. Agardh

REFERENCES: Linnaeus (1753: 1160), Setchell (1933: 209 - 210), Setchell (1933: 208), Silva *et al.* (1996: 657).

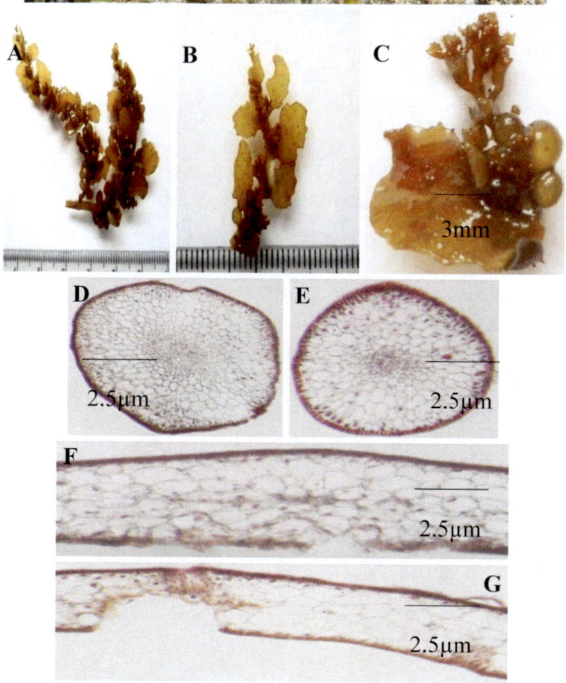

Sargassum C. Agardh

Sargassum acinarium (Linnaeus) Setchell

Description: Plants yellowish brown in colour, branches cylindrical, plant reaches upto 15 cm high. Leaves are in linear, oblong and ovate, cuneate bases, with wavy and irregular margin; upper leaves lanceolate, 1-1.5 cm long, 2-3 mm broad, dentate.Vesicles 4 mm short and spherical shape. receptacle branches occurring in cluster.

Rare: attached to rocks in lower intertidal region.

Distribution: Seeniappa dharga, Shingle island and Van island

A & B Habit of the plant

C. Receptacle & Vesicle

D. Transverse section of Stem

E. TS of *Receptacle*

F. TS of lamina

G. Longitudinal section of Leaf

Sargassum C. Agardh

REFERENCES: Misra (1967: 233), K. Srinivasan (1967: 137, fig. 33), S. Dixit (1968: 18), V. Krishnamurthy & H. Joshi (1970: 14), Untawale, Dhargalkar & Agadi (1983: 22), Silva *et al*. (1996: 662).

Sargassum C. Agardh

***Sargassum carpophyllum* J. Agardh**

***Description*:** Plants light brown in colour, plant reaches maximum of 35 cm high. Branches cylindrical or slightly flattened. Leaves long with 3- 7 cm in length and 1-2 cm in width, apices blunt or round, cuneate bases, wavy with marginal teeth along the entire margin; upper leaves lanceolate, Vesicles 4 mm long and spherical, receptacle 2-3 mm long, in cluster.

Rare: attached to rocks in lower intertidal region.

Distribution: Keelakarai

A. Habit of the plant

B. Leaf

C. Receptacle & Vesicle

D. Transverse section of Stem

E. TS of Receptacle

F. TS of lamina

Sargassum C. Agardh

REFERENCES: Kamat *et al.* (1991: 55), Kesava Rao *et al.* (1991: 283, 284, table 1), Raghukumar, Nagarkar & Raghukumar (1992: fig. 1, table 1), Ramaiah & Chandramohan (1992: 73 – 75), Sathe – Pathak *et al.* (1993: 159), Silva *et al.* (1996: 664).

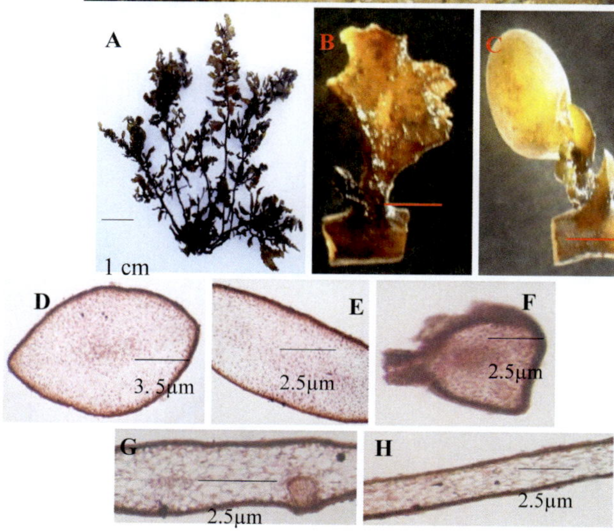

Sargassum C. Agardh

Sargassum cinereum **J. Agardh**

Description: Plants dark brown in colour, plant reaches maximum of 30 cm high. Leaves show different morphology; leaves at basal parts 1-2 cm long, 4-5 cm wide with rounded apices, cuneate bases, wavy with marginal teeth along the entire margin; upper leaves lanceolate, 1-1.5 cm long, 2-3 mm broad, dentate, partly curved. Vesicles 4 mm long and broad, receptacle single or in cluster.

Rare: attached to rocks in lower intertidal region.

Distribution: Nalupanai

A. Habit of the plant
B. Leaf & Receptacle
C. Vesicle
D. Transverse section of Stem
E. Longitudinal section of Stem
F. TS of Receptacle
G. TS of Leaf
H. LS of Leaf

Sargassum C. Agardh

REFERENCES: J. Agardh (1848: 310), Kutzing (1849: 620), Grunow (1867: 56), G. Martens (1868: 76), J. Agardh (1889: 119), K. Srinivasan (1967: 143, fig.51), Silva *et al.* (1996: 674).

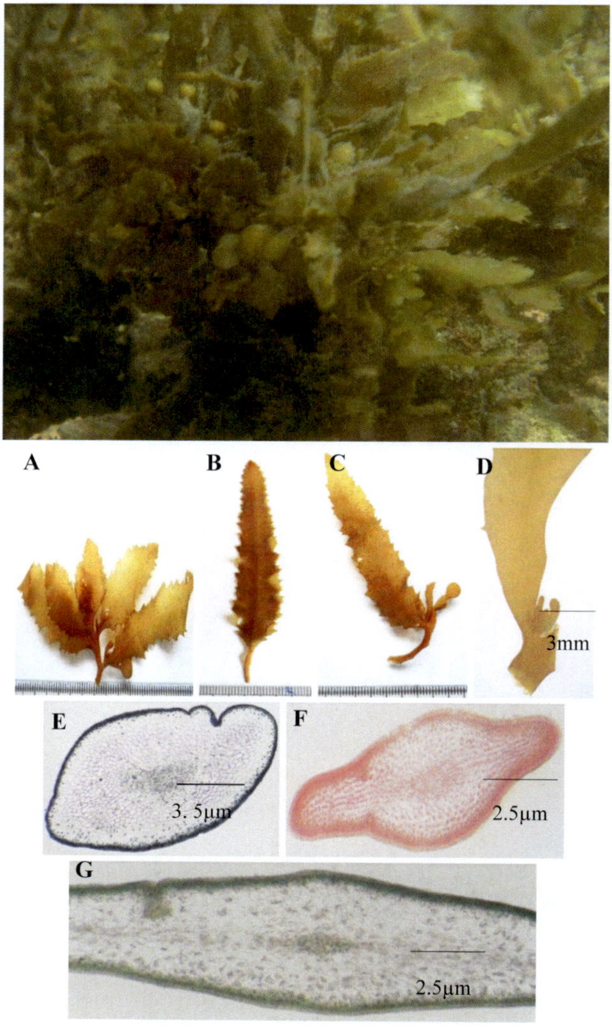

Sargassum C. Agardh

Sargassum gracile **J. Agardh**

***Description*:** Plant light yellowish brown in colour, grows upto 30 cm in height; Upper branches are cylindrical to club shaped. Leaf with serrated margins throughout, 3-6 cm long and 1-2 cm broad, broader in the middle and tapering toward apices; midrib conspicuous; vesicles with long stalk and spherical shape. Receptacles are small with spine. Crypto- stomata are dense.

Uncommon: occurring on coral stones in lower intertidal region.

Distribution: Keelakarai light house, Keelakarai, Bharathinagar, Ervadi, Valinokkam, Shingle island, Valai island and Anaipar island.

A. Habit of the plant

B. Leaf with serrated margin

C. *& D* Receptacle & Vesicle

E. Transverse section of Stem

F. TS of Receptacle

G. TS of Leaf

Sargassum C. Agardh

REFERENCES: Greville (1848: [1848 – 1849]: 274 – 275, pl. V: uppe figs. 1 - 7), Greville (1850: 87 – 88, pl. VII: uppe figs. 1 - 7), Sonder (1871: 43), Silva *et al*. (1996: 689).

Sargassum C. Agardh

Sargassum odontocarpum **Sonder**

Description: Plant yellowish brown in colour, grows upto 2 cm in height; Upper branches are cylindrical. Branches having small leaves of 2-5 mm long and 1-3 mm broad in lower side and in upper side; broader leaves of 2-5 cm length and 1-2 cm wide in middle part of the branches. Leaf with serrated margins throughout, oblong and lengthy shape. midrib conspicuous; vesicles with short stalk and spherical shape. Receptacles are small without spine. Crypto-stomata are scattered on either side.

Uncommon: occurring on coral stones in lower intertidal region.

Distribution: Nochiurani Shingle island, Hare island Appa island

A. Habit of the plant

B. Leaf

C. *& D* Receptacle & vesicle

E. Transverse section of *Stem*

F. TS of Receptacle

G. TS of Leaf

Sargassum C. Agardh

REFERENCES: A. Sreenivasa Rao, Umamaheswara Rao & Rattan Kumar (1998: 21), A. Sreenivasa Rao, Umamaheswara Rao & Rattan Kumar (2000: 16), Subbarangaiah & Apparao (2000: 18), A. Sreenivasa Rao & Umamaheswara Rao (2000b: 27), A. Sreenivasa Rao & Umamaheswara Rao (2000c: 39), A. Sreenivasa Rao & Umamaheswara Rao (2001: 33).

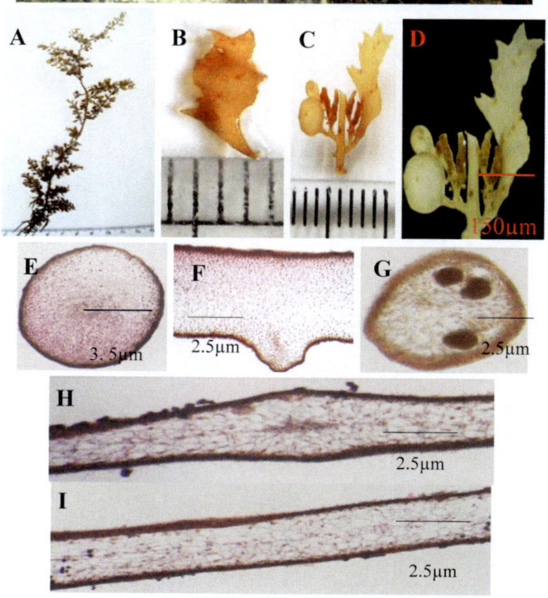

Sargassum C. Agardh

***Sargassum polycystum* C. Agardh**

Description: The plants are 1-2 m high. The erect branches have numerous spines on the stem. The creeping branches are smooth on the edges and form secondary branches with secondary holdfasts on the terminal portions. Leaves are lanceolate to oblong with serrations, and vesicles are spherical. Receptacles cylindrical.

Common: Form large communities on rocks in lower intertidal zones in relatively calm waters

Distribution: Pamban lighthouse, Nalupanai, Rameswaram Keelakarai light house, Keelakarai, Bharathinagar, Ervadi, Shingle island, Pullivsal island, Krusadai island and Mulli island, Valai island and Salli island.

A. Habit with spines on stem

B. Leaf

C.&D Receptacle & vesicle

E TS of Stem

F. LS of *Stem*

G. TS of Receptacle

H.TS of leaf

I. LS of leaf

Sargassum C. Agardh

REFERENCES: Pant (1990: 66, fig. 1 tables 5), Pandya, bapodra & parekh (1990: 557), Ramalakshmi & Chauhan (1990: 66, fig. 1 tables 1 – 3), Padmini Sreenivasa Rao, Sreenivasa Rao & Karmarkar (1991: 15 table 1), Silva *et al.* (1996: 700), Nilesh Joshi & Sridhar Murthy (2000: 52).

Sargassum C. Agardh

Sargassum swartzii **C. Agardh**

Description: Plant brown in colour, upto 25 cm high; attached by discoid holdfast; plant has primary flat and secondary branches; branches cylindrical below and flat above. leaves linear with 3- 8 cm long and 3-4 mm broad, cuneate at the basal parts; dentate distally, vesicles ovate or elliptical, rarely round with flat stalks and terete apical appendages; receptacles flattened, arranged in cymose clusters.

Common: attached on the rocks in the intertidal region upto 5 meter deep.

Distribution: Pamban light house, Thonithurai, Shingle island, Pullivasal island, Krusadai island, Manali island, Hare island, Mulli island, Appa island, Anaipar island, Valimunai island 13.Salli island and Koswari island

A. Habit of the plant
B. Leaf
C. Microscopic view of Receptacle & vesicle
D. Transverse section of *CS*
E. TS of Receptacle
F. TS of Leaf

Sargassum C. Agardh

REFERENCES: Tewari (1998: 1-8), Zaidi & Mairh (1997a: 113 – 130), Kaliaperumal *et al*. (1998: 141 – 151), Nilesh Joshi & Sridhar Murthy (2000: 52), Ramavat, A. Mehta & Mairh (2000: 76), Ramavat, Panikkar & Tewari (2000: 78).

Sargassum C. Agardh

***Sargassum tenerrimum* C. Agardh**

Description: Plants yellowish brown in colour, upto 30 cm tall, attached on substratum with a small conical discs; short terete primary axis; alternately branched; primary branches cylindrical, smooth with branches and branchlets; leaves thin, upto 6 cm long, 0.5 to 1.5 cm broad; broader below and narrower towards apices; margin dentate; vesicles more or less spherical, stalked; receptacles branched; anthredia and oogonia developing in different receptacles.

Rare: attached on rocks in intertidal region.

Distribution: Mully island

A. Habit of the plant

B. Microscopic view of Leaf

C. Microscopic view of Receptacle & vesicle

D. Transverse section of Stem *E*. Longitudinal sectionof *Stem*

F. TS of Receptacle

G. TS of Leaf

H. LS of Leaf

Sargassum C. Agardh

REFERENCES: Zaidi & Mairh (1997a: 113-130), Kaliaperumal *et al.* (1998: 141 – 151), A. Sreenivasa Rao & Umamaheswara Rao (1998: 11 – 17), Kaliaperumal & Sonia Sukumaran (1999: 15), M. Ganesan *et al.* (2000: 28), Velliyan *et al.* (2001: 23).

Sargassum C. Agardh

Sargassum wightii **Greville**

Description: Plant has richly branched thalli, radially organized; attached to the substratum with well developed discoidal holdfast. Axes flattened throughout. Leaves broad or narrow, oblong or linear, measuring 5-8 cm in length and 2-9 mm in breadth, margins a little sinuate, dentate or more or less entire; tapering at base and apex; midrib inconspicuous, with few crypto-stomata occurring on either sides of it. Vesicles large, spherical or ellipsoidal, being 5-8 mm long and 3-4 mm broad, stipe of the vesicles 5-7 mm long, seldom ending into a long tip. Receptacles cylindrical. Antheridial and oogonial conceptacles distinct, borne on separate receptacles

Common: attached on rocks in intertidal region.

Distribution: Munaikadu, .Pamban light house, Nalupanai, Villoondi theertham Ariyankundu, Thonithurai, Velankanninagar, Seeniappa dharga, Pudumadam, Thalathoppu, Sullimunthal, Kanchirankudi, Keelakarai, Bharathi nagar, Ervadi, Sadaimuniyanvalasai, Valinokkam, Manapad, Kuduthalai, Panjal, Leepuram Chinnamuttam, Vattakottai, Kanyakumari, Kovalam and all 20 islands of Gulf of Mannar.

A. Habit of the plant
B. Leaf
C. Microscopic view of Receptacle
D. Transverse section of Stem
E. Transverse section of Receptacle
F. Transverse section of Leaf

Fucales-Sargassaceae

Sargassum C. Agardh

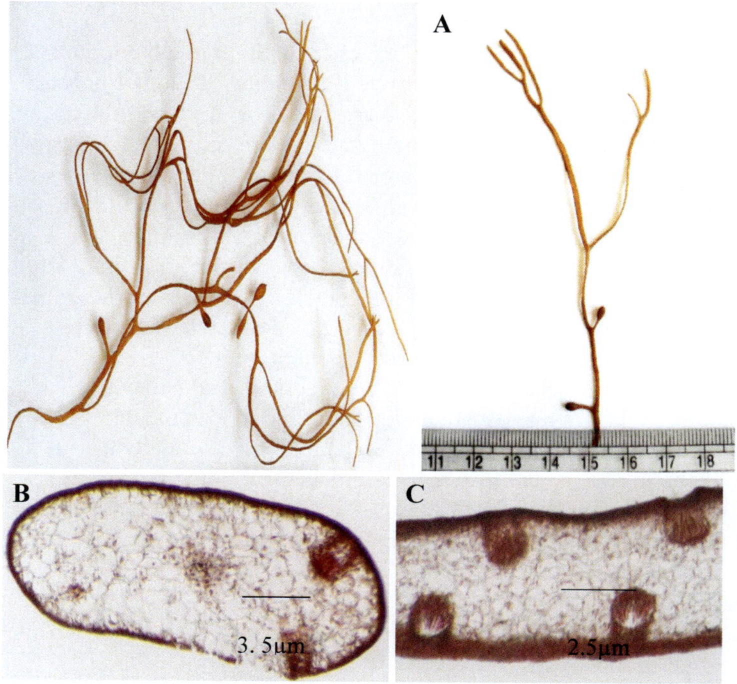

Sargassum sp.

Description: Thallus creeping narrow, branches regularly or irregularly dichotomous; narrow with 1 mm diameter. Vesicles small, 2-5 mm long and 1 mm wide, alternate. True lamina and receptacles are absent.

Rare: attached on small stones embedded in muddy sand.

Distribution: Kottaipattinam.

A. Habit of the plant with vesicle.

B. Transverse section of Leaf

C. Longitudinal section of Leaf

***Spatoglossum* Kützing**

Thallus dark brown, leaf shaped, irregularly branched ; attached to the substratum by a bunch of rhizoids produced from a short stalk; thallus not zonated as seen in many other Dictyotales; growth from apical cells not especially differentiated; structurally composed of several layers of larger medullary cells and a few layers of smaller cortical cells on each face. Reproductive organs scattered on the surface.

Key for the species

1. Thallus 20-35 cm in height, 226-268 µm thick, composed of superficial layers and 5 medullary layers ..... *S.asperum*
1. Thallus 10-15 cm in height, 146 µm thick; having superficial layers and 2-3 medullar layers .... *S. variabile*

Spatoglossum Kützing

REFERENCES: Maya & Nair (1992: 32), Subba Rao *et al*. (1992: 181, 182), V. Kumar (1993a: 139, 140), V. Kumar (1993b: 47), Rama Rao *et al*. (1993: 106), Silva *et al*. (1996: 608).

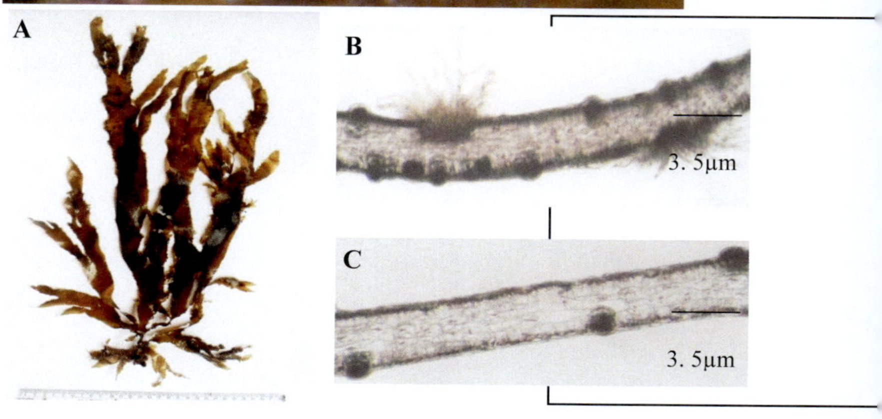

Spatoglossum Kützing

Spatoglossum asperum **J. Agardh**

Description: Plant dark brown in colour; 20-35 cm in height, 3-5 cm in breadth; thallus arising from a basal compact disc; irregularly, sub dichotomously divided. Margins sinuate, provided with small teeth, large and small proliferations arising from the margins. Superficial cells palisade like, richly provided with chromatophores. Reproductive bodies scattered all over both surfaces of the thallus.

Common: growing on rock in the mid littoral region.

Distribution: Keelakarai light house, Manapad, Panjal, Kanyakumari, Kovalam, Krusadai island, Hare island, Appa island, Anaipar island and Koswari island.

A. Habit of the plant

B. Transverse section showing reproductive bodies

C. Longitudinal section

Spatoglossum **Kützing**

REFERENCES: R. Parekh, Dave & E. Lewis (1980: table 1), Sreenivasa Rao & Parekh (1981: 578), Ohno & Mairh (1982: table 1), Untawale, Dhargalkar & Agadi (1983: 21), M. Dave & Chauhan (1985: 46, table 1), Silave *et al.* (1996: 609).

Spatoglossum Kützing

Spatoglossum variabile **Figari & De Notaris**

Description**:** Plants 10-15 cm in height and upto 5 cm in width; attenuated at the base; upper portion palmate; sub-dichotomously divided, margin dentate; in cross section thallus composed of superficial layer surrounding 2-3 layers of inner cells; superficial layer sub-quadrate; inner layer composed of polygonal cells. Tetrasporangia distributed over both surface of the thallus.

Common: grown on rocks in the mid littoral region.

Distribution: Keelakarai light house, Keelakarai, Bharathinagar, Chinnamuttam, Vattakottai, Krusadai island, Mully island, Upputhanni island, Van island and Koswari island.

A Habit of the plant.

B. Thallus with dentate margin

C. Transverse section showing polygonal cells

D. Longitudinal section with tetrasporangia.

***Stoechospermum* Kutzing**

Light brown, narrow blades; at the base narrowing into a more or less cylindrical stipe; attached by a discoid base; growth from a marginal row of in-rolled apical cells, as in *Padina*.

Key for the species

1. Lateral margin entire … *S.marginatum*
1. Lateral margin in- rolled … *S.polydioides*

***Stoechospermum* Kützing**

REFERENCES: V. Kumar (1993a: 139, 140), V. Kumar (1993b: 47), Venkateswarlu & Biabani (1995: 331), Wahidulla & De Souza (1995: 333), Silva *et al*. (1996: 610), Selvi *et al*. (2001: 23).

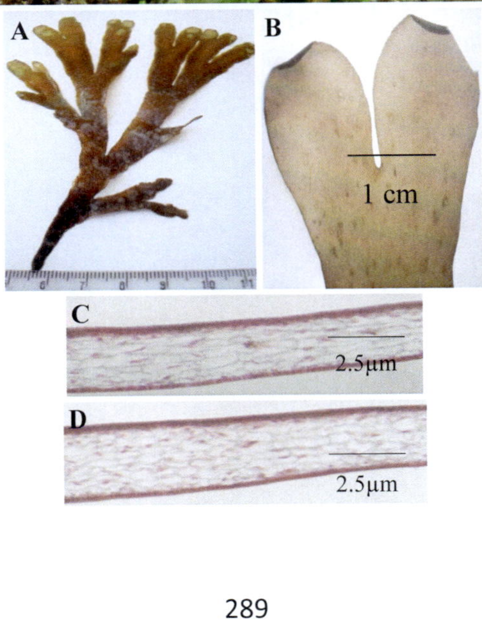

Stoechospermum Kützing

***Stoechospermum marginatum* (C. Agardh) Kützing**

Description: Plant yellowish brown in colour; 7-10 cm in height; 0.5 to 1.5 cm in width. Basal portion consists of irregular branches, short, intertwined together. Rhizoid arising from intertwined branches, fixing the plant to the substratum. Upper portion dichotomously branched. Lateral margin entire; apical margin involute. Thallus composed of superficial layer enclosing a middle layer in the younger region; middle portion becoming multilayered in mature. Superficial layer rectangular, rounded with chromatophores; middle layers large, rectangular. Reproductive bodies occurring in marginal sori.

Common: growing on rocks in the infra littoral region.

Distribution: Nalupanai, Thonithurai, Nochiurani, Pudumadam, Thalathoppu, Kanchirankudi, Keelakarai light house, Ervadi, Sadaimuniyan valasai, Manapad, Kuduthalai, Idinthakarai, Panjal,Leepuram, Chinnamuttam, Vattakottai, Kanyakumari, Kovalam and all 20 islands of Gulf of Mannar.

A. Habit of the plant

B. Upper portion of thallus showing dichotomous branch

C. Transverse section showing multilayered medulla.

D. Longitudinal section

Stoechospermum Kützing

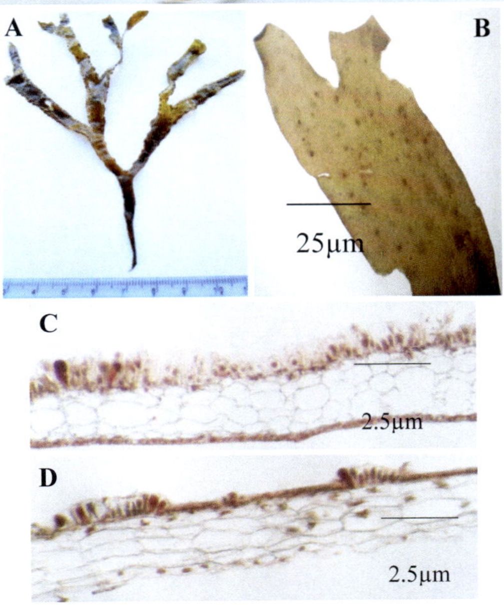

Stoechospermum Kützing

***Stoechospermum polydioides* (C. Agardh) Kützing**

Description: Plant pale yellowish brown in colour; 5 to 7 cm in height; 0.3 to 0.5 cm in width. Branches uniformly dichotomous, thin as compared to *S.marginatum*, lateral margin is serrated, apical margin inrolled. Thallus composed of superficial layer enclosing a middle layer in the younger region; middle portion becoming multilayered in mature. Superficial layer rectangular, rounded with chromatophores; middle layers large, rectangular. Reproductive bodies occurring in marginal sori.

Common: growing on rocks in the infra littoral region.

Distribution: Nalupanai, Thonithurai, Nochiurani, Pudumadam, Thalathoppu, Kanchirankudi, Keelakarai light house, Ervadi, Sadaimuniyan valasai, Manapad, Kuduthalai, Idinthakarai, Panjal,Leepuram, Chinnamuttam, Vattakottai, Kanyakumari, Kovalam and all 20 islands of Gulf of Mannar.

A. Habit of the plant

B. Upper portion of thallus showing dichotomous branch

C. Transverse section showing multilayered medulla.

D. Longitudinal section

Turbinaria **Lamouroux**

Tough brown alga with a prostrate system of fibrous branches and erect shoots; branching sparse. Erect branches are densely covered with short pyramidal club shaped branchlets also called leaves, sometimes including a gas vesicle. Cryptostomata are scattered on the surface of the short branchlets; structure as seen in Sargassum; reproductive organs oogonia and anthredia in pits on receptacular branchlets in the axils of the short branchlets.

Key for the species

1 Thallus simple 2

1 Thallus branched 3

2 Proximal part of the lateral without wing like extension ..... *T.ornata*

3 Proximal portion of the lateral obconical, distal portion broad and dentate .... *T.conoides*

3 Proximal portion of the lateral with angled margin distal portion peltate .... *T.decurrens*

Turbinaria Lamouroux

REFERENCES: Kaliaperumal & Kalimuthu (1993: 9, 11, fig 12), Menon *et al*. (1993:59), Kaliaperumal, Kalimuthu & Ramalingam (1996: 84 – 90), Silva *et al*. (1996: 710), Zaidi & Mairh (1997: 113 – 130), Saraswathi Jothi & Rengasamy (1998: 14).

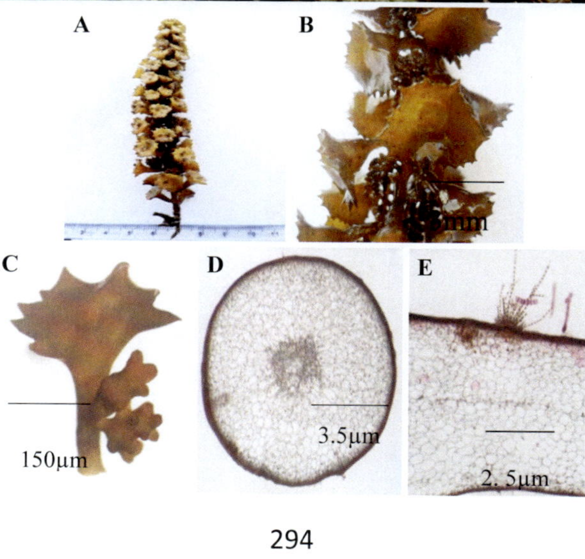

Turbinaria Lamouroux

***Turbinaria conoides* (J. Agardh) Kützing**

Description: The thallus is erect, 13 cm tall, dark brown in colour attached to substrate with branched hold fast; the erect axis is sub cylindrical supporting alternatively and polystichously arranged stalked turbinate leaves; leaves are 13 - 20 mm long with substrate stalk slightly tapering below and becoming expanded triangular above; distal marginal blade may be vesiculate or evesiculate 10 - 15 mm broad, concave at the centre which may be surrounded fully or partially by an inner crown of teeth, the outer margin is lined by sharp prominent teeth; receptacular cluster are racemose attached to the stalk at 1/4 distance from the main axis.

Common: on rocks and dead coral fragments in different environment in shallow areas on reef crust in strong wave action

Distribution: Pamban light house, Nalupanai, Thonithurai, Sullimunthal, Keelakarai light house, Keelakarai, Bharathi nagar, Ervadi and all 20 islands of Gulf of Mannar

A. Habit of the plant

B. Blade with prominent teeth.

C. Stalk bearing triangular blade

D. Transverse section.

E. Longitudinal section

Turbinaria Lamouroux

REFERENCES: Kamat (1991: 55), Jagtap (1992:57), Kalimuthu, Kaliaperumal & Ramalingam (1992: 122), Kaliaperumal & Kalimuthu (1993: 9, 11, fig. 14), Silva *et al*. (1996: 711), Zaidi & Mairh 1997: 113 – 130).

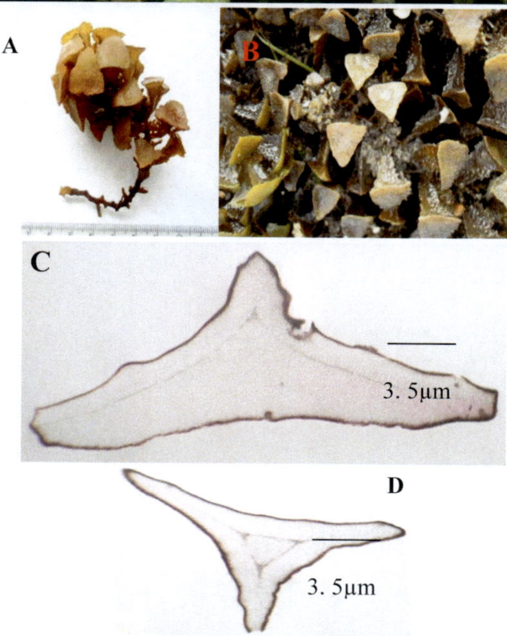

Turbinaria Lamouroux

Turbinaria decurrens **Bory de Siant-Vincent**

Description: Thalli erect, tough, dark to yellowish brown, with coarse, branched holdfasts; leaves fleshy, 11 to 17 mm long, pyramidal in lateral view. Leaves characteristically triangular in surface view, often with a concave surface; outer margin finely toothed to entire; cross-section of leaves triangular with or without a large central vesicle. Receptacular branches about 8 mm long, attached at the base of the leaf stalk. This species is distinct from other members of Turbinaria because of its triquetrous leaves.

Common: attached on rocks and dead coral in lower intertidal to subtidal zones along strongly wave exposed shorelines.

Distribution: Nalupanai, Thonithurai, Sullimunthal, Kanchirankudi, Keelakarai light house, Ervadi, Kanyakumari, Sadaimuniyanvalasai, Kovalam and all 20 islands of Gulf of Mannar.

A&B Habit showing triquetrous leaves

C. Triangular transverse section of leaves

D. Longitudinal section.

Turbinaria Lamouroux

REFERENCES: Umamaheswara Rao (1990: 324, tables 2, 4, 6), Kalimuthu, Kaliaperumal & Ramalingam (1992: 123), Kaliaperumal &Kalimuthu (1993: 9, 11, fig. 13), V. Kumar (1993a: 139, 140), Silva *et al*. (1996: 714), Renuka Bai *et al*. (1998: 91 – 95).

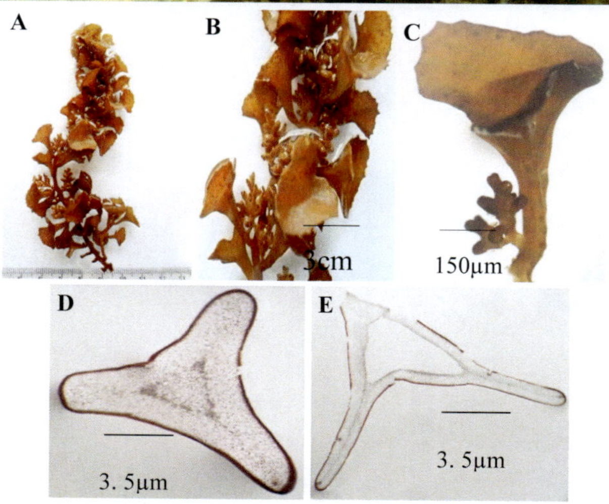

Turbinaria Lamouroux

Turbinaria ornata **(Turner) J. Agardh**

Description: Plant light yellowish brown to dark brown in colour. Stiff and erect plant, 2 -20 cm tall when reproductive. Blades conical, hard, thick, heart shaped with double row of stiff spines around the irregularly triangular margin of the blade when viewing from above. Holdfast bears one terete, erect portion and basal portion is conical or irregular, usually with several un branched or dichotomously branched root-like structures growing from basal area of the erect axes.

Common: on rocks and dead coral fragments in shallow areas on reef crust in strong wave action; lower intertidal to 30 m deep.

Distribution: Nalupanai, Thonithurai, Sullimunthal, Kanchirankudi, Keelakarai light house, Ervadi, Sadaimuniyanvalasai, Kanyakumari, Kovalam and all 20 islands of Gulf of Mannar

A. Habit of the plant

B & C. Conical and heart shaped blade

D. Transverse section

E. Longitudinal section

D. CS

E. LS

Zonaria C. Agardh

Zonaria C. Agardh

Thallus erect, stipe branched, and lamina divided into narrow division with tapering towards the base; below which it is covered with hairs. Segments fan shaped, their peripheral margin turned inside. Lateral segments obtuse with small irregular teeth or apparently serrated margin. Sorus not seen.

1. Thallus thick and fan shaped ....*Z.variegata*

Zonaria C. Agardh

Zonaria
(Lamouro
Descriptio
Plant ye
Fronds
terminally
overlappin
and 3-7 c
margin wi
structure.
of a super
walled ,
intermedia
walled is
medullary
large thin
cells.

Uncommo
rocks in t
region.
Distributi
Sethukarai
Bharathina
islands of
A. Habit o
B. Cross
thallus.
C. Longitu
thallus.

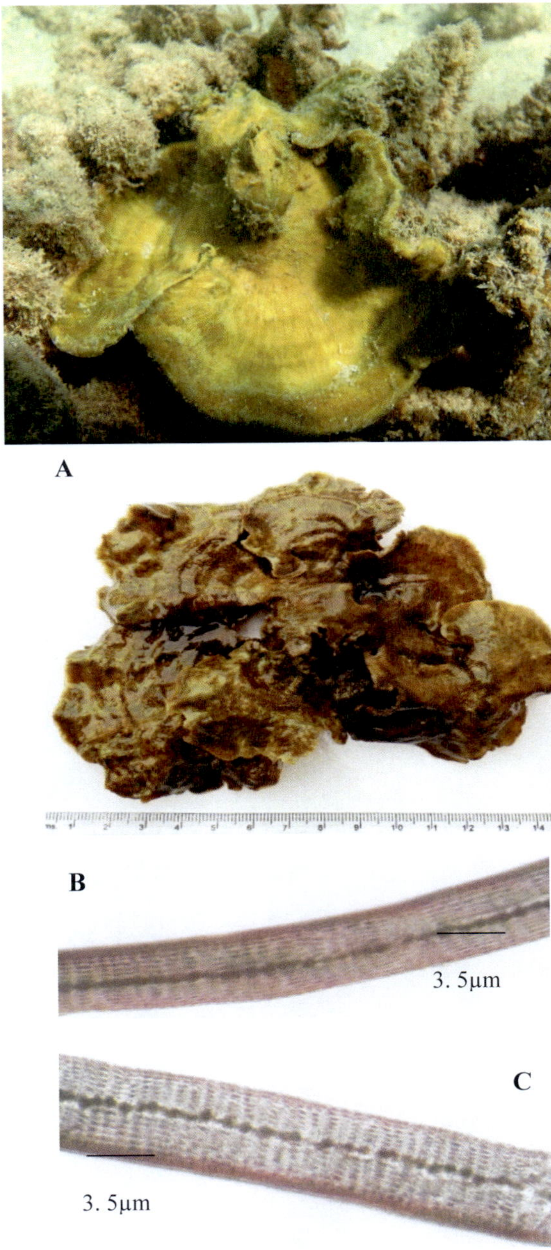

Zonaria C. Agardh

Zonaria variegata **(Lamouroux) C. Agardh**

Description:

Plant yellow in colour. Fronds very thick and terminally fan shaped, overlapping; 5-10cm across and 3-7 cm breadth; serrated margin with small teeth like structure. Thallus composed of a superficial layer of thin walled, three to four intermediary layers of thick walled isodiametric cells, a medullary layer consisting of large thin walled rectangular cells.

Uncommon: Attached on rocks in the lower intertidal region.

Distribution: Sullimunthal, Sethukarai, Keelakarai, Bharathinagar, Ervadi and islands of Gulf of Mannar.

A. Habit of the plant

B. Cross section of the thallus.

C. Longitudinal section of the thallus.

RHODOPHYTA

***Acanthophora* Lamouroux**

Thallus erect, 10cm or more in height with a cylindrical axis irregularly branched; branches covered wholly or partially or sometimes occasionally with spines. Growth by a single apical cell giving rise to an axial row of cells which is surrounded by 5 peri-central cells, covered by several rows of thick cortical cells. Trichoblasts are abundant close to the apical cells and can be seen on the tip of each branchlet. Cystocarps urn shaped born on short branches bearing one or two spines.

Key to the species:

1. Spines present on main axis 2
1. Spines absent on main axis 3
2. Both branches and branchlets containing spur like branches ..... *A. muscoides*
2. Tetrasporangial stichidia ovate with prolonged spines at apices ....*A. nayadiformis*
3. Only branchlets containing spur like branches short branches not constricted below ...... *A. spicifera*

Acanthophora Lamouroux

REFERENCES: Untawale, Reddy & Ambiye (1989: table 1), Krishnamurthy & Balasundaram (1990: fig. 5), Kalimuthu, Kaliaperumal & Ramalingam (1992: 124), Jeyasankar & Ramalingam (1993: 42), Silva *et al*. (1996: 469), Desikachary, V. Krishnamurthy & M. Balakrishnan (1998: 230)

Acanthophora Lamouroux

Acanthophora muscoides **(Linnaeus) Bory de Sant-Vincent**

Description: Thallus brittle, closely compact upto 16 cm high, color variable from brown-red, green, or yellow; branching irregularly radial. Branches cylindrical, 2.5 mm diameter, heavily corticated, possessing spur like spines. Branchlets uniformly abundant, short, also having spur like spines. Holdfast disc like; tetrasporangia in linear rows, tetrahedrally divided in short spine like branchlets. Cystocarps urn shaped, 0.5–1.0 mm diameter, solitary in axis of spines.

Common: On dead coral fragments, exposed rocks and on pebbles.

Distribution: Munaikadu, Pamban light house, Pudumadam, Thalathoppu, Kanchirankudi, Keelakarai Kuduthalai

A. Habit of the plant

B. Branch with spines

C. Transverse section of branch

D. Longitudinal section of branch

Acanthophora Lamouroux

REFERENCES: P. Subba Rao *et al*. (1985: 34), Krishnamurthy *et al*. (1990: fig. 5), Ragothaman & Haranatha Babu (1990: 91), Menon *et al*. (1993: 59), Silva *et al*. (1996: 469), Desikachary, V. Krishnamurthy & M. Balakrishnan (1998: 231).

Acanthophora Lamouroux

Acanthophora nayadiformis **(Delile) Papenfuss**

Description: Plants upto 15 cm tall, attached with substratum by rhizoids or basal discs; axes and branches covered sparingly with spines, spines pointing upward; short branches or ramuli with solitary spines.

Common: on solid exposed rocks and dead coral fragments.

Distribution: Nalupanai, Villoonditheeertham, Velankanni nagar, Seeniappadharga Nochiurani, Chinnamuttam, Van island, Koswari island.

A. Habit of the plant

B. Branches with upward spines

C. Transverse section of branch.

D. Longitudinal section of branch.

Acanthophora Lamouroux

REFERENCES: Jagtap (1993: table 1), Jeyasankar & Ramalingam (1993: 42), Nair *et al*. (1993: 186, 193), Silva et *et al*. (1996: 470), Kaliaperumal, Ezhilvalavan & Ramalingam (1997b: 38), Desikachary, V. Krishnamurthy & M. Balakrishnan (1998: 232).

Acanthophora Lamouroux

Acanthophora spicifera **(Vahl) Børgesen**

Description: Thallus brittle, 25 cm high, color white-pink, pale brown, green or yellow, branching regularly radial, cylindrical, smooth without spines, heavily corticated. Branches short, possessing spur like short spines. Holdfast irregularly lobed, disc like. Tetrasporangia in linear rows, 60–80 µm long and 40–50 µm diameter. Cystocarps urn shaped, 0.5–1.0 mm diameter, solitary in axes with spines.

Common: on rocks and coral stones upto 2 meter deep.

Distribution: Kottaipattinam Karankadu, Ponnagaram, Kodikkarai, Munaikadu, Pamban light house, Nalupanai, Villoonditheertham, Pamban, Thonithurai, Nochiurani, Pudumadam, Thalathoppu, Keelakarai light house, Ervadi, Valinokkam, Kuduthalai, Panjal, Kanyakumari Kovalam and all islands of Gulf of Mannar.

A. Habit of the plant with spines on branches

B. Transverse section of branch

C. Longitudinal section of branch

***Agardhiella* Schmitz**

Plants fleshy, slippery, usually 5- 30 cm height; rose, red or brown in colour. Branches repeatedly pinnately alternate or opposite. Main axis usually flattened, 2-15 mm wide. Branchlets pinnately opposite or alternate. Cortex 4- 6 cells thick. Inner cortex 2-4 layers; outer cortex 1-2 layers. Medullary cells loosely arranged. Tetrasporangia and cystocarps are spherical.

Key to the species:

Branching radially alternate; occasionally favouring one side,
main axis cylindrical ...... *A. subulata*

Agardhiella Schmitz

REFERENCES: C. Agardh, 1822 (1822-1823: 328), Kraft & Wynne (1979: 327 – 328), Kraft & Wynne, (1979: 329), Silva *et al.* (1996: 324).

Agardhiella Schmitz

Agardhiella subulata (C. Agardh) Kraft & Wynne

Description: Main axis cylindrical, 0.5–5.0 mm diameter, slightly pinched at base, tapering towards apices. Medullary filaments loosely tangled within mucilaginous gel; cortex 4–6 cells thick; inner cortex 2–4 layers, outer cortex 1–2 layers. Tetrasporangia and spermatangia are oval, 6–44 μm diameter. Cystocarps spherical, 1–2 mm diameter.

***Uncommon*:** On rocks and dead coral stones upto 10 meter deep.

Distribution: Manapad, Kuduthalai, Panjal, Chinnamuttam, Kanyakumari and Kovalam

A. Habit of the plant

B. Branches with tapering apices

C. Transverse section of branch.

D. Longitudinal section of branch.

Agardhiella
Agardh) Kra
Descriptio
cylindrical,
diameter, sli
base, taperin
Medullary f
tangled with
gel; cortex
inner cortex
cortex
Tetrasporangi
spermatangia
diameter. Cys
1–2 mm diam
***Uncommon*:**
dead coral sto
deep.
Distribution:
Kuduthalai,
Chinnamuttan
and Kovalam
A. Habit of th
B. Branches
apices
C. Transve
branch.
D. Longitud
branch.

***Ahnfeltiopsis* P. Silva & Decew**

Members of this algal genus have unisexual gametophyte and crustose sporophyte; gametophytes are dichotomous, sometimes proliferous; branches cylindrical or flattened; medulla pseudo-parenchymatous, cortex are small cells, carpogonial branches are three celled. Cortex increasing in thickness during the development of cystocarp with special pore through which carpospores are released. Spermatangia in sori elongate; Tetrasporangia cruciately divided.

Key to the species:

Thallus cylindrical below, compressed above, branches flabellate
…… *A. pygmaea*

Ahnfeltiopsis P. Silva & Decew

REFERENCES: Rajendran *et al.* (1991: 123), Kalimuthu, Kaliaperumal & Ramalingam (1992: 123), Maya & Nair (1992: 26), Nair *et al.* (1993: 186), Silva *et al.* (1996: 318), Desikachary, V. Krishnamurthy & M. Balakrishnan (1998: 130).

Ahnfeltiopsi
agardh) P. S
Description:
height, grega
isolated bus
well attache
discoid l
cylindrical at
compressed
mm in dia
dichotomous
single plane
Marginal pro
dark red in
psedo-parenc
all cells app
same size a
thick and co
arranged fil
cells.
Habitat: Ur
on rocks in in
Distribution:
Vattakottai,
Kovalam.
A. Habit of t
B. Branches
base and con
C. Transve
branch sh
parenchymat
D. Longitu
branch.

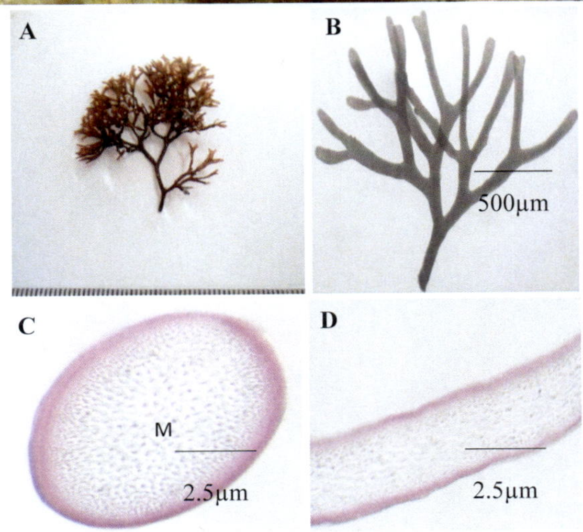

Ahnfeltiopsis P. Silva & Decew

***Ahnfeltiopsis pygmaea* (J. agardh) P. Silva & Decew**

Description: Plants 4–5 cm in height, gregarious, growing in isolated bushy clusters, very well attached by means of discoid holdfast; thalli cylindrical at the basis, slightly compressed higher up, 0.55 mm in diameter, repeatedly dichotomously branched in a single plane with wide axils. Marginal proliferations absent; dark red in colour. Medulla psedo-parenchymatous, with all cells approximately of the same size and shape; cortex thick and composed of radially arranged filaments of small cells.

Habitat: Uncommon: grown on rocks in intertidal zone.

Distribution: Panjal, Vattakottai, Kanyakumari and Kovalam.

A. Habit of the plant

B. Branches with cylindrical base and compressed above.

C. Transverse section of branch showing pseudo parenchymatous medulla (M)

D. Longitudinal section of branch.

Amphiroa Lamouroux

Amphiroa Lamouroux

Thallus articulated, forming erect and bushy or low dense tufts of branches with strong deposition of lime; branching dichotomous to irregular, with cylindrical to compressed branches; several uni porate and usually small conceptacles are scattered on the surface of the segments. Structurally formed by a few cortical layers of small cells and medulla of several layers of smaller, globose cells intercalated with one or a few layers of larger, cylindrical cells; articulations mostly with several layers of cells.

Key to the species:

1. Matured fragments cylindrical to sub-compressed …………2
2. Plants more than 2 cm high, branches dense and anastomosing ……*A. anastomosans*
3. Ends of the segments swollen and pad like ….*A. fragilissima*
3. Ends of the segments not swollen and paddle like ……*A. rigida*
4. Plants 5–10 cm high; conceptacles raised and tetrasporangia restricted to periphery of the conceptacle ……*A. anceps*
4. Plants less than 5 cm tall, conceptacles partly immersed, tetrasporangia found on the entire floor of the conceptacle …….*A. foliacea*

Amphiroa Lamouroux

REFERENCES: Untawale, Dhargalkar & Agadi (1983: 29), Subba Raovet *et al*. (1992: 182), Jagtap (1993: table 1), Kerkar (1994: table 1.) Silva *et al.* (1996: 219), Desikachary, V. Krishnamurthy & M. Balakrishnan (1998: 58).

Amphiroa Lamouroux

***Amphiroa anastomosans* Weber-van Bosse**

Description: Plants usually1–14 cm high, occasionally a little more; dense branching forming a compact tuft on the substratum; branching dichotomous; erect fronds cylindrical and articulated, branch showed a central medullary stand and a thin cortex; medullary strands consisting of long and short cell rows. Conceptacles prominent; hemispherical, scattered over the surface of the branches.

Uncommon: on rocks and other hard substratum in intertidal region up to 5 meter deep.

Distribution: Manapad and Mully island

A. Habit of the plant

B. Plant with cylindrical and dichotomous branching.

Amphiroa Lamouroux

REFERENCES: V. Kumar (1993b: 47), Nair *et al.* (1993: 185), Rama Rao *et al.* (1993: 106), Kaliaperumal *et al.* (1996: 73 – 77), Silva *et al.* (1996: 219), Desikachary, V. Krishnamurthy & M. Balakrishnan (1998: 61).

A

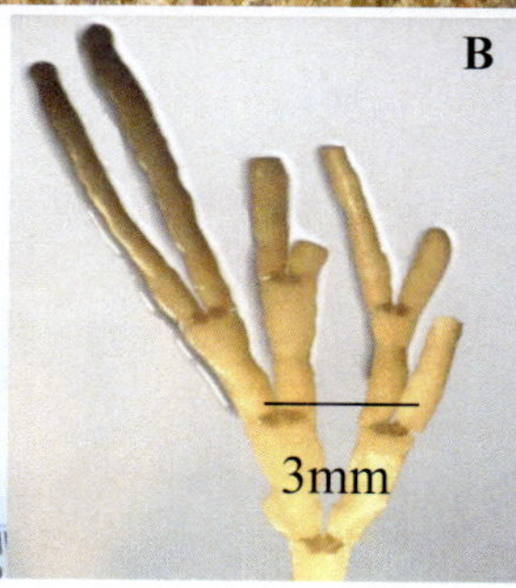

Amphiroa Lamouroux

Amphiroa anceps **(Lamarck) Decaisne**

Description: Plants 5–10 cm tall, red to bright purple in color; branching dichotomous, sub-trichotomous in the lower position; branches arising from the broadened top of the inter genicula; segments flattened and occasionally sub-terete in lower portions; conceptacle only on one side of the inter genicula or segments; occasionally on both sides.

Common: on rocks, dead coral pieces, pebbles, etc.

Distribution: Manapad, Kuduthalai, Panjal, Chinnamuttam, Kanyakumari, Kovalam and Koswari island.

A. Habit of the plant

B. Plants with sub-trichotomous branching

Amphiroa Lamouroux

REFERENCES: Rama Rao *et al*. (1993: 106), Kerkar (1994: table 1), Silva *et al*. (1996: 223 – 224), Vanillakumari & Subba Rangaiah (1997a: 22), Desikachary, V. Krishnamurthy & M. Balakrishnan (1998: 60), Subba Rangaiah & Vanillakumari (1999: 255 – 269).

A

B

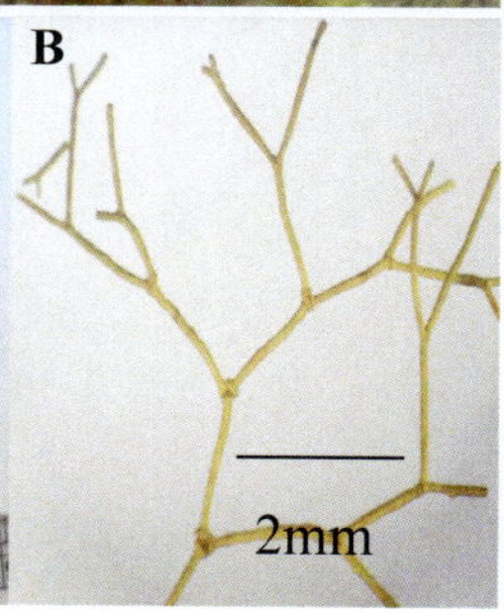

C

Amphiroa Lamouroux

Amphiroa fragilissima **(Linnaeus) Lamouroux**

Description: Plants 2–4 cm in height, fragile, calcified, calcareous in dense clumps or mats, 8 cm thick, yellow green to white pink in color; branching widely dichotomous, occasionally trichotomous, primarily at joints. Branches thin, brittle, cylindrical, often swollen at segment ends.

Common: on hard rocks and pebbles.

Distribution: Kottaipattinam, Arasaneripattinam, Karankadu, Ponnagaram, Thonithurai, Keelakarai light house, Manapad, Kanyakumari, Kovalam, Krusadai island, Valai island, Appa island and Koswari island.

A. Habit of the plant

B. Branches showing swollen at segment ends.

C. Branches showing pointed ends.

Amphiroa Lamouroux

REFERENCES: Quoy & Gaimard (1824: 628, pl. 93: figs. 2, 3), E.Ganesan (1968: 14 – 20, figs. 3, 4, 8 – 11, 21 – 31), V. Krishnamurthy & H. Joshi (1970: 19) Untawale, Dhargalkar & Agadi (1983: 29), Nair *et al*. (1986: 737), Kerkar (1994: table 1), Silva *et al*. (1996: 222).

Amphiroa Lamouroux

Amphiroa foliacea **Lamouroux**

Description: Plants sub-erect, 2–4.5 cm in tall, forming semi-rosette groups on rocks in the subtidal region; branches decumbent and more or less attached to substratum; branching irregularly dichotomous; often tetrachotomous in the upper portion. Inter-genicula or segment dimorphic; lower segment short cylindrical to sub cylindrical, 1–2 mm in width; upper branch much flattened and irregular in shape tending to the dorsi-ventral. Conceptacles scattered on the surface of segments.

Habitat: Uncommon: on rocks upto 5 meter deep.

Distribution: Manapad, Idinthakarai, Panjal, Leepuram, Chinnamuttam, Vattakottai Kanyakumari, Kovalam, Hare island and Koswari island

A. Habit of the plant

B. Branches showing flattened dorsi-ventral upper portion.

Amphiroa Lamouroux

REFERENCES: Gopinathan & Panigrahy (1983: 51), Untawale & Jagtap (1984: 101 table.1), Jagtap (1985a: 47), Jagtap (1993: table.1) Silva *et al.* (1996: 224), Desikachary, V. Krishnamurthy & Balakrishnan (1998: 60).

A

B

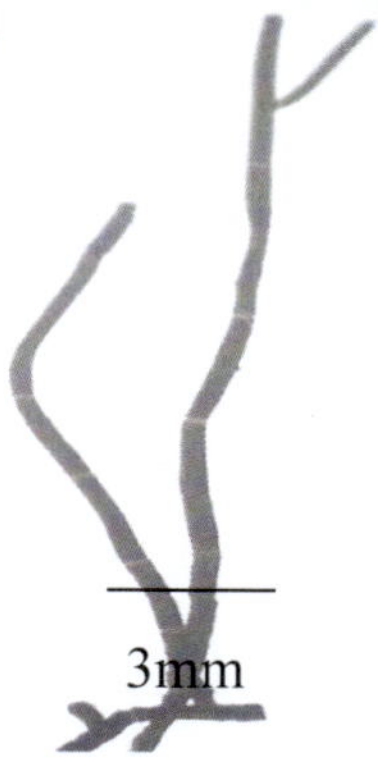

Amphiroa Lamouroux

Amphiroa rigida Lamouroux

Description: Plants 4–5 cm high; olive red in color, branching dichotomous or irregularly dichotomous; attachment to the substratum by a small basal crust; erect fronds cylindrical and clearly articulate; genicula bizonal or occasionally trizonal, upto 100 μm long; inter-genicula 3–8 mm long and 0.75–1 mm wide, cylindrical, multi zonal. Cortex well-developed, compact, medullary strand with one or two rows of long cells.

Rare: attached on rocks.

Distribution: Koswari island.

A. Habit of the plant

B. Plant with cylindrical and dichotomous branches

Asparagopsis Montagne

Plants 5- 15cm tall or more with creeping and erect axes attached by rhizoidal clusters arising from the prostrate system, erect axes bushy, densely covered with slender brush like branches and branchlets; branches sub-opposite, pinnate; thallus uniaxial, pseudo-parenchymatous; branchlets 30-40 µm in diameter, with 3 rows of cells; spermatangia in cylindrical clusters at the tip of the branchlet; cystocarps terminal on short branchlets, spherical or pyriform. The morphological dimorphism between sporophytic and gametophytic phases in this genus is remarkable. The difference is so striking that the phases have been identified as different genera in different families and orders. The tetrasporic phase is characterized by a filamentous and well branched thallus that resembles a member of the order Ceramiales, attached by several holdfast to other algae.

Key to the species:

Thallus with many stalks terminating in fine filamentous branchlets ….*A. taxiformis*

Asparagopsis Montagne

REFERENCES: Untawale, Reddy & Ambiye (1989: table 1), Desikachary, V. Krishnamurthy & Balakrishnan (1990: 189 – 190, fig. 38, pl.XXII), Pandya, Bapodra & Parekh (1990: 557), Jagtap (1993: table 1), P. Subba Rao & Ganesan (1995: 36), Silva *et al.* (1996: 183)

Asparagopsi
(Delile) Tre

Description: solitary, ext finely branc cm tall; ligl grey, lower l sparse; u compact. below, dens branchlets soft, repea divided. M solitary, lon 6 cells thick mass sprea runners (st 400–550 μm swellen, branchlets.

Common: o meter deep.

Distribution Manali isla Valimunai island, Upputhanni and Koswar

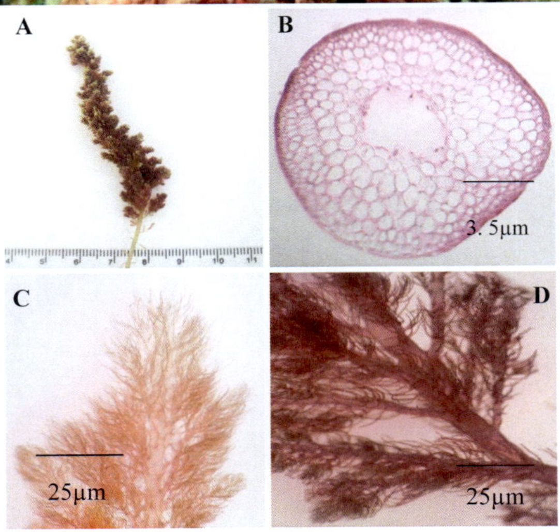

A. Habit wi upper portio
B. Transver branch
C. Upper po
D Middle pa

Asparagopsis Montagne

Asparagopsis taxiformis (Delile) Trevisan

Description: Thallus delicate, solitary, extremities fluffy and finely branched at apices; 20 cm tall; light pale red to red-grey, lower branching irregular; sparse; upper branching compact. Branches naked below, densely covered with branchlets above. Branchlets soft, repeatedly, alternately divided. Medullary filament solitary, longitudinal; cortex 3–6 cells thick. Hold fast fibrous mass spreading by prostrate runners (stolons). Cystocarps 400–550 μm diameter on short swellen, club shaped branchlets.

Common: on coral reefs to 20 meter deep.

Distribution: Krusadai island, Manali island, Appa island, Valimunai island, Anaipar island, Salli island, Upputhanni island, Van island and Koswari island.

A. Habit with densely covered upper portion
B. Transverse section of the branch
C. Upper portion of branch
D. Middle part of branch.

***Bostrychia* Montagne**

Thallus filamentous, 1-3 cm, partly dorsi-ventral, typically blackish; common in shaded crevices or on the roots and stems of mangroves. The structure is typically polysiphonous and branching mostly bilateral with in-rolled apices. Transverse division of pericentral cells occur in this genus. The tetrasporangia are produced in modified branches called strichidia.

Key to the species:

Ultimate branchlets monosiphonous ....*B. tenella*

Bostrychia Montagne

REFERENCES: Untawale, Dhargalkar & Agadi (1983: 38), Untawale, Reddy & Ambiye (1989: table 1.), G. Rao & Umamaheswara Rao (1991: 154 – 158, figs.1, 3), G. Rao (1995: 209), Silva *et al.* (1996: 475), Desikachary, V. Krishnamurthy & Balakrishnan (1998: 315).

Bostrychia Montagne

Bostrychia tenella **(Lamouroux) J. Agardh**
Description: Thallus prostrate, coarse, often forming mats of indeterminate shape to 5 cm high; black to dull purple in colour; branching mainly pinnate, feather like; apices incurled. Branchlets numerous, crowded, incurved, 25–300 μm diameter. Tetrasporangia 40–50 μm diameter. Tetrahedrally divided in four linear rows. Cystocarps oval to spherical, 700 μm diameter.
Uncommon: forming tightly adhering mats on rocks.
Distribution: Thonithurai
A. Habit of the plant
B. Plant showing crowded incurved branches.

Botryocladia (J. Agardh) Kylin

Plants macroscopic, erect, 3-20 cm tall or more, attached by a disc, with consistent system of stem like branches from where vesicular appendices resembling a bunch of grapes are produced; structure muti-faxial; vesicles filled with mucilage. Cystocarps in the walls of the vesicles; spermatangia as a small sori on the vesicles; tetrasporangia scattered and cruciately divided.

Key to the species:

Plants more than 20 cm tall, vesicles sub-spherical, 1–3 mm long, wall 2–3 layers; outer layer discontinuous ……….*B. leptopoda*

Botryocladia (J. Agardh) Kylin

REFERENCES: Parekh & Chauhan (1990: table 1), Kamat *et al.* (1991: 55), Bhakuni *et al.*(1992: 514), Rama Rao *et al.*(1993: 106), Silva *et al.* (1996: 353), Desikachary, V. Krishnamurthy & Balakrishnan (1990: 172).

Botryoclai
Agardh) I
Descriptio
Plants erec
cylindrical
thick de
branching
dichotomo
central tis
50–130
surroundec
small cells
vesicles s
diameter,
minutely s
scattered o
vesicular
four on
hemispher
Uncommi
rocks and
10 meter d
Distributio
Kovalam
A. Habit o
B. Plant
dichotomo
C. Transve
D. Longitu

Botryocladia (J. Agardh) Kylin

Botryocladia leptopoda (J. Agardh) Kylin

Description:

Plants erect, upto 60 cm high; cylindrical stem; 2–2.5 mm thick densely branched; branching irregularly dichotomous, solid with central tissue of large cells, 50–130 µm diameter, surrounded by a cortex of small cells 4–6 µm diameter; vesicles small, 1–5 mm in diameter, sub-spherical and minutely stalked. Cystocarps scattered on the surface of the vesicular branchlets, one to four on each vesicles, hemispherical, prominent.

Uncommon: attached to rocks and coral stones upto 10 meter depth.

Distribution: Kanyakumari, Kovalam and Appa island.

A. Habit of the plant

B. Plant showing irregular dichotomous branch

C. Transverse section

D. Longitudinal section

***Bryocladia* Schmitz**

Plants generally upto 8 cm in height. Thallus is creeping interwoven filaments bearing erect axes, well developed, cylindrical polysiphonous filaments with 6-12 pericentral cells in each segment; growth by means of transversely or obliquely dividing apical cell; erect branches un branched below but profusely branched in upper portion; branches spirally arranged. Cystocarps ovate to urn shaped.

Key to the species:

Thallus creeping, branches spirally arranged …….*B. thwaitesii*

Bryocladia Schmitz

Bryocladia Schmitz

***Bryocladia thwaitesii* (Harvey ex J. Agardh) De Toni**

Description: Thallus erect, 5–7 cm in height, arising from basal rhizoidal plexus; basal filament decumbent, fixed by mono-siphonous rhizoids with discoids tips; erect filaments of uniform thickness without laterals at basal portions, densely branched above; primary branches 6–7 peripheral cells; lower portion with 10–12 peripheral cells; twice as long as broad.

Rare: attached on rocks.

Distribution: Muttukadu.

A. Habit of the plant

B. Plant with densely branched upper part

C. Middle part of plant

D. Transverse section of lower portion of branch with 10 peripheral cells (P)

E. Longitudinal section.

***Carpopeltis* Schmitz**

Plant always erect, 10 cm tall, fronds flat, sub-flabellate, dichotomously to pinnately divided with distinct midrib in the lower part; 5 mm broad, inner cortical cells irregular, outer cortical cells smaller. Tetrasporangia in regular patches, cystocarps small, embedded in medulla.

Key to the species:

Fronds dichotomously to pinnately divided with distinct midrib. ……. *C.maillardii*

Carpopeltis Schmitz

REFERENCES: Montagne & Millardet (1862: 8 – 9, PL. XXIV), Chiang (1970: 68 – 69, PL. 5C), Desikachary, V. Krishnamurthy & Balakrishnan (1990: 264 – 265, pl.XXXIX), Silva *et al.* (1996: 188).

Carpopelti
(Montagn
Chiang
Descriptio
creeping
4–9 cm
complanat
cartilagino
but flex
attached
holdfast;
present e
specimens
like, 0.5–
proximal
undulated
curved
dichotomo
Medulla
packed;
thick.
Rare : on
Distributio
A. Habit o
B. Plant
dichotomo
C. Trans
thallus
D. Longit
densely pa

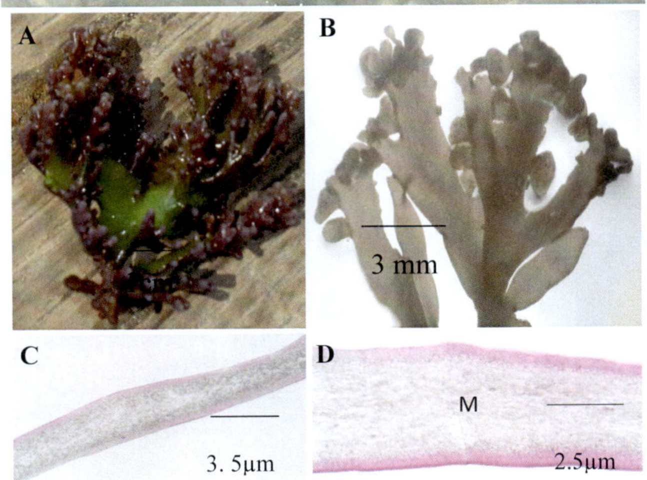

Carpopeltis Schmitz

Carpopeltis maillardii (Montagne & Millardet) Chiang

Description: Plants with creeping and erect portions; 4–9 cm high, composed of complanate fronds, cartilaginous, stiff and tough but flexible; very well attached by a discoid holdfast; dark, a short stipe present especially in older specimens; upper parts strap like, 0.5–4 mm wide with proximal midrib; thin undulated and rounded curved apices; branching dichotomous to irregular. Medulla narrow, densely packed; cortex 4–8 cells thick.

Rare: on small rocks.

Distribution: Kovalam

A. Habit of the plant

B. Plant with irregular dichotomous branch.

C. Transverse section of thallus

D. Longitudinal section with densely packed medulla(M)

Centroceras Kützing

Centroceras Kützing

Thallus usually erect, sometimes found in prostrate mats; cylindrical, macroscopic with erect branches typically organized in segments separated by nodes; the cells of the central filaments elongated, becoming macroscopic, covered with a cortex of small rectangular cells developing from a ring of large nodal cells, a ring of 1-3 celled spines present at the nodes, cystocarps arising at the nodes. Tetrasporangia at a ring at the node.

Key to the species:

Spines present at joints …… *C.clavulatum*

Centroceras Kützing

REFERENCES: Maya & Nair (1992: 26), Raghukumar, Ngarkar & Raghukumar (1992: fig. 1), Jagtap (1993: table I.), Menon *et al.* (1993: 186, 192, 193), Silva *et al.* (1996: 387), Desikachary, V. Krishnamurthy & Balakrishnan (1998: 215).

Centroceras Kützing

Centroceras clavulatum (C. Agardh) Montagne

Description: Thallus filamentous, upto 20 cm long; dark brown-maroon in color; branching narrowly dichotomous; apices forked, slightly incurved. Branches 50–150 µm diameter, heavily pigmented, dark pigmentation at joints. Joints whorled with spines; corticating cells in two layers. Tetrasporangia oval to spherical. Segments 300–500 µm long, cortical cells in one layer. Spermatangia in terminal clusters. Carposporophyte formed laterally at joints.

Common: grown on hard rocks and coral stones mostly in intertidal region.

Distribution: Muttukadu, Mahabalipuram, Pondicherry, Velankanni nagar, Seeniappadharga, Pudumadam, Thalathoppu, Kanchirankudi, Keelakarai, Bharathi nagar, Ervadi, Panjal, Leepuram, Vattakottai, Pullivasal island and Valai island.

A. Habit of the plant

B. Plant with dichotomous branch

C. Branch with spines at joints.

Ceramium Roth

Thallus erect, mostly delicate, regularly bifurcated, typically banded in nodes and internodes. Branching pseudo-dichotomous or alternate, sometimes irregular. A remarkable elongation of the central cells makes up the internodes whereas the nodes consist of a band of smaller cells that surround the point of contact between two central cells. In some species the cortication of small cells may extend until it completely covers the internodal cells making the algae resemble Centroceras.

Key to the species:

1. Number of segments between dichotomous 2–4 ……..*C. cruciatum*
1. Number of segments between dichotomous 9–13 ……*C. deslongchampii*
2. Fronds not compressed, apices of branches incurved, branch dichotomous or sub dichotomous ………..*C. subdichotomum*
2. Thallus more than 1.5 cm tall; branch evenly Dichotomous ………..*C. brevizonatu*

Ceramium Roth

Ceramium
Peterson
Descriptio
forming fil
in small ta
red to
sometimes
branching
pincer like
incurved o
segments
diameter, h
rhizoids f
mats. Tetra
embedded.
Rare: on
corals.
Distributio
Upputhanni
A. Habit o
B. Plan
dichotom

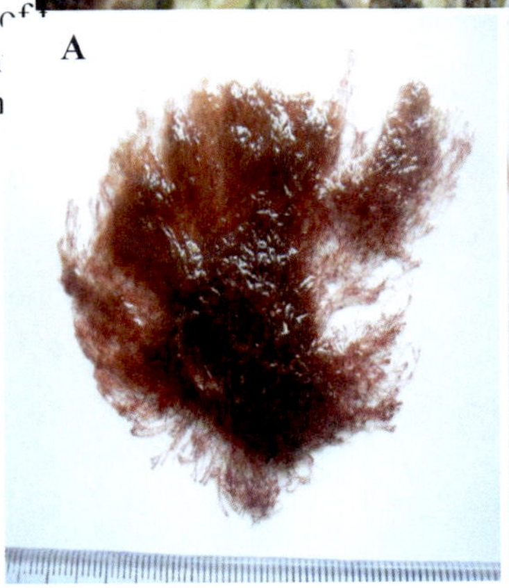

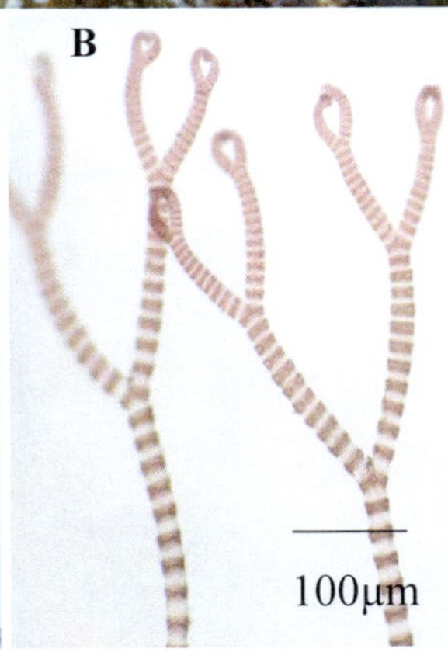

Ceramium Roth

Ceramium brevizonatum **H. Peterson**
Description: Thallus soft, forming filamentous tufts or in small tangled clumps; rose red to bright red and sometimes pink in colour; branching dichotomous, pincer like at apices; laterals incurved or spur like at joints; segments 80–120 μm diameter, heavily pigmented; rhizoids forming tight knit mats. Tetrasporangia partially embedded.
Rare: on rocks and dead corals.
Distribution: Thonithurai and Upputhanni island
A. Habit of the plant
B. Plant showing evenly dichotomous branches

Ceramium Roth

REFERENCES: V. Krishnamurthy *et al.* (1990: figs. 4, 8), Covelong & B. Subramanian (1990: 89, 90, figs. 1 – 13), Thomas & Subbaramaiah (1991: 205, table 2), Kalimuthu, Kaliaperumal & Ramalingam (1992: 124), Silva *et al.* (1996: 393), Desikachary, V. Krishnamurthy & M. Balakrishnan (1998: 207).

Ceramium c
& Harvey
Description:
as small de
high; light p
colour; bra
dichotomous
incurved; se
diameter w
longer belo
μm dia
pigmented;
bands; lowe
with largest
cells sm
numerous,
originating
Tetrasporang
diameter and
μm diameter
Common: c
stones mostl
intertidal wi
action.
Distribution
Nalupanai, V
Ariyankundu
Velankannir
dharga, Pum
Manapad, Pa
Chinnamutta
Kanyakumar
island and S
A. Hal

A

Ceramium Roth

Ceramium cruciatum **Collins & Harvey**

Description: Thallus creeping as small dense tufts to 15 cm high; light pink to rose red in colour; branching unequally dichotomous; apices slightly incurved; segments 80–120 µm diameter with short above, longer below. Joints 120–140 µm diameter, heavily pigmented; cells in several bands; lower or middle band with largest cells; other band cells smaller. Rhizoids numerous, colourless, originating from joint cells. Tetrasporangia oval, 35–50 µm diameter and spermatangia 5–8 µm diameter.

Common: on rocks and coral stones mostly located in intertidal with heavy wave action.

Distribution: Munaikadu, Nalupanai, Villoonditheertham, Ariyankundu, Velankanninagar, Seeniappa dharga, Pumadam Valinokkam, Manapad, Panjal, Leepuram, Chinnamuttam, Vattakottai, Kanyakumari, Kovalam, Hare island and Salli island.

A. Habit of the plant

Ceramium Roth

REFERENCES: Chauvin ex Duby (1830: 967), V. Kumar (1993b: 47), Silva *et al.* (1996: 394), Desikachary, V. Krishnamurthy & M. Balakrishnan (1998: 112).

Ceramium
Chauvin e
Descriptio
Plants
irregularly
adventious
present, s
60–75 µı
Number
dichotomo
cystocarps
by 4
Tetraspora
around the
the cortex.
Rare: on
Distributi
A. Habit o
B. Plant w
branches
C. Branch
of segmen
dichotomy

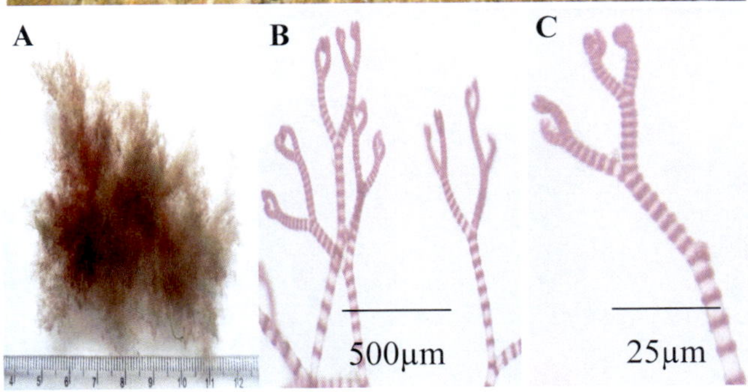

Ceramium Roth

Ceramium deslongchampii Chauvin ex Duby

Description: Plants 6–7 cm high; irregularly dichotomous; adventious branches often present, small. Axial cells 60–75 µm in diameter. Number of rows between dichotomous 9–13; cystocarps lateral, surrounded by 4 small ramuli. Tetrasporangia dispersed around the node; immersed in the cortex.

Rare: on dead coral pieces.

Distribution: Nalupanai.

A. Habit of the plant

B. Plant with dichotomous branches

C. Branch with more number of segments between dichotomy.

Ceramium Roth

REFERENCES: Mahadevan & R. Ganesan (1995: 12), S. Dixit (1966: 18), V. krishnamurthy & H. Joshi (1970: 23), Untawale, Dhargalkar & Agadi (1983: 34), Silva *et al.* (1996: 403), Silva *et al.* (1996: 213).

Ceramium
Weber va
Descriptio
1–12 cm
broad;
procumbe
rhizoids;
dichotomo
curved; a
apical po
broad in
portion u
Cortical f
upwards
internode
portion, l
base.
Rare: on c
Distributi
Manali isl
island.
A. Habit o
B. Pla
dichotomo
C. Bran
segments l

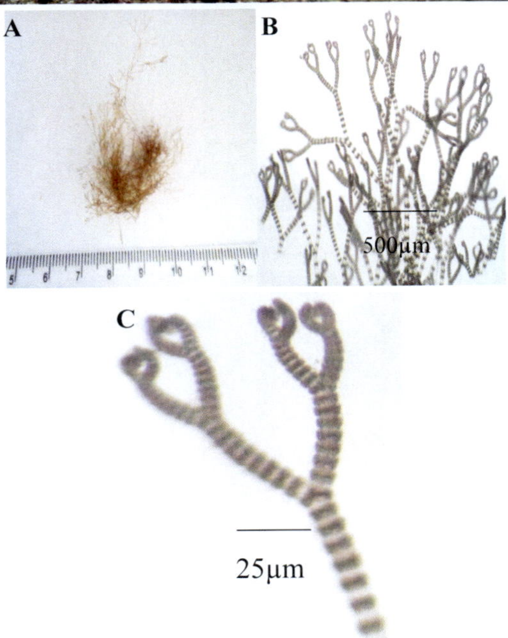

Ceramium Roth

Ceramium subdichotomum **Weber van Bosse**

Description: Fronds small, 1–12 cm high; 100–250 μm broad; primary axis procumbent, attached by rhizoids; branching sub-dichotomous; apices inwardly curved; articulating short at apical portion; as long as broad in older parts; nodal portion uniformly corticated. Cortical filaments extending upwards and dowmwards; internode small in upper portion, long towards the base.

Rare: on coral rocks.

Distribution: Thonithurai, Manali island, Upputhanni island.

A. Habit of the plant

B. Plant with sub dichotomous branches

C. Branch with many segments between dichotomy

***Champia* Desvaux**

Thallus macroscopic, mostly erect, 3- 12 cm in height, cylindrical to flattened with discoid holdfast. The thallus is hollow, typically segmented, with hollow segments separated by diaphragm; growth multiaxial with interrupted cortical layers of small cells and few layers of larger subcortical cells; vertical filaments surround the central cavity and glandular cells are present. Tetrasporangia tetrahedrally divided. Cystocarps in the form of prominent globose.

Key to the species:

1. Plants over 10 cm high …..2
1. Plants less than 10 cm high ….3
2. Plants upto 20 cm high; segments twice as long ……..*C. somalensis*

3. Segments 1–2 mm diameter, 1–5 mm diameter long…..*C. parvula*
3. Segments 1–4 mm diameter, as long as wide ….*C. salicornioides*
4. Cystocarps globular …………*C. globulifera*
4. Cystocarps not globular …………5
5. Plants 2–5 cm high; cystocarps conical ……..*C. compressa*

Champia Desvaux

REFERENCES: Untawale, Dhargalkar & Agadi (1983: 34), Chaugule & Deodhar (1985: 63 – 64, figs. 1 - 3), M. Dave *et al.* (1987: table 1), Maya & Nair (1992: 26), Silva *et al.* (1996: 345), Desikachary, V. Krishnamurthy & M. Balakrishnan (1998: 187).

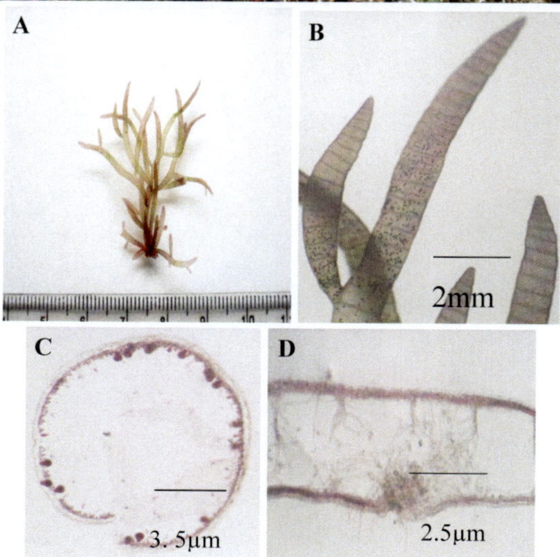

Champia Desvaux

Champia compressa **Harvey**

Description: Plants small, 2–5 cm in height; red purple in color; gelatinous; fronds compressed to flattened 3–4 mm wide; branches alternate at point of origin and expanded above; Tetrasporangia aggregated on main axis, sparse on branches. Cystocarps conical, upto 5 numbers on each segments.

Rare: on coral rocks upto 10 meter deep.

Distribution: Van island

A. Habit of the plant

B. Upper portion of branch

C. Transverse section of branch.

D. Longitudinal section of the branch showing tetrasporangial sori.

Champia Desvaux

REFERENCES: Chauhan (1978: table 4), Chauhan & Mairh (1978: 35), Untawale, Dhargalkar & Agadi (1983: 34), Kalimuthu, Kaliaperumal & Ramalingam (1992: 123), Silva *et al.* (1996: 345), Desikachary, V. Krishnamurthy & Balakrishnan (1998: 188).

Champia Desvaux

Champia globulifera **Børgesen**

***Description**:* Plants upto 6 cm in height, violet red in color; attached by rhizoids; well-developed basal discs from which several erect fronds arise; decumbent branches also produced from basal portion; these forming new attachment discs when contact with rock. Branches opposite or alternate, simple or branched again; fronds broadest at the middle, tapering gradually to the extremities; apices obtuse.

Uncommon: on rocks and dead corals.

Distribution: Panjal, Kanyakumari, Kovalam, Hare island, Appa island, Anaipar island and Koswari island.

A. Habit of the plant

B. Plant with opposite branches

C. Branches with cystocarp (c)

D. Transverse section of branch.

E. TS of cystocarp

Champia Desvaux

REFERENCES: Untawale, Reddy & Deshmukhe (1989: table 1), V. Krishnamurthy *et al*. (1990: 271, figs.4, 5, 7, 8), Kalimuthu, Kaliaperumal & Ramalingam (1992: 124), Jagtap (1993: table 1), Silva *et al.* (1996: 346), Desikachary, V. Krishnamurthy & Balakrishnan (1998: 184).

A

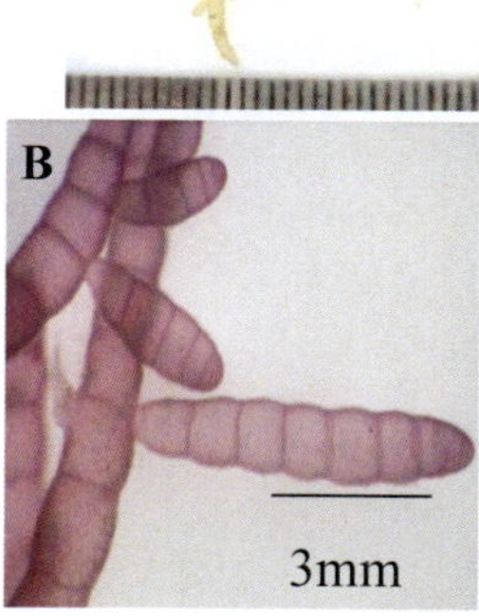

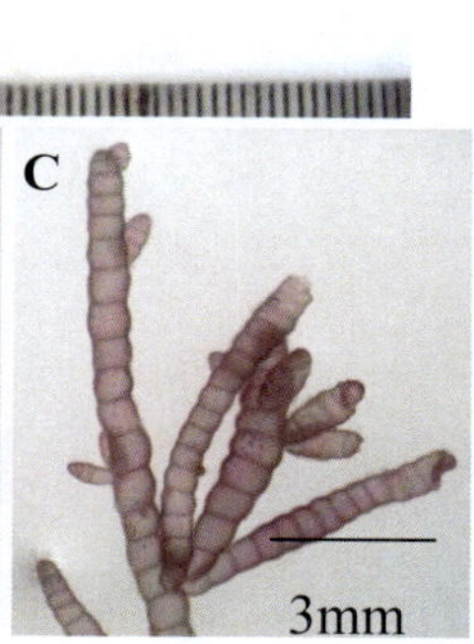

Champia Desvaux

***Champia parvula* (C. Agardh) Harvey**

Description: Thallus gelatinous, forming soft intestinal clumps to 10 cm high; translucent yellow to pale red in colour; branches variable, mostly alternate. Branches cylindrical to slightly flattened; apices tapering, bluntly pointed segments swollen, barrel shaped; 08–20 mm diameter and 1–5 mm diameter long. Joints slightly constricted; partitions thin.

Common: on sand-covered rocks; as epiphytes on seagrasses.

Distribution: Velankanni nagar, Seeniappadharga, Nochiurani, Pudumadam, Thalathoppu, Kanchirankudi, Keelakarai, Valinokkam, Mannapad, Valai island and Upputhanni island.

A. Habit of the plant

B. Middle part of filament with barrel shape segments.

C. Upper part of filament.

***Champia* Desvaux**

Champia Desvaux

Champia salicornioides **Harvey**

Description: Thallus gelatinous, forming soft robust tufts upto 12 cm high; translucent pale red in colour; branching irregular, somewhat clustered. Branches cylindrical, apices tapering, bluntly pointed. Segments short barrel, swollen, 1–4 mm diameter as long as wide; joints constricted; partitions thin; tetrasporangia spherical to 90 μm diameter; cystocarps urn shaped, scattered.

Rare: on rocks and small dead coral pieces scattered in intertidal region.

Distribution: Thonithurai and Upputhanni island.

A. Habit of the plant

B & C. Branches bearing urn shape cystocarps (c)

D. Cystocarp (c)

E. Transverse section of thallus

F. Longitudinal section of thallus

Champia Desvaux

REFERENCES: E. Lewis & Gonzalves (1962: 301, tables 1 – 3), S. Dixit (1996: 17), V. Krishnamurthy & H. Joshi (1970: 23), Untawale, Dhargalkar & Agadi (1983: 34), Silva *et al*. (1996: 348), Desikachary, V. Krishnamurthy & Balakrishnan (1998: 183).

Champia Desvaux

Champia somalensis Hauck

Description: Plants forming pyramidal tufts; upto 20 cm in height; dark red in color; tough and carteloginous in consistency; attachment to the substratum by means of rhizoids; basal half of plants practically devoid of branches; branching irregularly pinnate; branches alternate or opposite; main axis 2–3 mm wide, side branches slightly less. Cystocarps scattered, 800–900 µm in diameter and upto 900 µm high.

Rare: on rocks.

Distribution: Thonithurai

A. Habit of the plant

B. Branch bearing cystocarp (C)

C. Transverse section of thallus

D. Longitudinal section of thallus

Cheilosporum (Decaisne) Zanardini

***Cheilosporum* (Decaisne) Zanardini**

Thallus erect, formed by regular calcareous segments united by non calcified articulations; segments cylindrical at the base, becoming flattened, heart-like or winged at top. Structurally, segments are composed of a cortex with smaller cells and a medulla with alternating rows of long and short cells; uni-porate conceptacles immersed on the lateral expansion of the segments.

Key to the species:

1. Branches slender, cylindrical ......*C. acutilobum*
1. Branches marked flattened ....*C. spectabile*

Cheilosporum (Decaisne) Zanardini

REFERENCES: Decaisne (1842: 125), De Toni (1833), Boergesen (1943: 19 -21, fig. 5), Borgesen (1950: 7 – 8, fig. 1), Jagtap (1993: table I), Silva *et al*. (1996: 227).

Cheilospo
(Decaisne
Descripti
high; ere
chalky
branching
dichotomo
cylindrical
proliferous
0.5 mm
wide; con
portion. 1.
lower port
mid rib an
Rare: on
and subtid
Distributi
dharga anc
A. Habit o
B. Branc
lower pa
upper part
C.Transve
D. Longi
thalli

Cheilosporum (Decaisne) Zanardini

Cheilosporum acutilobum **(Decaisne) Piccone**
Description: Plant 4–7.5 cm high; erect and articulate; chalky red in color; branching regularly dichotomous; intergenicula cylindrical, slender and proliferous in the basal parts; 0.5 mm long and 0.5 mm wide; compressed in upper portion. 1.5 to 3 mm wide in lower portion with prominent mid rib and lateral wings.
Rare: on rocks in intertidal and subtidal regions.
Distribution: Seeniappa dharga and Kuduthalai
A. Habit of the plant
B. Branch with cylindrical lower part and flattened upper part
C. Transverse section of thalli
D. Longitudinal section of thalli

Cheilosporum (Decaisne) Zanardini

REFERENCES: V. Krishnamurthy & Balasundaram (1990: 7, fig. 5), Krishnamurthy *et al.* (1990: figs. 4 – 7), D. Rao *et al.* (1991: table 1, 2), Kalimuthu, Kaliaperumal & Ramalingam (1992: 32), Nair *et al.* (1993: 186), Kerkar (1994: table 1).

Cheilosporum (Decaisne) Zanardini

Cheilosporum spectabile **Harvey ex Grunow**

Description: Plants small and delicate, erect and articulated, dense in the littorals region; branch regularly dichotomous. Genicula narrow and line like. Intergenicula flattened; 0.5–1 mm long and 0.5–3.5 mm wide, with prominent mid rib and wings; the base of wings broadly inserted.

Common: on rocks and boulders

Distribution: Nochiurani, Pudumadam, Manapad, Kuduthalai, Idinthakarai, Leepuram, Vattakottai, Kanyakumari and Kovalam.

A. Habit of the plant

B. Plant with dichotomous branches.

C. Branches showing inserted wings.

D. Transverse section of thallus

E. Longitudinal section.

Chondria C. Agardh

Chondria C. Agardh
Thallus macroscopic, upto 25 cm in height, branches cylindrical, rarely flattened and well branched; last order branches usually spindle shaped with a constriction and pointed apices; apices sometimes blunt to club shaped in which case the apical cell may be sunken in an apical depression. A tuft deciduous and sometimes inconspicuous trichoblast is commonly found at the apices. Central axis distinct; 5 pericentral cells; inner cortex large; outer cortex small. Cystocarps ovate; tetrasporangia on the upper parts of the branches, spirally arranged, arising from the pericentrals, tetrahedral.

Key to the species:

1. Apex of ramuli depressed …….2
1. Apex of ramuli protruding ………3
2. Branching regular or irregular; tetrasporangia in clustered ramuli …..*C. dasyphylla*

2a.Branching with irregular ramuli …….. *C.dasyphylla var. stellata*

3. Branching regular, often geminate tetrasporangia in geminate ramuli ……..*C. capillaris*
4. Cystocarp single on ramuli ……5
4. Cystocarp several on ramuli ………..*C. seticulosa*
5. Cystocarp urceolate without a lateral spine ………*C. armata*

Chondria C. Agardh

REFERENCES: V. Krishnamurthy *et al*. (1990: fig. 4), R. Parekh & Chauhan (1990: table 1), Kamat *et al*. (1992: table 1), Kumar *et al*. (1992: table 1), Nair *et al*. (1993: 193), Silva *et al*. (1996: 479), Desikachary, V. Krishnamurthy & M. Balakrishnan (1998: 334).

Chondria C. Agardh

Chondria armata **(Kützing) Okamura**
Description: Plants violet red to pinkish red in color; 4–10 cm tall, with a rhizotomous prostrate portion attached to the rocky substratum by conical holdfast and cluster of rhizoids; erect fronds cylindrical, pinnately branched, ramuli dichotomously arranged; 3–8 mm long, cylindrical, up to 400 µm thick. Cystocarps on ramuli in the middle part.
Uncommon: attached on big rocks and boulders.
Distribution: Pamban, Pullivasal island, Hare island, and Mulli island
A. Habit of the plant
B. Plant with cylindrical and pinnately branched filament
C. Transverse section of the thallus
D. Longitudinal section of the thallus

Chondria C. Agardh

REFERENCES: Hudson (1778: 571), Wynne (1991: 317), Silva *et al*. (1996: 480), Desikachary, V. Krishnamurthy & M. Balakrishnan (1998: 337).

Chondria C. Agardh

Chondria capillaris **(Hudson) Wynne**
Description: Plants brownish violet in color; 5–20 cm tall; pyramidly arranging branches; ramuli 4–10 mm long, cylindrical when young, clavate when old. Cystocarps in modified branches, sessile, ovate to sub globose. Tetrasporangia densely scattered in young strichedia.
Rare: on dead coral pieces
Distribution: Nalupanai.
A. Habit of the plant
B. Branches with cylindrical ramuli
C. Branch with pyramid shape
D. Transverse section of thalli
E. Longitudinal section of thalli.

Chondria C. Agardh

REFERENCES: Untawale & Jagtap (1984: 101, table 1), M. Dave *et al*. (1987: table 1), Jagtap (1987: table 1), Krishnamurthy et al. (1990: figs. 4,5,7), Silva *et al*. (1996: 481), Desikachary, V. Krishnamurthy & M. Balakrishnan (1998: 335).

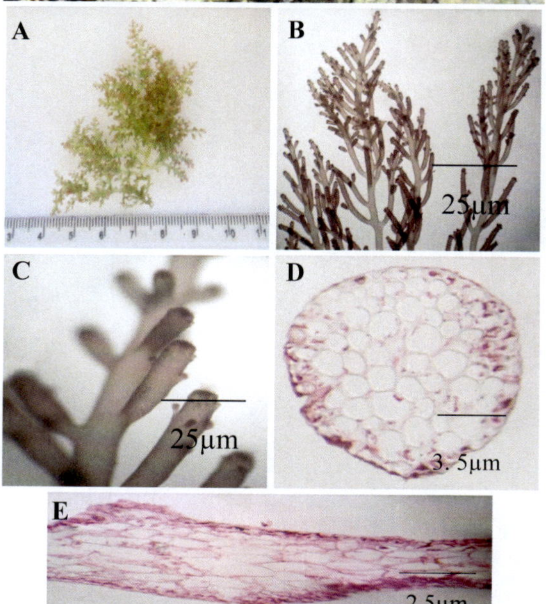

Chondria C. Agardh

***Chondria dasyphylla* (Woodward) C. Agardh**

Description: Thallus soft, sparse, bushy, 5–10 cm high; light yellow to dark red in color; branching irregular, cylindrical, 1.0–1.5 mm diameter. Branchlets numerous, cylindrical to cone shaped; opposite or spirally arranged, constricted at base. Tetrasporangia spherical, 40–170 μm diameter. Cystocarps spherical to oval, 1 mm diameter.

Common: attached on rocks, small pebbles, coral pieces

Distribution: Kottaipattinam, Pamban light house, Ariyankundu, Nochiurani, Thalathoppu, Manapad, Pullivasal island and Upputhanni island.

A. Habit of the plant
B. Plant showing opposite branches
C. Branch with cone shaped branchlets
D. Transverse section of thallus
E. Longitudinal section of thallus

Chondria C. Agardh

REFERENCES: Borgesen (1934: 47), K. Srinivasan (1965: 248), S. Dixit (1966: 20), V. Krishnamurthy & H. Joshi (1970: 25), Silva *et al*. (1996: 482), Desikachary, V. Krishnamurthy & M. Balakrishnan (1998: 336).

Chondria C. Agardh

***Chondria dasyphylla* (Woodward) C. Agardh var. *stellata* Børgesen**

Description: This variety is distinguished from type species by irregular ramification of the branches. The ramuli may be produced both on main axis and on branches and branchlets. A cross section of the main axis show five large pericentral cells surrounding an axial cell; inner cortex of larger cells and outer cortex of small cells in one or two layers.

Rare: on rocks

Distribution: Villoondi theertham

A. Habit of the plant

B. Branches showing small ramuli.

C. Transverse section showing 5 pericentral cells (P) surrounding an axial cell.

D. Longitudinal section of thallus

Chondria C. Agardh

REFERENCES: S. Dixit (1980: 76), V. Krishnamurthy (1980: 55), Untawale, Dhargalkar & Agadi (1983:37), P. Subba Rao *et al.* (1992: 181), Silva *et al.* (1996: 484), Desikachary, V. Krishnamurthy & M. Balakrishnan (1998: 336).

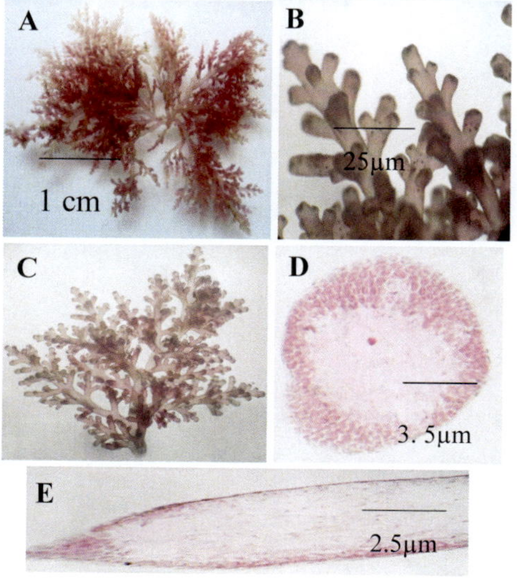

Chondria C. Agardh

Chondria seticulosa **(Forsskål) C. Agardh**

Description: Plants dark red in color; prostrate, slender, cylindrical, 3–8 cm in height, 0.5 mm thick; more branches and branching irregularly dichotomous and sometimes pinnate, slightly alternate at base. Thallus composed of filamentous medulla and cortex of anticlinal rows of cells. Cystocarps in ultimate branchlets; sessile, sometimes solitary.

Habitat: Common: on rocks.

Distribution: Seeniappadharga, Manapad, Panjal, Chinnamuttam, Kanyakumari, Kovalam and Koswari island.

A. Habit of the plant

B. & C. Plant showing irregular dichotomous branches.

D. Transverse section of thallus.

E. Longitudinal section.

***Chondrocanthus* Kützing**

Thallus erect, foliose, compressed or cylindrical; usually becomes cartilaginous when drying. Structure multiaxial with several layers of small cells in the cortex and filamentous medulla. Cystocarps prominent or immersed in the thallus, carposporangia mixed with filaments, tetrasporangia cruciately divided, produced in immersed sori.

Key to the species:

Plant irregularly dichotomous with spinous branches

…….*C.acicularis*

Chondrocanthus Kützing

REFERENCES: G. Rao & Umamaheswara Rao (1986: 103), Chennubhotla, Kaliaperumal & Kalimuthu (1987: 108 – 1009, fig. 13 D), Vidyavathi & Sridhar (1991: fig. 2) Kalimuthu, Kaliaperumal & Ramalingam (1992: 123), Silva *et al.* (1996: 290), Desikachary, V. Krishnamurthy & M. Balakrishnan (1998: 128).

Chondroc
(Roth) Fr
Descriptio
in color;
cylindrical
0.5 mm th
and bran
dichotomo
pinnate, s
base. Tha
filamentou
cortex of
cells. Cyst
branchlets;
solitary.
Common:
Distributio
Seeniappa
Panjal, Ch
Kanyakum
Koswari is
A. Habit o
irregular d
branches.
B. Transve
showing fi
(M)and co
C. Longitu

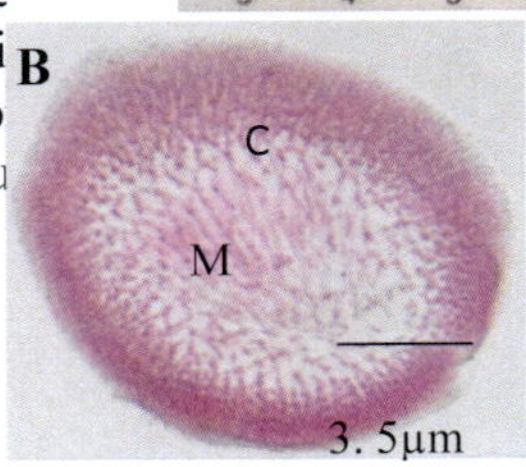

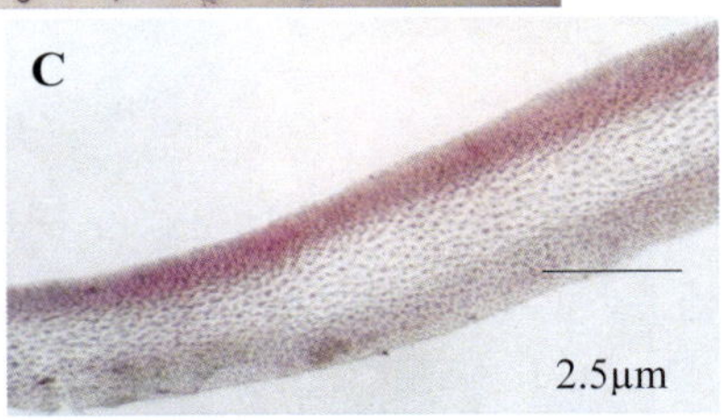

Chondrocanthus Kützing

***Chondrocanthus acicularis* (Roth) Fredericq**

Description: Plants dark red in color; prostrate, slender, cylindrical, 3–8 cm in height, 0.5 mm thick; more branches and branching irregularly dichotomous and sometimes pinnate, slightly alternate at base. Thallus composed of filamentous medulla and cortex of anticlinal rows of cells. Cystocarps in ultimate branchlets; sessile, sometimes solitary.

Common: on rocks.

Distribution: Seeniappadharga, Manapad, Panjal, Chinnamuttam, Kanyakumari, Kovalam and Koswari island.

A. Habit of plant showing irregular dichotomous branches.

B. Transverse section showing filamentous medulla (M)and cortex (C)

C. Longitudinal section.

***Claudea* Lamouroux**

Thallus has stipe continuing into a stout axis bearing a net like structure formed 3-4 order of narrow blade like segments distinguishable into primary, secondary and tertiary blades; each order of blade arising from midrib of blades of preceding order; cystocarp dome shaped, sessile. Tetrasporangia in sori on the short blades.

Key to the species:

Plants small; fronds with primary and secondary blades unilaterally produced ……*C. elegans*

Claudea Lamouroux

REFERENCES: Kylin (1956: 437, fig 348C), V. May (1965: 395), S. Dixit (1996: 19), R. Norris (1987: 311 – 315, figs. 1 – 6), Silva *et al*. (1996: 453), Desikachary, V. Krishnamurthy & M. Balakrishnan (1998: 263).

Claudea Lamouroux

Claudea elegans **Lamouroux**

Description: Plants 1–3 cm in height; fronds composed of primary branches arising from cylindrical main axis; blade like, parallelly arranged; secondary branches produced on primary branches, often connecting primary branches; cystocarps usually on apex of mid portion of primary and secondary blades.

Rare: attached on coral stones in 10 meter deep.

Distribution: Mully island.

A. Habit of the plant

B. & C. Plant showing net like branches.

***Coelathrum* Borgesen**

Thallus erect, 4-10 cm in height, delicate and hollow, cylindrical, regularly bifurcated, with constriction between bifurcations with 1-2 layers of large cells facing the internal cavity and 1-2 layers of small chloroplast rich cells on the surface; gland cells produced from the inner cells. Cystocarps prominent, scattered on the surface of the thallus; tetrasporangia scattered and cruciately divided.

Key to the species:

1. Plants large, upto 25 cm high; segments small with multilayered wall. ....*C. muelleri*

1. Plants smaller, upto 10 cm; segments large with 3 layered wall. ....*C. opuntia*

Coelarthrum Børgesen

REFERENCES: V. Krishnamurthy & H. Joshi (1970: 22), Gopalakrishnan (1971: table I, II), M. Dave *et al*. (1987: table 1), R. Parekh & V.D. Chauhan (1990: table 1), Silva *et al*. (1996: 358), Desikachary, V. Krishnamurthy & M. Balakrishnan (1998: 169).

Coelarthr
(Sonder) I
Descripti
Plants b
erect; a
substratum
basal por
distinctly
trichotomc
segments l
ovoid, 15-
3–6 mm v
mm long a
Wall of
layered.
hemispher
the segme
Uncommo
and rocks i
Distributi
A. Habit o
trichotomc
B. Branch
cylindrical
segments.
C. Hollow
section.

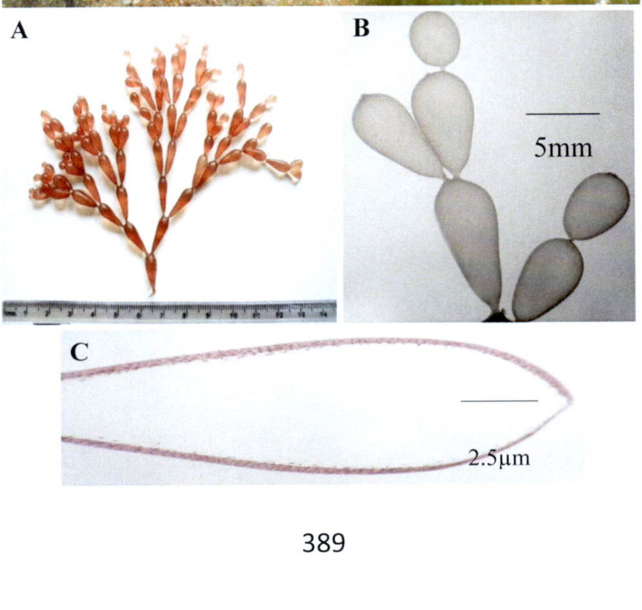

Coelarthrum Børgesen

Coelarthrum muelleri (Sonder) Børgesen

***Description*:**

Plants bright red in color, erect; attached to the substratum by a stoloniferous basal portion; erect fronds distinctly articulated; di or trichotomously branched; segments hollow; cylindrical, ovoid, 15–24 mm long and 3–6 mm wide; joints upto 2 mm long and 0.5 mm broad. Wall of segments 4–6 layered. Cystocarps hemispherical, scattered over the segments.

Uncommon: on coral stones and rocks in 10 meter deep.

Distribution: Koswari island

A. Habit of the plant showing trichotomous branches

B. Branches showing cylindrical and ovoid segments.

C. Hollow longitudinal section.

Coelarthrum Børgesen

REFERENCES: Ohno & Mairh (1982: table 1), Untawale, Dhargalkar & Agadi (1983: 33), Chennubhotla, Kaliaperumal & Kalimuthu (1987: 108 - 109), Silva *et al*. (1996: 358), Desikachary, V. Krishnamurthy & M. Balakrishnan (1998: 170), Monika Ram *et al*. (2000a: 45 – 50).

Coelarthru
(Endlicher
Descriptio
Plants bri
attached b
creeping sy
8–10 cm l
wide; ob
segments
Cystocarps
surface of s
Rare: on
lower inter
deep.
Distributio
Mully islan
A. Habit of
oblong seg
B. Transv
segments
C. Longitu
segment w
wall.

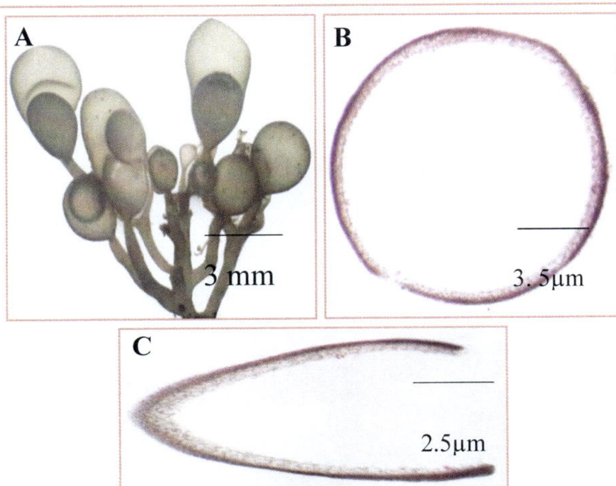

Coelarthrum Børgesen

Coelarthrum opuntia **(Endlicher) Børgesen**

Description

Plants bright red in color, attached by a stoloniferous creeping system. Erect fronds 8–10 cm long and 6–8 mm wide; oblong, walls of segments three layered. Cystocarps scattered on the surface of segments.

Rare: on coral stones in lower intertidal to 10 meter deep.

Distribution: Manapad and Mully island.

A. Habit of the plant showing oblong segments.

B. Transverse section of segments

C. Longitudinal section of the segment with three layered wall.

Corynomorpha J. Agardh

***Corynomorpha* J. Agardh**

Plants erect, cylindrical, clavate, upto 15cm tall, dark red in colour. Thallus club shaped, mostly unbranched and bifurcated once, firm, attach to the substratum by a discoidal base. A cross section of the thallus shows a cortical portion of bead like cells and medulla of loose and anastomosing star shaped cells. Tetrasporangia cruciately divided.

Key to the species:

Fronds cylindrical, simple, rarely dichotomously divided
…..*C.prismatica*

Corynomorpha J. Agardh

REFERENCES: V. Krishnamurthy & H. Joshi (1970: 20), V. Krishnamurthy (1980: 54), Untawale, Dhargalkar & Agadi (1983: 31), Nair *et al*. (1986: 738), Desikachary, V. Krishnamurthy & M. Balakrishnan (1990: 228 – 230, fig. 44.pl.XXVII), Silva *et al*. (1996: 185).

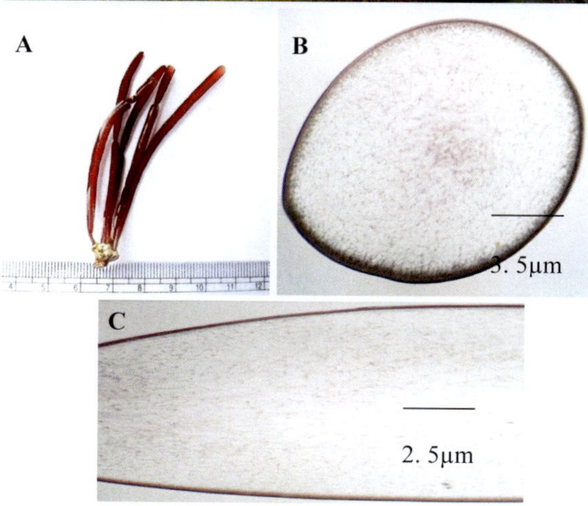

Corynomorpha J. Agardh

Corynomorpha prismatica **(J. Agardh) J. Agardh**

***Description*:**

Plants 8–15 cm tall; attached to rocks by small holdfast; erect axes 1–1.5 mm thick at basal portion, 3–4 mm in diameter above, clavate with rounded tufts. Simple, or sometimes sub-dichotomously branched. Thallus composed of filamentous medulla and compact cortex. Cystocarps at swollen apices.

Common: attached on rocks in shaded places.

Distribution: Manapad, Mully island, Appa island and Koswari island.

A. Habit of the plant showing simple filaments with rounded apex

B. Transverse section of filament showing filamentous medulla (M)

C. Longitudinal section of the filament

***Cryptonemia* J. Agardh**

Thallus foliose, branching mostly in one plane, sometimes with marginal proliferations; thallus stiff with stipe at the base; a midrib more or less developed, present in most species. Small, well pigmented cells in anticlinal rows form the cortex; ganglionic cells can be seen in the sub cortex and filamentous medulla. Cystocarp immersed in the thallus, tetrasporangia cruciate.

Key to the species:

1. Erect fronds flabellate, apices broadly rounded, basal parts of thallus often worn down to the mid rib

   ………*C. lomation*

2. Plants more than 15 cm tall, fronds palmately or irregularly divided; proliferations marginal

   ………*C. coriacea*

Cryptonemia J. Agardh

REFERENCES: V. Krishnamurthy & H. Joshi (1970: 19), Untawale, Dhargalkar & Agadi (1983: 30), P.V. Subba Rao *et al.* (1985: 33), Desikachary, V. Krishnamurthy & M. Balakrishnan (1990: 261, pl. XL), P.V. Subba Rao *et al.* (1992: 182), Silva *et al.* (1996:190).

Cryptonen
Schmitz
Descriptio
Plants lar;
cm tall, ai
palmately
broad; le:
margin
subtidal pl
substratum
Rare: on
region.
Distributi
A. Habit o
palmate br
B. Thallus
proliferate
C. Transve
thallus
D. Longitu
thallus sho
medulla (M

Cryptonemia J. Agardh

Cryptonemia coriacea **Schmitz**

***Description*:**

Plants larger, more than 15 cm tall, and thicker. Thallus palmately lobed upto 2 cm broad; leathery in texture; margin proliferated. A subtidal plant found on rocky substratum and small caves.

Rare: on rocks in subtidal region.

Distribution: Manapad

A. Habit of the plant showing palmate branches

B. Thallus showing proliferated margin.

C. Transverse section of thallus

D. Longitudinal section of the thallus showing filamentous medulla (M)

Cryptonemia J. Agardh

REFERENCES: Borgesen (1937: 324), Chacko, Mahadevan & Ganesan (1955: 12), S. Dixit (1966: 14), Umamaheswara Rao (1969: 44), V. Krishnamurthy & M. Balakrishnan (1990: 259), Silva *et al*. (1996: 191).

Cryptonen
(Bertoloni
Descriptio
Plants dar
15 cm
pinnately
broadly r
fronds smc
Uncommo
and on cor
Distributic
Pudumada
Keelakarai
Kanyakum
Koswari is
A. Habit o
B. Fronds
margin
C. Trans
frond
D. Longi
frond v
medulla (M

Cryptonemia J. Agardh

***Cryptonemia lomation* (Bertoloni) J. Agardh**

***Description*:**

Plants dark red in color; upto 15 cm tall; branches pinnately arranged; apices broadly round. Margin of fronds smooth.

Uncommon: in subtidal rocks and on corals.

Distribution: Nalupanai Pudumadam, Kanchirankudi, Keelakarai, Kuduthalai, Kanyakumari Kovalam and Koswari island.

A. Habit of the plant

B. Fronds showing smooth margin

C. Transverse section of frond

D. Longitudinal section of frond with filamentous medulla (M)

***Dasya* J. Agardh**

***Dasya* C. Agardh**

Plants with a stout main branch, covered with spirally disposed, uniseriate bifurcated branchlets. Main branches with 5 or rarely 4 pericentral cells, usually covered by a variable amount of rhizoidal cortication. Carposporangia in branching series, tetrasporangia in stalked strichedia, on pseudolateral tetrahedral.

Key to the species:

Pseudo laterals sub dichotomously branched, not curved upwards *...D.flagellifera*

Dasya J

REFERENCES: V. Krishnamurthy & H. Joshi (1970: 27) & Mairh (1978: 35), Untawale, Dhargalkar & Agadi (198 Krishnamurthy & Vardarajan (1991a: 48, figs. 11 - 13), S (1996: 436), Desikachary, V. Krishnamurthy & M. Ba (1998: 278).

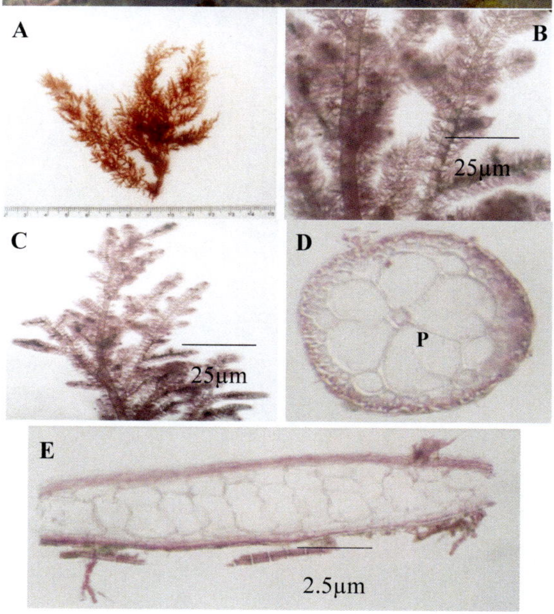

Dasya J. Agardh

Dasya flagellifera **Børgesen**

Description

Plants about 14 cm in height; branching irregular with frequent adventious branches arising on older branches; primary axis about 1 mm in thickness with an axial cell surrounded by 5 pericentral cells; pseudolaterals flagella like, curve upwards. Tetrasporangia in strichedia, which are ovate-oblong.

Rare: attached on coral stones in 5 meter deep.

Distribution: Koswari island.

A. Habit of the plant

B. & C. Plant showing flagella like branches

D. Transverse section showing an axial cell surrounded by 5 pericentral cells (P)

E. Longitudinal section of frond.

Dictyurus Bory de Saint-Vicent

Dictyurus Bory de Saint-Vicent

Plants erect, radially organized upto 60cm in height; bushy and cylindrical bifurcated main axis from which an orderly fusion uniseriate branches produce a sort of net give the thallus a prismatic outline. Cortication seen on axis. Plants attached by discoid holdfast.

Key to the species:

Axes with central axial cells and 5 pericentral cells ….*D.purpurascens*

Dictyurus Bory de Saint-Vicent

REFERENCES: Jagtap (1987: table 1), Untawale, Reddy & Ambiye (1989: table 1), Krishnamurthy et al. (1990: 271), V. Krishnamurthy & Varadarajan (1992: 141, figs. 20 – 25), Silva *et al*. (1996: 438), Desikachary, V. Krishnamurthy & M. Balakrishnan (1998: 438).

Dictyurus
Bory de Sa
Descriptio
Plants atta holdfast w bearing a blades; bl laterals succession; axial cells cells, spe like; cyst protruding laterals.
Uncommo
Distributio
Kanyakum Koswari isl
A. Habit of net like bra
B. Transve frond
C. Longitu frond.

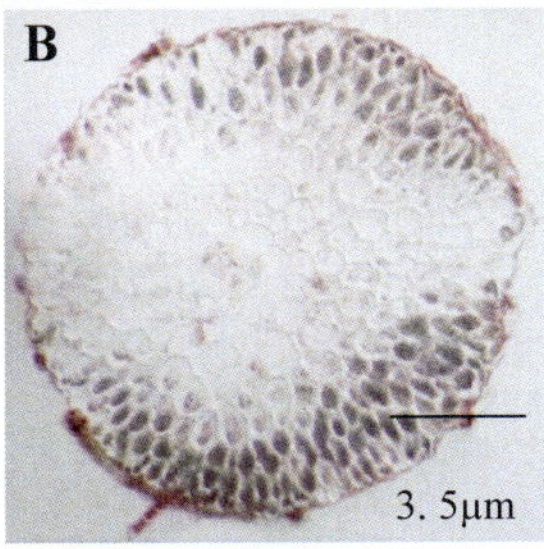

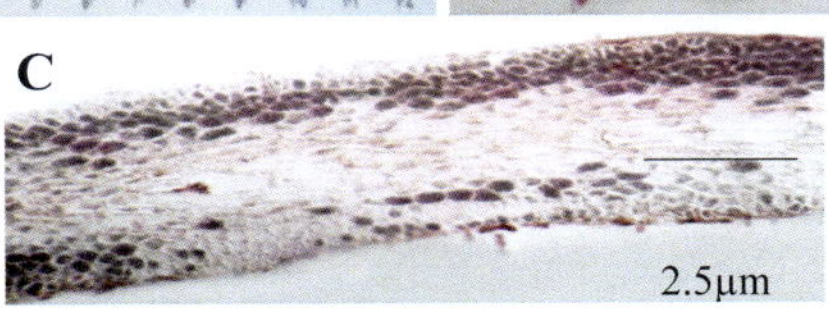

Dictyurus Bory de Saint-Vicent

Dictyurus purpurascens Bory de Saint-Vicent

Description:

Plants attached by a disc like holdfast with an erect axis, bearing a series of net-like blades; blades initiated as laterals as acropetal succession; axis with row of axial cells and 5 pericentral cells, spermatangia is net like; cystocarps spherical, protruding out of the network laterals.

Uncommon: on rocks.

Distribution: Manapad, Kanyakumari, Kovalam and Koswari island.

A. Habit of the plant showing net like branch.

B. Transverse section of the frond

C. Longitudinal section of the frond.

Digenea J. Agardh

Digenea J. Agardh

Plants light to dark red, tough, bushy and erect, 5-25cm tall. The erect branches are cylindrical, dichotomously to laterally branched, sparsely bifurcated and beset with short, unbranched branchelts with an apical cell and trichoblast. The polysiphonous nature of the thallus is obscured by regular cell division in the main branches, but can be seen in the short laterals. Cystocarps oval, at the tip or in mid portion of detrimental branch. Tetrasporangia in linear row in upper part of the branches.

Key to the species:

Branches pinnate, branchlets with pointed apices

…. *D.simplex*

Digenea J. Agardh

REFERENCES: Borgesen (1945: 39 - 40), Borgesen (1953: 57 - 58), Jagtap (1992: 57), Jagtap (1993: table 1), Silva *et al*. (1996: 490), Desikachary, V. Krishnamurthy & M. Balakrishnan (1998: 309), Oza & Ramavat (2000: 22).

Digenea J. Agardh

Digenea simplex **(Wulfen) C. Agardh**
***Description*:**
Plants dark red to brownish red in color; attached by a basal crust; 5–25 cm or more in height; upto 5 cm thickness; pinnately branched above; branches bearing alternating branchlets; tapering to acute apices; branches corticated at base only. Cystocarps ovoid to spherical in shape occurring on determinate branches, 330 µm long and 250 µm wide.
Uncommon: on rocks and stones upto 10 meter deep.
Distribution: Manapad, Pullivasal island and Koswari island.
A. Habit of the plant
B. Thallus showing pinnately branched fronds
C. Branch with cystocarp (C)
D. Transverse section
E. Longitudinal section showing tube like cylinder of surface cells.

Enantiocladia Falkenberg

***Enantiocladia* Falkenberg**

Plant erect, 5-10 cm in height, ribbon like, slightly concave thallus; dark red, turns black when drying; Thallus pinnately branched; with an evident of mid rib formed by a central row of cells, five pericentral cells and the cortical layer; the lateral pericentral cells expand to from a wing on each side of the mid rib; with a central layer of large cells and one or more layers of cortical cells on each face; wings tooth like or serrated along the margins. Tips of branches inrolled. Cystocarps global with stout stalk; Tetrasporangial strichidia formed by transformation of ultimate branches

Key to the species:

Thallus ribbon like, branches short with incurved apices ...*E.prolifera*

Enantiocladia Falkenberg

REFERENCES: Sobha & Nair (1985: 23), Nair *et al.* (1986: 738), Chennubhotla, Kaliaperumal & Kalimuthu (1987: 108 – 109), V. Krishnamurthy *et al.* (1990: 271), Silva *et al.* (1996: 493), Desikachary, V. Krishnamurthy & M. Balakrishnan (1998: 326).

Enantiocla
Falkenber
Descriptio
Thallus er
height; ribl
branched
branchlets
incurved;
pericentral
axes deve
extensions;
branchlets
like. Cysto
stout stal
formed by
ultimate br
Uncommo
shaded pla
Distributio
Manapad,
Kovalam.

A. Habit of
B. Transve
frond.
C. Longitu
frond.

Enantiocladia Falkenberg

Enantiocladia prolifera Falkenberg

***Description*:**

Thallus erect, 5–10 cm in height; ribbon like; sparingly branched along the margin; branchlets short, apices incurved; axis with five pericentral cells, the lateral axes developing wing like extensions; ultimate branchlets re- curved, hook like. Cystocarps globose with stout stalk; tetrasporangia formed by transformation of ultimate branches.

Uncommon: on rocks in shaded place

Distribution: Valinokkam, Manapad, Kanyakumari and Kovalam.

A. Habit of the plant

B. Transverse section of the frond.

C. Longitudinal section of the frond.

***Euptilota* Kützing**

Plants erect, flat, subdichotomously or pinnately branched in one plane. Cells of the primary axis and branches covered by cortical layer formed by downwardly growing rhizoids; spermatangia globose or irregular clusters on ultimate branches, cystocarp terminal.

Key to the species:

Main axis and branches except ultimate ones corticated; tetrasporangia with cruciate tetraspores

…….*E. articulata*

Euptilota Kützing

REFERENCES: Lucas & Perrin (1947: 338, fig. 164), Umamaheswara Rao (1969: 47), Umamaheswara Rao (1974e: 65, figs. 12 - 14), S. Dixit (1980: 75), Silva *et al.* (1996: 409), Desikachary, V. Krishnamurthy & M. Balakrishnan (1998: 249).

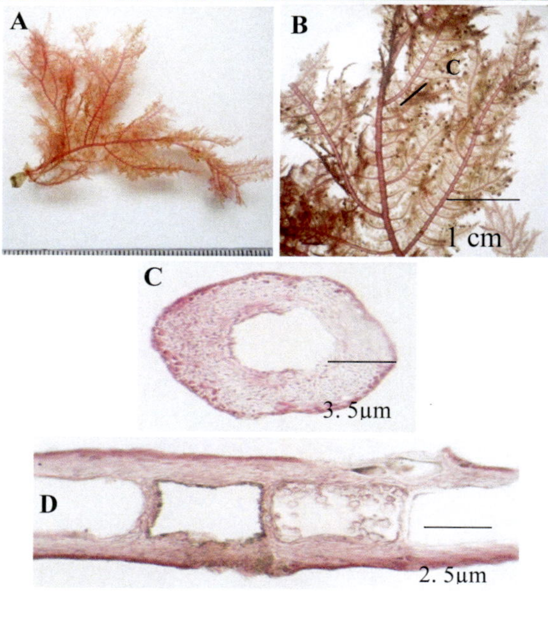

Euptilota Kützing

***Euptilota articulata* (Kützing) Kützing**

***Description*:**

Thallus pink to red in color; feather like; with monosiphonous axis as well as branches expect the ultimate branchlets covered by rhizoidal cortex. Cystocarps terminal on ultimate branches surrounded by involucral filaments. Tetrasporangia borne at the apices of ultimate branchlets.

Rare: on coral stone and small coral pieces at 10 meter deep.

Distribution: Mully island

A. Habit of the plant

B. Feather like branch with cystocarp (C) on tip of the branch.

C. Transverse section

D. Longitudinal section

***Galaxaura* Lamouroux**

Plants usually about 10 centimeters long. Thallus cylindrical or ribbon shaped, regularly bifurcated and with a variable degree of lime deposition; branches with a smooth surface or with delicate pigmented hairs. Cross section of a branch shows a medulla composed of slender, colourless filaments surrounded by radial filaments with larger, chloroplast-rich cells; in thallus expanded cells are closely packed. In some species, the morphology of the sexual phase is different from the morphology of the sporophytic phase. In the past, the two phases were classified as different species and this still complicates the taxonomy of the genus.

Key to the species:

Both medulla and cortex 3 layered ......*G.rugosa*

Nemaliales-Galaxauraceae

Galaxaura Lamouroux

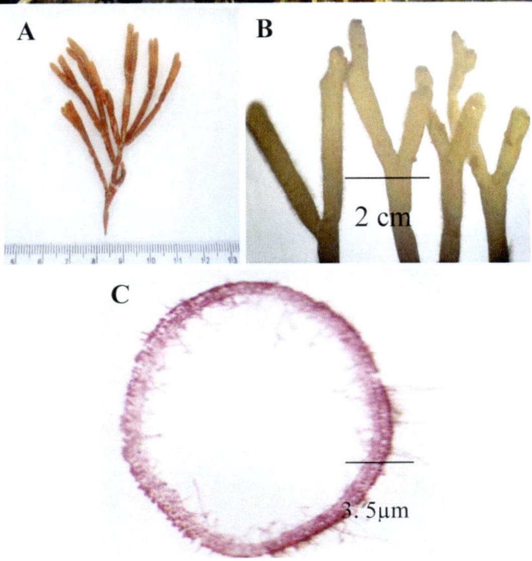

Galaxaura Lamouroux

***Galaxaura rugosa* (Ellis & Solander) Lamouroux**
***Description*:** Thallus bushy, stiff, compact, dome shaped to 12 cm height; branching dichotomous; calcification heavy; branches cylindrical, 1–2 mm diameter. Joints Medullary filaments 8–20 μm diameter, colorless, cortex 3–4 cells thick. Holdfast inconspicuous.
Common: on coral fragment, rocks in 1 m deep.
Distribution: Munaikadu, Pamban light house, Sadaimuniyanvalasai, Manapad, Kuduthalai, Idinthakarai, Kanyakumari and all 20 islands of Gulf of Mannar.
A. Habit of the plant
B. Plant showing dichotomous branches.
C. Transverse section of the branch.

Gelidiella J. Feldmann & G. Hamel

Gelidiella J. Feldmann & G. Hamel

Plants cylindrical to flattened with erect branches produced from creeping branches attached to the substratum; a single apical cell is present on each branch. The thallus is formed by a medulla of well packed longitudinal filaments with thick cell walls, covered by a few layers of globose cells with chloroplast. The genus is morphologically similar to *Gelidium* and *Pterocladia* but does not have thick walled rhizoidal cells.

Key to the species

Plant regularly pinnate, tetrasporangia in swollen strichedia, irregularly arranged ......*G.acerosa*

Gelidiella J. Feldmann & G. Hamel

REFERENCES: P. V. Subba Rao (1997: 43), Kaliaperumal *et al.* (1998: 141 – 151), Mairh *et al.* (1998: 110 – 126), A. Srinivasa Rao (1998: 22), Zaidi & Mairh (1999a: 28), Redekar & Raje (2000: 41 – 43).

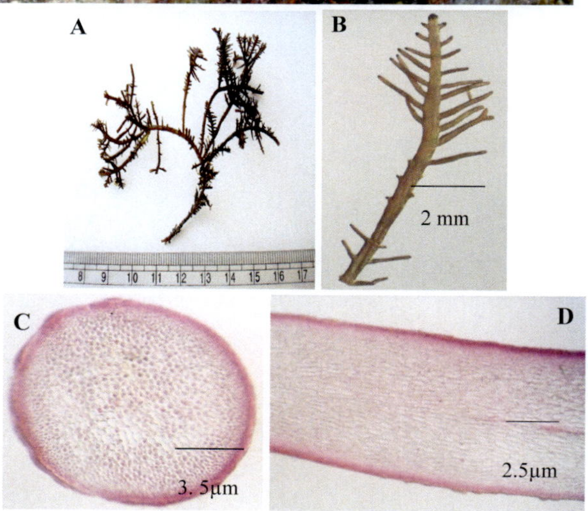

Gelidiella J. Feldmann & G. Hamel

Gelidiella acerosa **(Forsskål) J. Feldmann & G. Hamel**
Description: Plants growing from a stoloniferous holdfast, uprights 4–7 cms long forming tufts of tough and flexible axes, brownish black in shaded places to greenish straw colored when directly exposed to sunlight. Erect axis cylindrical in basal parts to compress in upper parts; 1 mm in diameter; mostly pinnately branched, rarely radially or unilaterally branched. Main axis frequently recurved; 2–6 mm long; gradually shorter towards the branch apices. Internal structure consisting of a small ovoid-darkly pigmented cells, centrally grading to larger, irregularly shaped medullary cells.
Common: on coral stones and rocks upto 5 meter deep.
Distribution: Pamban light house, Nalupanai, Villoonditheertham, Ariyankundu, Rameshwaram, Pamban, Thonithurai, Velankanni nagar, Nochiurani, Pudumadam, Sallimunthal, Kanchirankudi, Keelakarai light house, Bharathi nagar, Ervadi, Sadaimunyian valasai, Valinokkam, Manapad, Leepuram, Kanyakumari, Kovalam and all 20 islands of Gulf of Mannar.
A. Habit of the plant showing pinnately branched thallus
B. Branch tapering towards apex
C. Transverse section of frond
D. Longitudinal section

***Gelidiopsis* Schmit**

The thallus is erect, 3-10cm tall, irregularly branched with a very firm, wire like consistence. Branches cylindrical to compressed. A cortical layer of small cells, a subcortex of larger cells and solid medulla of small cells with thick cell walls. Cystocarps are oval in shape developing singly or in groups at the tips of the branches. Tetrasporangia in large groups in swollen branches; mostly tetrahedrally divided.

Key to the species:

Fronds irregularly or sparsely branched, branches simple with obtuse apex ....*G. variabilis*

Gelidiopsis Schmit

REFERENCES: Umamaheswara Rao (1990: 324, table 2, 5, 7, 8), Kalimuthu, Kaliaperumal & Ramalingam (1992: 123 - 126), Maya & Nair (1992: 26), Nair *et al.* (1993: 186, 192, 193) Silva *et al.* (1996: 362), Desikachary, V. Krishnamurthy & M. Balakrishnan (1998: 176).

Gelidiopsis
Agardh) S
Descriptio
Plants ver
well branc
stolons; er
high; dark
tough but
cylindrical
slightly co
upper por
apex; in
consisting
pigmented
composed
cells grac
angular in
Medulla co
cells of mi
Common:
overhangs
region to 1
Distributio
.Velankann
Seeniappad
Nochiurani
Thalathopp
Kanchirank
light hous
Ervadi,
valasai,
Manapad,
Kanyakum
all 20 isl
Mannar.

A. Habit of
B. Plant
branches

***Gelidiopsis variabilis* (J. Agardh) Schmit**

***Description*:** Plants very well attached by well branched thin pinkish stolons; erect fronds 5–8 cm high; dark red in colour; tough but flexible; all axis cylindrical and very thin; slightly compressed in the upper portion with obtuse apex; internal structure consisting of darkly pigmented outer cortex composed of 4 layers of small cells grading into larger angular inner cortical cells. Medulla composed of hyaline cells of mixed size.

Common: on under rock overhangs in lower intertidal region to 1 m deep.

Distribution: Thonithurai Velankanni nagar, Seeniappadharga, Nochiurani, Pudumadam, Thalathoppu, Sallimunthal, Kanchirankudi, Keelakarai light house, Bharathi nagar, Ervadi, Sadaimuniyan valasai, Valinokkam, Manapad, Leepuram, Kanyakumari, Kovalam and all 20 islands of Gulf of Mannar.

A. Habit of the plant

B. Plant with irregular branches

C. Upper part of branch with obtuse apex.

D. Transverse section of branch

E. Longitudinal section of branch

Gelidium Lamouroux

Gelidium Lamouroux
Thallus firm and erect, cylindrical to flattened; branching pinnate or irregular. Attached to the substratum by disks produced from prostrate branches. Cystocarp bilocular with carpospoangia produced on both sides of the central placenta.

Key to the species:

1. Plants more than 5.0 cm tall; 0.5–2 mm broad with blunt apices and sunken growing points ....*G. micropterum*
1. Erect fronds upto 1.5 cm high; simple, irregularly branched; 0.2–0.75 mm wide; cylindrical below and flattened above with rounded or attenuate tips ....*G. pusillum*

Gelidium Lamouroux

REFERENCES: Nair *et al.* (1986: 736), Untawale, Reddy & Deshmukhe (1989: table 1), Desikachary, V. Krishnamurthy & M. Balakrishnan (1990: 195 – 197, figs. 41 A-D, pl.XXIII), Maya & Nair (1992: 26), Sindhu & Panikkar (1992: 122, figs. 1 – 11), Silva *et al.* (1996: 139).

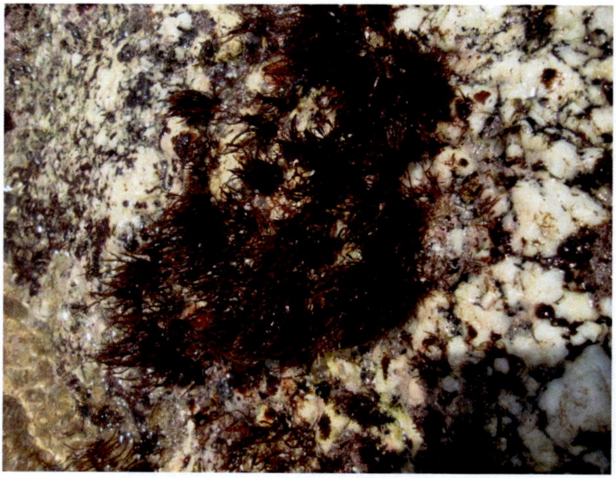

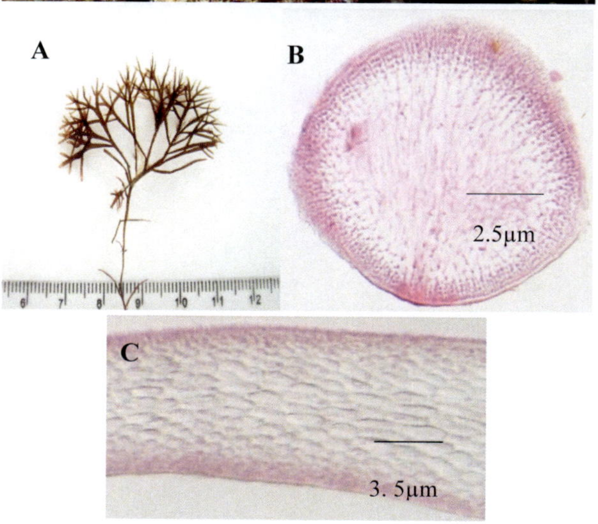

Gelidium Lamouroux

Gelidium micropterum **Kützing**

Description**:**

Plants dark red in colour, upto 5 cm tall; growing as tufts or patches; flattened, 0.5–2 mm broad with blunt apices and sunken growing points; branching bi or tri pinnate or irregular sometimes; tetrasporangia confined to short swollen fertile tips with rounded ends, scattered all over the surface of the fertile portion; cystocarp swollen, sub-spherical.

Uncommon: on rocks in shaded place.

Distribution: Sullimunthal, Keelakarai, Panjal, Kanyakumari, Kovalam, Krusadai island and Manali island

A. Habit of the plant showing pinnate branches

B. Transverse section of plant

C. Longitudinal section of plant

Gelidium Lamouroux

REFERENCES: Vidyavathi & Sridhar (1991: fig. 2), Kaliaperumal & Ramalingam (1992: 123), Raghukumar, Nagarkar & Raghukumar (1992: fig. 1, table1), Ramaiah & Chandramohan (1992: 74, 75), Sindhu & Panikkar (1992: 122, 125, figs. 12 - 20), Silva *et al.* (1996: 140).

Gelidium Lamouroux

Gelidium pusillum **(Stackhouse) Le Jolis**
Description: Thallus erect or creeping, tough, wiry, 1–2 cm high; dark marron black; growth from a single apical cell. Blades strap shaped above, cylindrical below to 1.5 mm wide, 110–170 µm thick; branchlets opposite, pinnate to regular. Holdfast in conspicuous pad like; irregularly spaced. Cystocarps with 2 chambers, embedded.
Common: on rocks or other hard surfaces in wave exposed areas often with high wave shock intertidal to 8 m deep.
Distribution: Pamban light house, Villoonditheertham, Ariyankundu, Pamban Seeniappadharga, Nochiurani, Thalathoppu, Kanchirankudi, Valinokkam, Leepuram Chinnamuttam, Kanyakumari, Kovalam and all 20 islands of Gulf of Mannar
A. Habit of the plant showing pinnate branches
B. & C. Branches showing strap shape
D. Transverse section
E. Longitudinal section

Gigartina Stackhouse

Thallus cylindrical or compressed, flat or foliose, branched laterally or forked; plant mostly cartilaginous, rarely membranous. Growth is by apical meristematic cells terminating as multi- axial structure. Medulla and cortex are net like structure. The outer cortex is compact, small cells.

Key to the species:

Plant cylindrical, branches dichotomous or pinnate .... *G.acicularis*

Gigartina Stackhouse

Gigartina
(Wulfen) I
Descriptio
Fı
cylindrical
0.5mm t
elongated,
compresse
irregularly
sometimes
Cystocarps
branchlets,
tetrasporar
sori, found
of the bran
Rare: Grc
Distributic
A. Habit
pinnate brı
B. Branc
branchlets.
C. Trans
thallus
D. Longi
thallus.

Gigartina Stackhouse

Gigartina acicularis **(Wulfen) Lamour**

Description: Frond slender, cylindrical, 3-8cm in height, 0.5cm thick, branches elongated, sometimes compressed, branches irregularly dichotomous or sometimes pinnate. Cystocarps in ultimate branchlets, sessile tetrasporangia collected in sori, found on the lower parts of the branchlets.

Rare: Grown on rocks.

Distribution: Manapad

A. Habit of the plant with pinnate branches.

B. Branch with opposite branchlets.

C. Transverse section of thallus

D. Longitudinal section of thallus

***Gracilaria* Greville**

A very large genus with broad geographical distribution and very variable morphology not only among the many species but also with same species which makes the taxonomy difficult. In cross section, the thallus is composed of a variable number of large medullary cells with few chloroplasts and lots of starch granules; the cells decreases in size towards the surface and become richer in chloroplasts. Cystocarps are salient on the thallus surface and very conspicuous looking like dark warts. Some species of this genus supply most of the raw material for world agar production.

Key to the species:

1. Thallus erect, cylindrical or compressed. 2
1. Thallus cylindrical 3
2. Thallus compressed or flat 15
2. Thalli succulent; 3-5cm thick with many shoots arising from the base forming cushions or patches .... *G. salicornia*
3. Thallus consisting of articulated cylindrical segments 4
3. Thallus not articulated 5
4. Segments short cylindrical, constriction at the base without depression at the apex ....*G.opuntia*
4. Plant up to 10cm tall ....*G.canaliculata*
5. Plants thick, several shoots arising from the base 6
5. Plants less than 2mm thick 12
6. Fronds short, branches curing downwards constricted ....*G.crassa*
7. Fronds with long cylindrical axis unbranched to about 2/3 of height; dichotomously branched above ; ultimate branches compress .... *G.fergusonii*
7. Fronds dichotomously branched in the upper parts as crowns; upper part slightly compressed ....*G.rostrata*
8. Erect fronds much branched almost from the base 9
9. Segments short ....*G.debilis*
9. Segments longer 10
10. Spermatangial cavity shallow ....*G.disticha*
10. Spermatangial cavity deep 11
11. Frond 1- 1.5 mm thick ....*G.dura*
12. Branches without constriction 13

Gracilaria Greville

13. Branches alternate, constricted at base ....*G.arcuata*
14. Branches uniformly thin throughout ....*G.verrucosa*
14. Branches thin at base, becoming thicker and attenuating towards the apex ....*G.edulis*
15. Fronds flat; thallus upto 2-3 cm broad, margin of branches proliforms ....*G.textorii*
15. Thallus more than 3cm wide; base constriction ....*G.curtissiae*
16. Plants 10 – 14 cm tall, branching at basal parts usually dichotomous, upper parts Irregularly branched proliferation absent .... *G.pudumadamensis*
17. Frond regularly, closely, dichotomously branched thick and cartilaginous Proliferation sparse ......*G.corticata*
17. (a) Fronds irregularly branched at upper parts, flat below, Cylindrical above with spinous apices ....*G.corticata var. cylindrica*
18. Fronds thin; profusely branched, branching dichotomous or polychotomous, alternate or pinnate, proliferation abundant ....*G.folifera*

Gracilariales-Gracilariaceae

Gracilaria Greville

REFERENCES: V. Krishnamurthy *et al.* (1990: fig. 6), Kalimuthu, Kaliaperumal & Ramalingam (1992: 191,23), Jagtap (1993: table 1), Silva *et al.* (1996: 155), Desikachary, V. Krishnamurthy & M. Balakrishnan (1998: 108).

***Gracilaria*
Zanardini**
Descriptio
Plants 1
several fr
discoid l
cylindrical,
diameter,
irregularly
branches
slightly co
cortex one
medulla lar
diameter;
cortex to
cystocarp
globose.
Rare: At
stones and
Distributio
A. Habit of
B. Plant sl
fronds
C. Transv
frond
D. Longi
frond

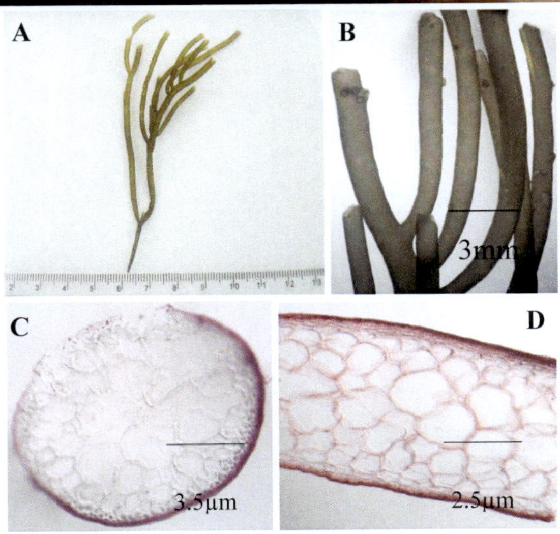

Gracilaria Greville

***Gracilaria arcuata* Zanardini**

***Description*:** Plants 10-12 cm height; several fronds arising from discoid holdfast. Fronds cylindrical, 2-3 mm in diameter, alternately or irregularly branched, branches and branchlets slightly constricted at base; cortex one cell in thickness; medulla large, 340-820μm in diameter; transition from cortex to medulla abrupt; cystocarp prominent, globose.

Rare: Attached on small stones and dead corals

Distribution: Valai island

A. Habit of the plant

B. Plant showing cylindrical fronds

C. Transverse section of frond

D. Longitudinal section of frond

Gracilaria Greville

REFERENCES: Kutzing (1868: 29, pl. 82: figs. D. E.), Sonder (1871: 56), Umamaheswara Rao (1974b: 677, figs. Ig – j, pl. I: A, B), S. Dixit (1980: 72), Untawale, Dhargalkar & Agadi (1983: 32).

Gracilaria
Sonder
Descriptio
Plants
decumben
succulent,
and brittl
pinkish
greenish;
holdfast. A
3 mm
decumben
dichotomo
internal st
1-2 layer
cortical c
large
medullary
spherical;
towards pe

Rare: On
substratum
Distributi

A. Habit of
B. Plant s
frond
C. Trans
frond
D. Longi
frond.

Gracilaria Greville

Gracilaria canaliculata **Sonder**

Description**:** Plants gregarious, decumbent, up to 10cm tall, succulent, stiff cartilaginous and brittle, bright to dark pinkish red, sometimes greenish; attached by a holdfast. Axes cylindrical, 2-3 mm in diameter, decumbent, 4-6 times dichotomously branched; internal structure consists of 1-2 layers of small outer cortical cells surrounding a large celled medulla; medullary cells polygonal to spherical; decreasing the size towards periphery.

Rare: On rocks and hard substratum.

Distribution: Valai island

A. Habit of the plant

B. Plant showing cylindrical frond

C. Transverse section of frond

D. Longitudinal section of frond.

Gracilaria Greville

REFERENCES: Oza & Gorasia (1998: 12), A. Srinivasa Rao (1998: 22), Nageswara Rao *et al.* (1999: 129 – 138), Kaliaperumal *et al.* (1998: 141 – 151), Zaidi *et al.* (1999: 66 – 82), Redekar & Raje (2000: 41 – 43).

Gracilaria
Agardh) J. A
Description
Plants
yellowish red
9-17cm in
repeatedly
branched
segments;
occasionally
tips of segme
1-2 layerec
radially el
Medulla lar
Cystocarps
elevated and l
Common: O
hard substra
action is more
Distribution:
Thonithurai,
nagar,
Nochiurani,
Thalathoppu,
Vaalinokkam
Kuduthalai,
Leepuram,
Kovalam, an
Gulf of Mann

A. Habit of t
repeatedly
branches.
B. Transve
thallus
C. Longitud
thallus

Gracilaria Greville

Gracilaria corticata **(J. Agardh) J. Agardh**
Description: Plants dark red to yellowish red in colour, bushy; 9-17cm in height. Fronds repeatedly dichotomously branched with yellow segments; 1-2mm wide; occasionally up to 2.5-30mm; tips of segments acute. Cortex 1-2 layered; outer layer radially elongated cells. Medulla large, ovoid shape. Cystocarps hemi spherical, elevated and1.0-1.5mm wide.
Common: On rocks and other hard substratum where tidal action is more.
Distribution: Muttukadu Thonithurai, Velankanni nagar, Seeniappadharga, Nochiurani, Pudumadam Thalathoppu, Kanchirankudi, Vaalinokkam, Manapad, Kuduthalai, Idinthakarai, Leepuram, Kanyakumari, Kovalam, and all 20 islands of Gulf of Mannar.
A. Habit of the plant showing repeatedly dichotomous branches.
B. Transverse section of thallus
C. Longitudinal section of thallus

Gracilaria Greville

REFERENCES: Gopinathan & Panigrahy (1983: 50, pl. I; F), Kaliaperumal & Chennubhotla *et al*. (1986: 107 – 110, figs. 1 – 3, tables 1, 2), Kaliaperumal, Kalimuthu, & Ramalingam (1990: table 1), Kaliaperumal, Kalimuthu, & Ramalingam (1992: 191), Jeyasankar & Ramalingam (1993: 42), Silva *et al*. (1996: 162).

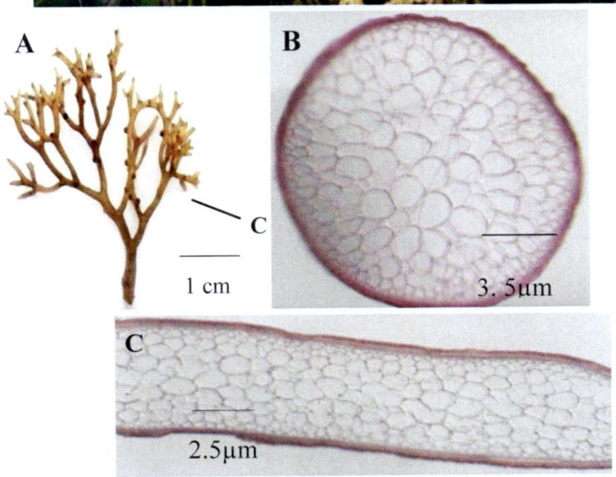

Gracilaria Greville

***Gracilaria corticata* (J. Agardh) J. Agardh var. *cylindrica* Umamaheshwara Rao**

Description: Plant dark red to yellowish red in colour; bushy, 10cm tall; more rigid, often with variegated spots on fronds; fronds dichotomously, alternately or irregularly branched; compressed in lower parts; sub-terete to cylindrical above with pointed or spinous apices. Cystocarp globose, scattered over the thallus.

Common: On boulders, rocks and other hard substratum.

Distribution: Velankanni nagar, Seeniyappadharga, Kanchirankudi, Ervadi, Valinokkam, Manapad, Kanyakumari, Kovalam and all 20 islands of Gulf of Mannar.

A. Habit of the plant showing cystocarp (C) on the branch

B. Transverse section

C. Longitudinal section

Gracilaria Greville

Gracilaria
Description
Thallus
about 5cm
several fro
discoid
lending
secondarily
of a singl
two layers
radially elo
inner
compressed
larger cell
central
surrounded
outer
compressed

Uncommo
rocks and o
Distributio
Ervadi
A. Habit of
B. Pl
dichotomou
C. Transv
frond
D. Longit
frond

Gracilaria Greville

Gracilaria crassa **Harvey**

Description: Thallus thick,cylindrical, about 5cm tall,3.5mm thick, several fronds arising from a discoid holdfast, some lending downwards and secondarily attached; cortex of a single or occasionally two layers of cells, outer radially elongated 10 × 8μm; inner cells radially compressed. Medulla of larger cells in the center; central cells oblong; surrounded by oval cells; outer cells radially compressed.

Uncommon: on coral stones, rocks and other substratum.

Distribution: Thonithurai and Ervadi

A. Habit of the plant

B. Plant showing dichotomous branches

C. Transverse section of frond

D. Longitudinal section of frond

***Gracilaria* Greville**

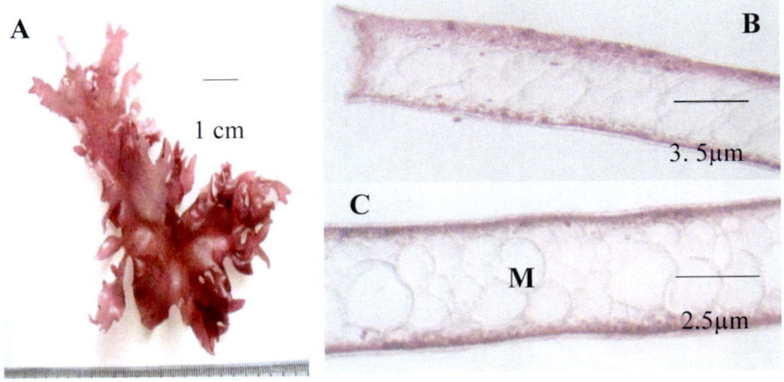

Gracilaria Greville

Gracilaria curtissiae **J. Agardh**

Description**:** Thallus as elongated clumps of flat blades, up to 40cm height; pink to straw yellow in colour. Blade strap shaped, 3-22cm long; 1-3.5 mm thick, constricted at base; margin smooth; branchlets marginal; irregularly placed; apices bluntly rounded. Medullary cells irregularly spherical; cortex 1-3 layer thick. Cystocarp hemi spherical.

Rare: On hard substratum and rocks.

Distribution: Muttukadu

A. Habit of the plant

B. Transverse section

C. Longitudinal section showing irregularly spherical medulla (M).

Gracilariales–Gracilariaceae

Gracilaria Greville

REFERENCES: Untawale, Dhargalkar & Agadi (1983: [32]), V. Krishnamurthy & Rajendran (1987a: 10, 14, figs. 8, tables 1), V. Krishnamurthy (1989: figs. 2C, 6A), V. Krishnamurthy (1991: table V), Silva *et al*. (1996: 163), Desikachary, V. Krishnamurthy & M. Balakrishnan (1998: 103).

Gracilaria d
Børgesen
***Description*:**
Plants cylind
thickness; c
colour; up tc
; sub
branched at t
the fronds;
and ramified
ends. Corte
outer layer o
and inner l
compressed.
large cells; c
shaped scat
mature parts
Rare: On ha
Distribution:
Hare island, N
A. Habit of th
B. Plan
subdichotom
C. Trans
showing la
cells (M).
D. Longitud
frond.

Gracilaria Greville

Gracilaria debilis **(Forsskål) Børgesen**
Description: Plants cylindrical, 1-2 mm in thickness; dark green in colour; up to 10cm in height; di or sub dichotomously branched at the upper ends of the fronds; sub- fastigiated and ramified at the distal ends. Cortex 1-2 layered, outer layer of elongated cells and inner layer of radially compressed. Medulla with large cells; cystocarps dome shaped scattered over the mature parts of the thalli.
Rare: On hard substratum.
Distribution: Shingle island, Hare island, Mully island.
A. Habit of the plant
B. Plant showing subdichotomous branch
C. Transverse section showing larger medullar y cells (M).
D. Longitudinal section of frond.

Gracilaria Greville

REFERENCES: Nair *et al.* (1986: 737), Krishnamurthy & Rajendran (1987a: 10, 14, figs. 2, 3, tables 1), V. Krishnamurthy (1989: figs. 2A, 6D, 8B), Nair *et al.* (1993: 186, 193), Silva *et al.* (1996: 165), Desikachary, V. Krishnamurthy & M. Balakrishnan (1998: 105).

Gracilaria
Agardh) J.
Description
Thallus atta
holdfast. M
up to 13
cylindrical,
branched.
branchlets
base; abou
Cortex sing
elongated c
2-3 layers
cells; inner
oblong, oft
cystocarps c
1.5mm in
over the sur
Rare: On tl
Distributior
A. Habit of
cylindrical
branching.
B. Transver
showing lar
cortex (C).
C. Longitu
axis.

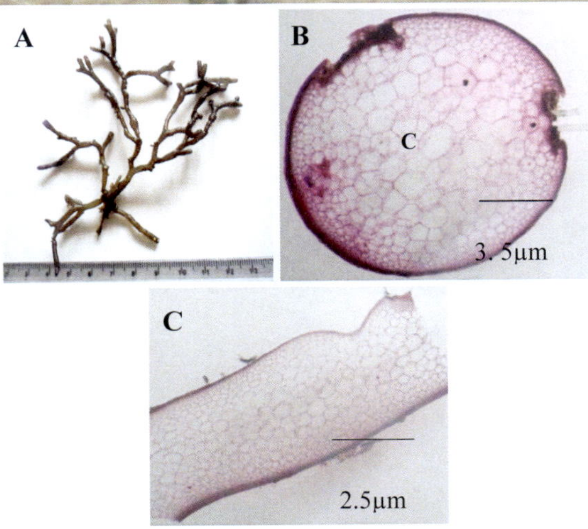

Gracilaria Greville

***Gracilaria disticha* (J. Agardh) J. Agardh**

***Description*:** Thallus attached by a discoid holdfast. Main axis several, up to 13 cm in height, cylindrical, and irregularly branched. Branches and branchlets not constricted at base; about 1-5mm thick. Cortex single layered of small elongated cells; outer medulla 2-3 layers of small ovoid cells; inner cortex of large, oblong, often angular cells; cystocarps dome shaped, about 1.5mm in width, scattered over the surface of the thallus.

Rare: On the coral stones.

Distribution: Kovalam

A. Habit of the plants showing cylindrical axis with irregular branching.

B. Transverse section of axis showing larger cells of inner cortex (C).

C. Longitudinal section of axis.

Gracilaria Greville

REFERENCES: Subba Rao *et al.* (1992: 182), Jagtap (1993: table I), Silva *et al.* (1996: 165), Siddhanta *et al.* (1997: 95 – 99), Desikachary, V. Krishnamurthy & M. Balakrishnan (1998: 107), Siddhanta *et al.* (1995: 17)

Gracilaria
J. Agardh
Descriptio
green to y
colour. 1
1mm
cylindrical
branched;
by discoid
often
branchlets
branchlets
at base,
sometimes
single laye
elongated;
1-2 layers
cells; in
layers of
Cystocarps
scattered c
Distributi
A. Habit o
B. Plant
covered w
C. Trans
frond.
D. Longi
frond.

Gracilaria Greville

Gracilaria dura (J. Agardh) J. Agardh

Description: Plants dark green to yellowish green in colour. 11-18cm in height; 1mm in thickness; cylindrical, irregularly branched; attached to rocks by discoid holdfast; branches often covered with branchlets; branches and branchlets slightly constricted at base, pointed at apex; sometimes forked. Cortex single layer of cells; narrowly elongated; outer medulla of 1-2 layers of small oblong cells; inner medulla 7-8 layers of large ovoid cells. Cystocarps dome shaped scattered on thallus.

Distribution: Cultivated.

A. Habit of the plant

B. Plant showing branches covered with branchlets.

C. Transverse section of frond.

D. Longitudinal section of frond.

Gracilaria Greville

REFERENCES: Mairh *et al*. (1998: 110 - 126), Pon Siraimeetan & Selvaraj (1998: 20), Kaliaperumal *et al*. (1998: 20), Kannupandi *et al*. (1998: 20), Kaliaperumal *et al*. (1998: 141 – 151), Kaladharan *et al*. (2000: 61 – 65).

Gracilaria
Gmelin) P.
Descriptior
subtidal; ε
stones;
cylindrical
branched, 2
1.5mm in tl
attenuating
Branching
fourth ord
layer; mec
ovoid, l
elevated aı
the thallus.
Common:
and coral
region.
Distributio
Arasanerip
Ammapatt
Kodikkara
Rameshwa
islands of (
A. Habit oı
B. Plant
constricted
C. Tr
showing cy
D. Long
showing cy
E. Transv
cystocarp
F. Cystoca
release.

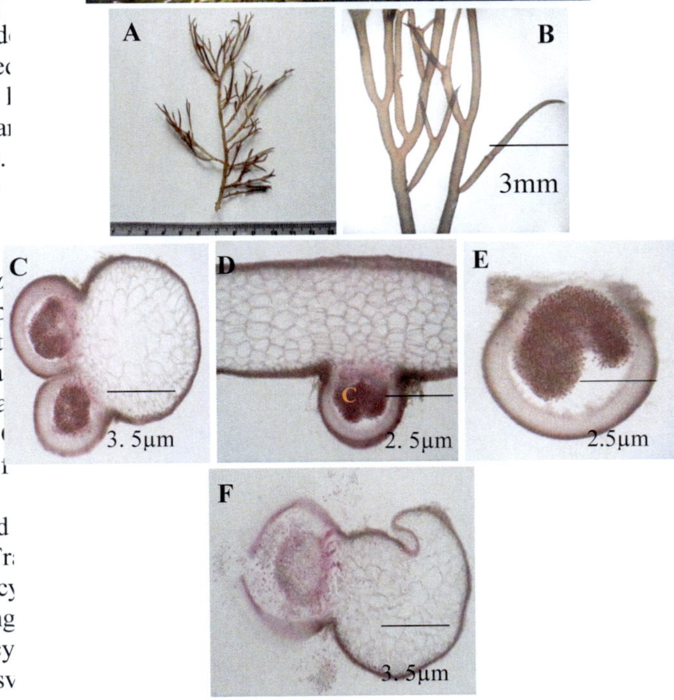

Gracilaria Greville

Gracilaria edulis (S. Gmelin) P. Sliva

Description: Plants subtidal; attached to coral stones; frond erect, cylindrical and highly branched, 2mm thick in base; 1.5mm in the middle and then attenuating to the apex. Branching upto third or even fourth order. Cortex single layer; medulla 4-6 layers, ovoid, large cystocarp elevated and scattered over the thallus.

Common: On shaded rocks and coral stones in subtidal region.

Distribution: Kottaipattinam, Arasaneripattinam, Ammapattinam, Karankadu, Kodikkarai, Ponnagaram, Rameshwaram and all 20 islands of Gulf of Mannar.

A. Habit of the plant
B. Plant showing branch constricted at base
C. Transverse section showing cystocarp
D. Longitudinal section showing cystocarp
E. Transverse section of cystocarp
F. Cystocarp showing spore release.

Gracilaria Greville

Gracilaria
Agardh
Descriptio
9cm tall.
up to abou
and t
dichotomo
form a
ultimate
compresse
thick,
tapering t
extensive
cortex sin
cells.
Common:
exposed to
Distributi
Valinokka
Kuduthala

A. Habit o
B. Upper
with repe
branching
C. Trans
frond
D. Longi
frond

Gracilaria Greville

***Gracilaria fergusonii* J. Agardh**

Description: Thallus about 9cm tall. Frond un branched up to about one half its height and then repeatedly dichotomously branched to form a flabellate crown; ultimate branches often compressed. Main axes 3mm thick, below gradually tapering to 1.5mm. Medulla extensive with 9-12 layers; cortex single layer with small cells.

Common: On rocks in exposed to high tidal splash.

Distribution: Kanchirankudi, Valinokkam, Manapad, Kuduthalai, Leepuram.

A. Habit of the plant
B. Upper portion of plant with repeated dichotomous branching
C. Transverse section of frond
D. Longitudinal section of frond

Gracilaria Greville

REFERENCES: Premnathan *et al*. (1992: table I), Jayasankar & Ramalingam (1993:42), Kaliaperumal & Kalimuthu. (1993: 9, 11, fig. 6), Nair *et al*. (1993: 193), Silva *et al*. (1996: 168), Desikachary, V. Krishnamurthy & M. Balakrishnan (1998: 107).

Gracilaria fc
Børgesen
Description:
up to 16cm i
by a discoid
stones and
intertidal or
Fronds
gradually nar
with lacinate
dichotomous
dichotomous.
layer of ra
cells. Med
composed c
Transition f
medulla ab
elevated; sub
over the fron

Common:
substratum,
pebbles.

Distribution:
Arasaneripatt
Ammapattina
Thonithurai,
Kovalam, Sh
island, Ana
island, Uppu
Koswari islar

A. Habit of tl
B. Plant s
dichotomous
C. Transvers
frond.
D. Longi
showing ar
cells (M)

Gracilaria Greville

Gracilaria foliifera (Forsskål) Børgesen

Description: Plant flat, foliose, up to 16cm in height, attached by a discoid holdfast to coral stones and rocks at lower intertidal or subtidal regions. Fronds broad at base, gradually narrow towards apex with lacinate ends; branching dichotomous or sub dichotomous. Cortex single layer of radially elongated cells. Medulla of large composed of angular cells. Transition from cortex to medulla abrupt. Cystocarps elevated; subspherical scattered over the fronds.

Common: On hard substratum, rocks and small pebbles.

Distribution: Kottaipattinam, Arasaneripattinam, Ammapattinam, Ponnagaram, Thonithurai, Kanyakumari Kovalam, Shingle island, Hare island, Anaipar island, Salli island, Upputhanni island and Koswari island

A. Habit of the plant

B. Plant showing irregular dichotomous branch

C. Transverse section of the frond.

D. Longitudinal section showing angular medullary cells (M)

Gracilaria Greville

REFERENCES: Durairatnam (1962: 11 – 12, fig. 3), Umamaheswara Rao (1974b: 688, 690, fig. 4k, pl. III: D), Silva *et al*. (1996: 173), Desikachary, V. Krishnamurthy & M. Balakrishnan (1998: 100).

Gracilaria
Durairatnan
Description
Thallus succ
height, cylin
mm thickn
branched; b
constricted a
towards the
constrictions
segments in
shows an ir
large cells
somewhat ra
an outer med
of radially c
and single
small cell
clavated, her
Uncommon:
other hard su
Distribution
Krusadai isla
Valai island
island

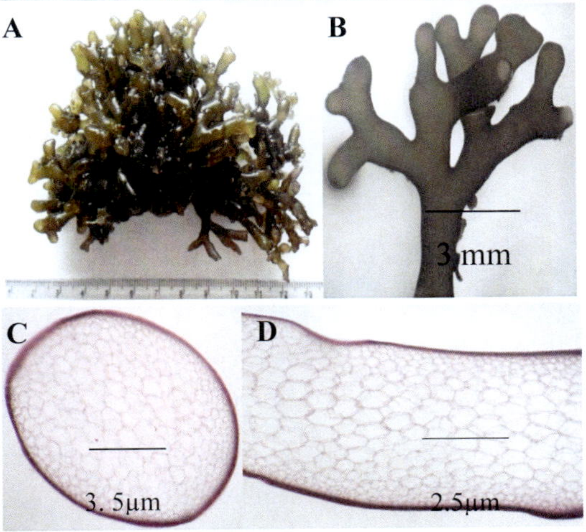

A. Habit of th
B. Branches
apex.
C. Transve
frond.
D. Longituc
frond.

Gracilaria Greville

Gracilaria opuntia Durairatnam
Description: Thallus succulent; 6-8 cm in height, cylindrical, 2.5 to 3.0 mm thickness, irregularly branched; branches slightly constricted at base, swollen towards the apex; intercalary constrictions seen on some segments in section, thallus shows an inner medulla of large cells in 8-10 layers, somewhat radially elongated; an outer medulla of 1-2 layers of radially compressed cells and single layer cortex of small cells. Cystocarps clavated, hemi spherical.
Uncommon: On rocks and other hard substratum.
Distribution: Thonithurai, Krusadai island, Appa island, Valai island and Valimunai island
A. Habit of the plant
B. Branches with swollen apex.
C. Transverse section of frond.
D. Longitudinal section of frond.

Gracilaria Greville

REFERENCES: V. Krishnamurthy & Rajendran (1987b: 69), V. Krishnamurthy (1989: fig. 6E), Silva *et al*. (1996: 174), Reeta Jayasankar (1997: 42), Desikachary, V. Krishnamurthy & M. Balakrishnan (1998: 122).

Gracilaria
V. Kri
Rajendra
Descriptio
in height,
first di
irregular.
layer of c
cortex
progressiv
ovoid to
globoid,
thallus.
Uncomm
intertidal r
Distributi
A. Habit o
B. Plant
dichotomc
C. Tra
showing
transition
medulla.
D. Longit

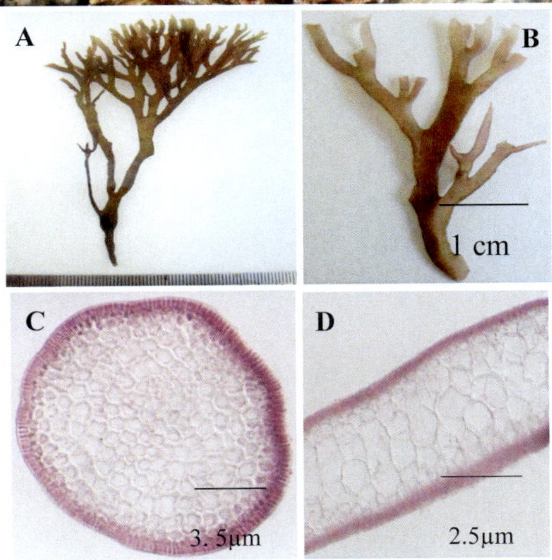

Gracilaria Greville

Gracilaria pudumadamensis **V. Krishnamurthy & Rajendran**

Description: Plants 10-14cm in height, foliose, branching first dichotomous, later irregular. Cortex of a single layer of cells; transition from cortex to medulla progressive. Medulla of large ovoid to oblong. Cystocarp globoid, scattered over the thallus.

Uncommon: On rocks in intertidal region.

Distribution: Pudumadam.

A. Habit of the plant

B. Plant showing irregular dichotomous branches.

C. Transverse section showing progressive transition from cortex to medulla.

D. Longitudinal section

Gracilaria Greville

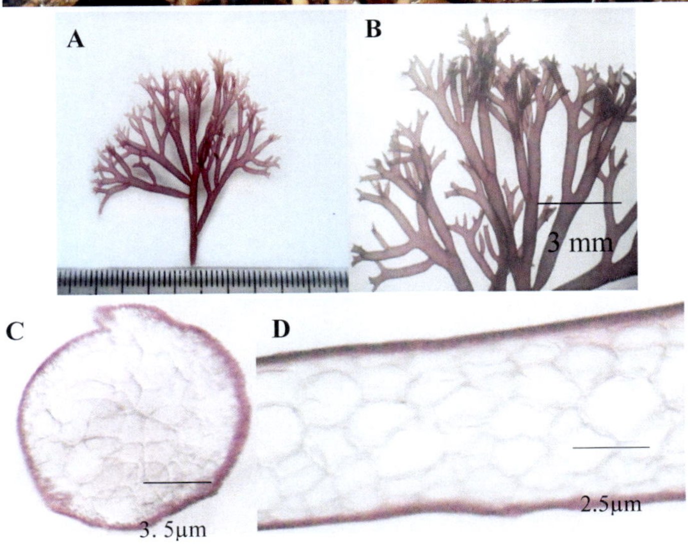

Gracilaria Greville

Gracilaria rostrata **Uma maheswara Rao**

***Description*:** Fronds dichotomously branched in upper parts; as flabellate thalli; ultimate branches with pointed or blunt apices; cystocarps rostrate; single or in clusters.

Uncommon: On rocks.

Distribution: Muttukadu

A. Habit of the plant

B. Branches with pointed apices.

C. Transverse section of frond.

D. Longitudinal section of frond

Gracilaria Greville

REFERENCES: C. Agardh (1820a), Dawson (1954a: 4), Jagtap (1992: 57), Jagtap (1993: table 1), Silva *et al.* (1996: 175).

Gracilari
Agardh)
Descripti
Plant
yellowish
upto 16
attached
thallus b
branched,
lower bra
not atten
upper br
below, e
and swoll
showing
cortex 1-
with large
Uncomm
rocks and
Distribut
Kottaipat
Arasaneri
Ponnagar
island
A. Habit
B. Branch
depressio
C. Trans
frond
D. Longi
frond.

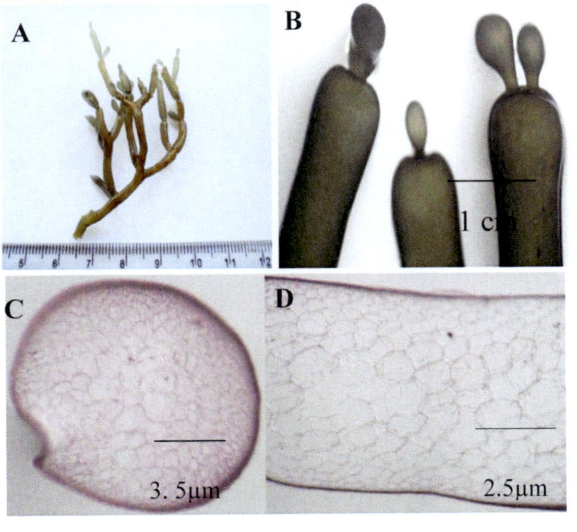

Gracilaria Greville

Gracilaria salicornia **(C. Agardh) Dawson**

***Description*:** Plant brownish to yellowish red in colour; upto 16cm in height; attached by a small disc; thallus bushy, irregularly branched, cylindrical, lower branches cylindrical, not attenuated at the base; upper branches attenuated below, elongate, clavate, and swollen at the apex and showing apical depression; cortex 1-2 layers, medulla with large central cells.

Uncommon: On small rocks and coral pieces.

Distribution: Kottaipattinam, Arasaneripattinam, Ponnagaram and Anaipar island

A. Habit of the plant

B. Branches showing apical depression

C. Transverse section of frond

D. Longitudinal section of frond.

Gracilaria Greville

REFERENCES: Umamaheswara Rao (1991: table V), Kaliaperumal *et al.* (1992: 13), Kalimuthu, Kaliaperumal & Ramalingam (1992: 123), Subba Rao *et al.* (1992: 182), Silva *et al.* (1996: 177), Desikachary, V. Krishnamurthy & M. Balakrishnan (1998: 117).

Gracilaria
(Suringar) I
Description:
foliose,
dichotomous
mature plant
47mm
attenuated a
entire or pro
usually one l
large. Transi
to medulla al
prominent,
globose in st
Common:
other hard su
Distribution:
Arasaneripat
Ammapattina
Ponnagaram,
Manapad Up
Koswari isla
A. Habit of t
B. Surface
cystocarps
C. Trans
cystocarp
D. Transve
frond
E. Longi
frond

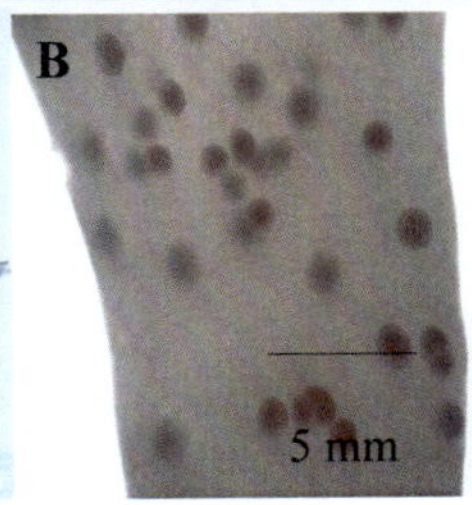

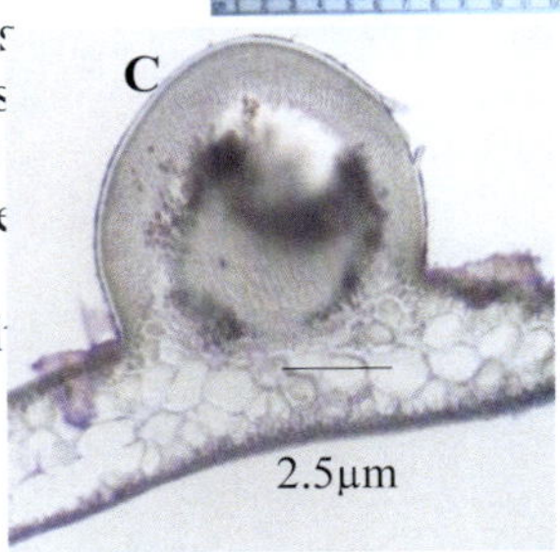

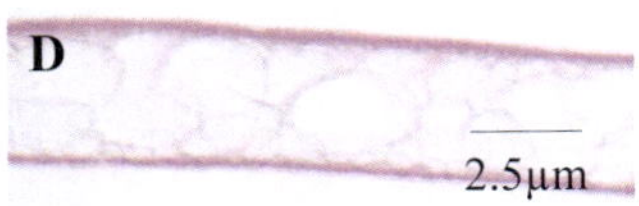

Gracilaria Greville

Gracilaria textorii **(Suringar) De Toni**
Description: Thallus flat, foliose, branching dichotomous or irregular; mature plants 9-35cm tall, 6-47mm broad with attenuated apices; margins entire or proliferous. Cortex usually one layered; medulla large. Transition from cortex to medulla abrupt. Cystocarp prominent, emergent, globose in structure.
Common: On rocks and other hard substratum.
Distribution: Kottaipattinam Arasaneripattinam, Ammapattinam, Ponnagaram, Valinokkam, Manapad Upputhanni island Koswari island.
A. Habit of the plant
B. Surface of frond with cystocarps
C. Transverse section of cystocarp showing spores.
D. Transverse section of frond
E. Longitudinal section of frond

Gracilaria Greville

REFERENCES: Kaliaperumal & Kalimuthu (1993: 9, 11), Silva *et al*. (1996: 345), Kannupandi *et al*. (1998: 20), Desikachary, V. Krishnamurthy & M. Balakrishnan (1998: 110), Mairh *et al*. (1998: 110 – 126), Zaidi, Oza & Mairh (1998: 15).

Gracilaria Greville

***Gracilaria verrucosa* (Hudson) Papenfuss**

Description: Plant yellowish to brownish red in colour; sub tidal plant up to about a meter in length. Thallus erect, cylindrical, highly branched and attached by a small disc; frond with single main axis, uniformly thin throughout; branching lateral, opposite or sub opposite in arrangement; branches upto third or even fourth order; cortex of 2-4 layers; outer most radially elongated. Medulla of 3-5 layers, large, spherical or ovoid. Cystocarps sub spherical, elevated and scattered over the thallus.

Uncommon: On small coral stones and attached in sand

Distribution: Kottaipattinam, Arasaneripattinam, Karankadu, Ammapattinam, Kodikkarai, Nalupanai and Anaipar island.

A. Habit of the plant with cystocarps on thallus

B. Plant showing sub opposite branches

C. Transverse section of frond

D. Longitudinal section of frond with big medullary cells (M)

Grateloupia C. Agardh

Plant erect, leathery, sometimes gelatinous with many fronds from discoid holdfast. Fronds compressed or foliose, divided dichotomously or pinnately or palmately. Medulla has lobed cells and filamentous rhizoids, cortex many cells thick, reproductive elements scattered on the thallus, cystocarps immersed.

Key to the species:

1. Fronds divided or branched 2
2. Plants upto 15cm tall, fronds narrow,0.25-2cm broad ....*G.lithophila*
3. Blades palmately divided, proliferations very few ....*G.wattii*
4. Blades pinnately branched often pinnate or cirrhose ....*G.filicina*
    a) Fronds upto 1.00mm wide, with dense proliferation ....*G.horrida*
    b) Fronds narrow upto 1mm wide many flat proliferations arising from the edge of fronds ....*G.pectinata*
    c) Fronds filiform, irregularly branched with many proliferous branches and branchlets arising from the main axis ....*G.cirrhosa*

Grateloupia C. Agardh

REFERENCES: Rajendran *et al*. (1991: table 1, 2), Vidyavathi & Sridhar (1991: fig. 2), Kalimuthu, Kaliaperumal & Ramalingam (1992: 123), Nair *et al.* (1993: 185, 186, 193), Silva *et al*. (1996: 194), Subramanian *et al*. (1998: 17).

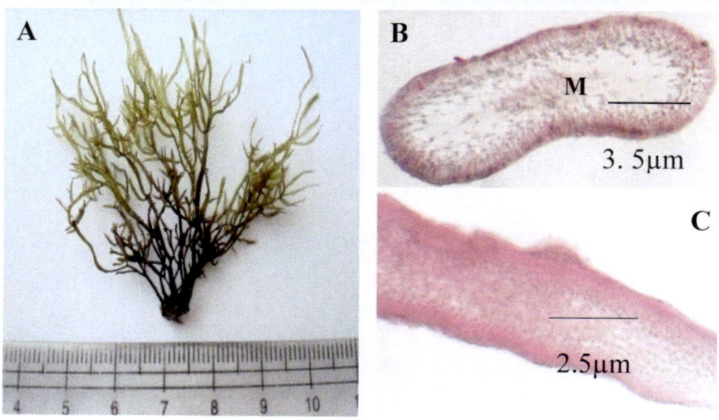

Grateloupia C. Agardh

***Grateloupia filicina* (Lamouroux) C. Agardh**

Description: Plant brownish green to blackish red in colour; branching near base, irregular to opposite. Blades flat, 0.5-2.5mm wide, branchlets irregularly opposite from margin, branchlets numerous; apices bluntly pointed. Medullary filaments cylindrical, loosely arranged in mucilaginous gel. Cortex 2-3 cells thick. Cystocarp spherical.

Common: On coral stones and small pebbles.

Distribution: Kasimedu, Muttukadu, Kovalam, Ponnagaram, Nochiurani, Pudumadam, Manapad, Krusadai island, Upputhanni island and Koswari island.

A. Habit of the plant showing irregular opposite branches

B. Transverse section of frond showing loosely arranged medulla (M)

C. Longitudinal section of frond

Grateloupia C. Agardh

REFERENCES: Børgesen (1935: 53 – 54), K. Srinivasan (1973: 13), Silva *et al*. (1996: 194).

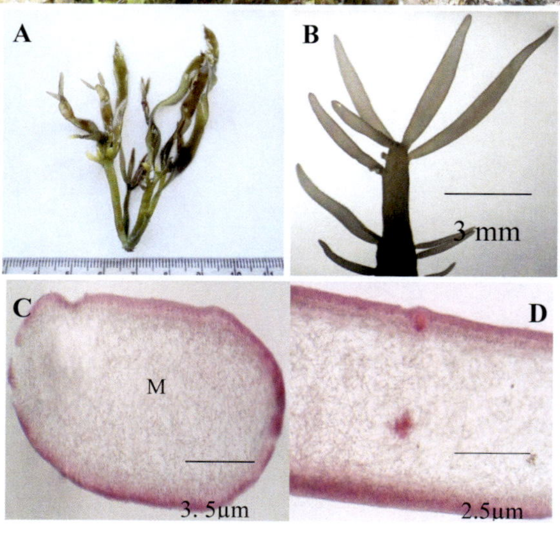

Grateloupia C. Agardh

***Grateloupia filicina* (Lamouroux) C. Agardh var. *filicina* forma *cirrhosa* Børgesen**

***Description*:** Plant brownish green in colour, fronds pinnately branched. Blades flat with dense proliferations. Ultimate branchlets show tendril like appearance. Medullary filaments cylindrical, loosely arranged. Cortex 2-3 layered. Cystocarp spherical.

Rare: Attached on rocks in high wave action region.

Distribution: Nochiurani and Valinokkam

A. Habit of the plant

B. Plant showing tendril like branchlets

C. Transverse section of frond showing cylindrical medulla (M)

D. Longitudinal section of frond.

Grateloupia C. Agardh

REFERENCES: Kutzing (1843: 397, pl.76. I), Borgesen (1935: 53 – 54), K. Srinivasan (1973: 13), Desikachary, V. Krishnamurthy & M. Balakrishnan (1990: 241 - 242), Silva *et al*. (1996: 194).

Grateloupia C. Agardh

***Grateloupia filicina* (Lamouroux) C. Agardh var. *filicina* forma *horrida* (Kützing) Børgesen**

Description: Plant brownish green in colour. Fronds flat upto 5 mm wide with numerous dense proliferations. Cortex 1-2 layers of elongated cells, medullary filaments cylindrical, loosely arranged. Cystocarp spherical.

Uncommon: attached on rocks in intertidal region.

Distribution: Muttukadu, Arasaneripattinam, Manapad, Kanyakumari and Kovalam

A. Habit of the plant

B. Plant showing dense proliferations.

C. Transverse section of frond.

Grateloupia C. Agardh

REFERENCES: Borgesen (1935: 53 – 54), Borgesen (1935: 83) K. Srinivasan (1973: 13), Silva *et al*. (1996: 194).

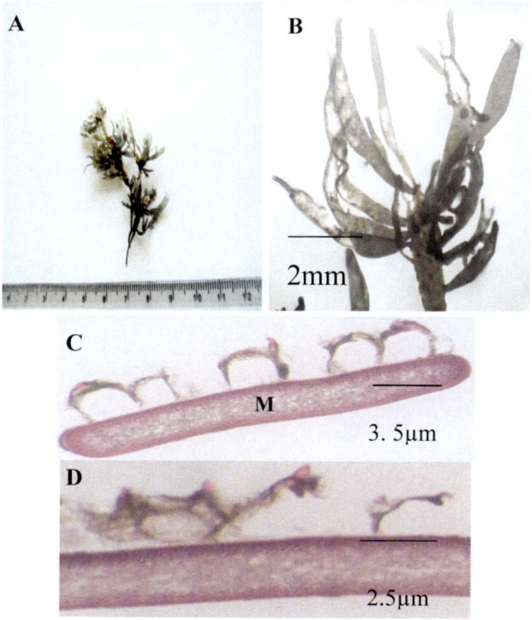

Grateloupia C. Agardh

Grateloupia filicina (Lamouroux) C. Agardh var. _filicina_ forma _pectinata_ Børgesen

Description: Plant greenish colour. Frond flat, narrow with 1 mm wide, many flat small branches arising from the edge of the fronds and giving pectinate appearance. Cortex 1-2 layers and medullary filaments are cylindrical and compactly arranged.

Rare: attached to rocks in lower intertidal region

Distribution: Kovalam

A. Habit of the plant

B. Branch showing pectinate appearance.

C. Transverse section showing compact medulla (M)

D. Longitudinal section of frond.

Grateloupia C. Agardh

REFERENCES: Kalimuthu, Kaliaperumal & Ramalingam (1992: 123), Maya & Nair (1992: 32), Premnathan *et al*. (1992: table I), Nair *et al*. (1993: 185, 193), Suresh *et al*. (1993: 113), Silva *et al*. (1996: 196).

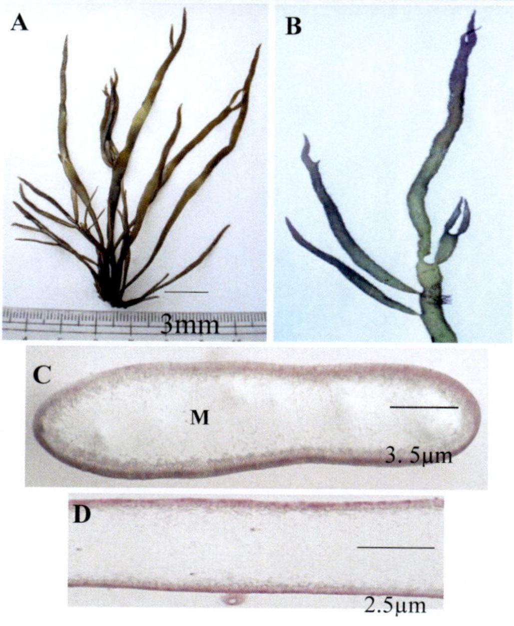

Grateloupia C. Agardh

Grateloupia lithophila **Børgesen**

Description: Plants 5-10cm long; yellowish green to purplish red; attached by a small basal disc; fronds basal cylindrical, flattering upward; tapering proximately; dichotomous branching near the basis; apices acute. Medulla cells loosely arranged, outer cortical cells compact.

Common: On rocks in high wave action.

Distribution: Ennore, Kasimedu, Muttukadu, Kovalam, Mahabalipuram, Pondicherry, Ammapattinam Villoonditheertham, Nochiurani, Pudumadam, Sallimunthal, Keelakarai, Valinokkam, Manapad, Leepuram Vattakottai, Kanyakumari, Kovalam, Shingle island, Krusadai island and Koswari island.

A. Habit of the plant

B. Branches showing tapering end.

C. Transverse section with loosely arranged medulla(M)

D. Longitudinal section

Grateloupia C. Agardh

REFERENCES: Holmes (1896: 351), De Toni (1905: 1569 – 1570), K. Srinivasan (1965: 245), S. Dixit (1970: 119), V. Krishnamurthy & H. Joshi (1970: 20), Untawale, Dhargalkar & Agadi (1983: 30), Desikachary, V. Krishnamurthy & M. Balakrishnan (1990: 239 - 240), Silva *et al.* (1996: 196).

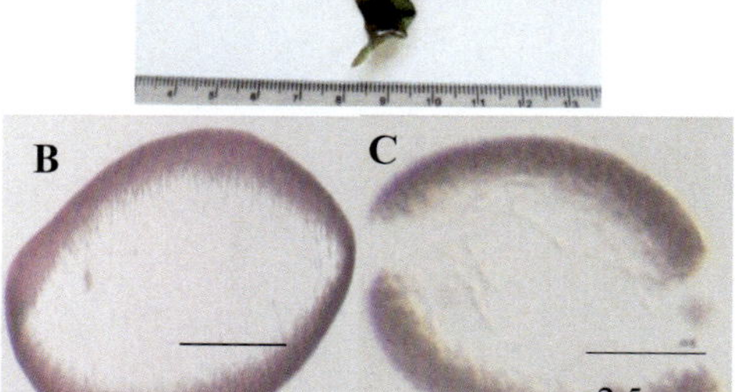

Grateloupia C. Agardh

Grateloupia wattii **Holmes**

Description: Plants up to 25cm in height; with a basal holdfast; the single erect frond expanding rather abruptly divided into the broad blade; lamina divided palmately; the lobes oblong, lanceolate and imbricate, hardly tapered at the base and obtuse apices. Cystocarps small rather densely scattered all over the thallus.

Uncommon: On rocks in upper intertidal region where high wave action.

Distribution: Pondicherry

A. Habit of the plant showing palmate branches.

B. Transverse section of frond

C. Longitudinal section of frond

Griffithsia C. Agardh

Griffithsia C. Agardh

Plants erect, filamentous, delicate, freely dichotomously or irregularly branched; cells of filaments usually globular but also cylindrical macroscopic; branching alternate or bifurcated. Cystocarp developing on 3 celled fertile branches; tetrasporangia naked or involucrate.

Key to the species:

Plants less than 5 cm tall, filaments oblong, cylindrical, incurved.
....*G. opuntioides*

Griffithsia C. Agardh

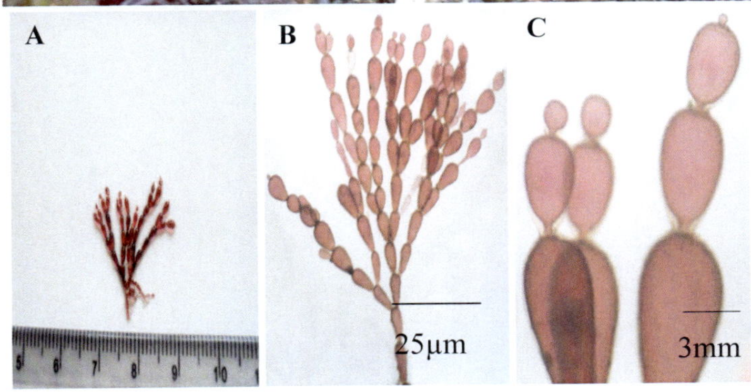

Griffithsia C. Agardh

Griffithsia opuntioides **C. Agardh**

Description: Fronds light red to brownish red in colour; 2-5cm in height; regularly dichotomous. Fastigate, flabellate with clariform cells, vesicles surrounded by a ring of involucrate cells, tetrasporangia in verticiliate.

Rare: Attached on rocks.

Distribution: Panjal.

A. Habit of the plant

B. Plant showing dichotomous branches.

C. Vesicle

Halichrysis (J. Agardh) Schmitz

***Halichrysis* (J.Agardh) Schmitz**

Plants about 3cm tall with cylindrical and contorted stipes of 2mm dia.; branches ending with lobed membranous blade; blade 2-4cm broad about 300µm thick; medulla with 2-4 layers of thin walled cells and cortex with small cells; cystocarps scattered on the blades, sub-spherical, rostrate.

Key to the species:

Branches ending with a lobed membranous blade

…….*H. thivyae*

Halichrysis (J. Agardh) Schmitz

REFERENCES: Dawson (1963: 448 – 449, pl. 81: fig. 5), K. Srinivasan (1965: 247), V. Krishnamurthy & H. Joshi (1970: 22), Untawale, Dhargalkar & Agadi (1983: 33), Silva *et al*. (1996: 365), Desikachary, V. Krishnamurthy & M. Balakrishnan (1998: 177).

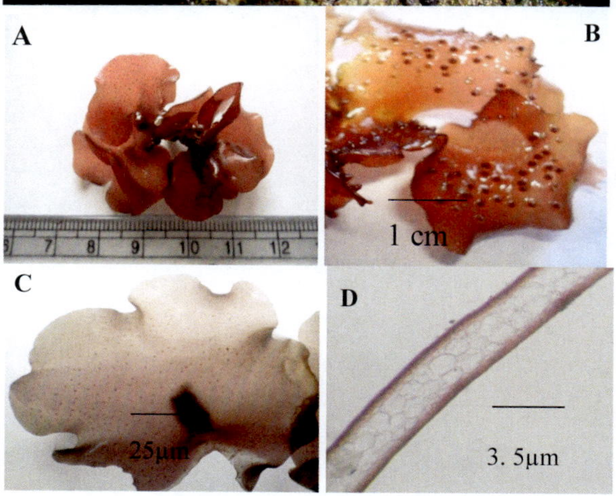

Halichrysis (J. Agardh) Schmitz

Halichrysis thivyae **(Dawson) Eisenman & Moe**

Description: Plant dark red in colour; grown prostrate; fronds soft, arising from unbranched short cylindrical stipe. Blades 2-4cm broad, membranous, margin often recurving towards the substratum. Blades narrow at the point of contact, expended distally and frequently overlapping. Internally, the thallus composed of 2 rows of cortical cells; 5 layers of medulla.

Uncommon**:** On coral stones and rocks upto 20m deep.

Distribution: Keelakarai light house, Bharathi nagar, Ervadi, Appa island, Valimunai island, Anaipar island and Koswari island

A. Habit of the plant

B. Thallus with reproductive parts

C. Thallus with minute cystocarps.

D. Longitudinal section of thallus.

***Haloplegma* Montagne**

Plants spongy, compressed, fronds irregularly lobed and divided, composed of reticulately joined monosiphonous filaments; inner cells of the fronds larger; outer smaller; peripheral cells free, small cells. Spermatangia rare, forming cylindrical masses on short filaments; cystocarps hemispherical; tetrasporangia terminal on peripheral filaments.

Key to the species:

1. Fronds flabelliform, indistinctly lobed; ultimate ramuli straight ………*H.duperreyi*
1. Frond pinnately lobed; ultimate ramuli recurved ……….*H.preissii*

Haloplegma Montagne

REFERENCES: Untawale, Dhargalkar & Agadi (1983: 35), Subramanian (1985: 72, 80), M. Dave *et al.* (1987: table 1), Subba Rao (1992: 182), Silva *et al.* (1996: 414), Desikachary, V. Krishnamurthy & M. Balakrishnan (1998: 231).

Haloplegm
Montagn
Descript
in colou
height,
below,
flabellifo
distinctly
irregularl
upto 2 c
fimbriate
filaments
as broad,
smaller to
Uncomm
habitats o
15 cm de
Distribut
and Kosw
A. Habit
B. Tran
thallus
C. Lon
thallus

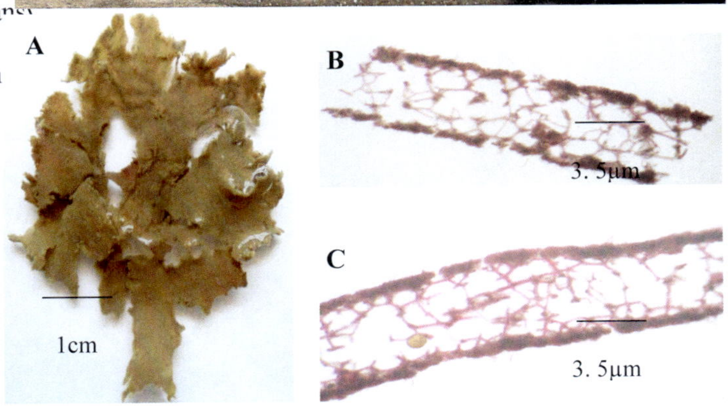

Haloplegma Montagne

Haloplegma duperreyi **Montagne**

Description: Plants dark red in colour, upto 12 cm in height, spongy; stipulate below, expanded into a flabelliform frond above; distinctly lobed, and irregularly divided; segments upto 2 cm in width; margin fimbriate; peripheral filaments short; cells as long as broad, gradually becoming smaller towards apex.

Uncommon: in low light habitats on hard surfaces upto 15 cm deep.

Distribution: Mully island and Koswari island

A. Habit of the plant

B. Transverse section of thallus

C. Longitudinal section of thallus

Haloplegma Montagne

REFERENCES: W. Hooker (1844: pl. DCXIII), Montagne (1845: 149), S. Dixit (1970: 123), Silva *et al.* (1996: 415), Desikachary, V. Krishnamurthy & M. Balakrishnan (1998: 232).

Haloplegm
(Harvey) N
Descriptio
flabellum,
lobes oblo
unequally
less en
filaments
segments t
half times l
Rare: On
upto 15m d
Distributio
Koswari isl
A. Habit of
B. Transve
thallus
C. Longitu
thalus

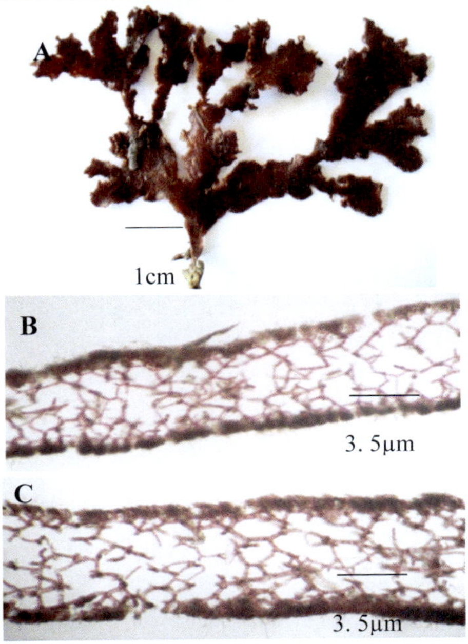

Haloplegma Montagne

***Haloplegma preissii* (Harvey) Montagne**
Description: Fronds sub flabellum, pinnately lobed, lobes oblong, with in margin unequally developed more or less entire; peripheral filaments elongated with segments three to three and a half times long as broad.
Rare: On hard substratum upto 15m depth.
Distribution: Mully island, Koswari island.
A. Habit of the plant
B. Transverse section of thallus
C. Longitudinal section of thalus

Halymenia C. Agardh

***Halymenia* C. Agardh**

Thallus usually foliose, well branched with a delicate slippery and gelatinous texture; discoid holdfast and inconspicuous stipe; in cross section the structure shows a cortex of small cells, a subcortex of larger cells and a filamentous medulla produced from star shaped cells at the center of the thallus. Cystocarps immersed in the outer medulla; tetrasporangia scattered among the cortical cells, cruciately divided.

Key to the species:

1. Fronds sub entire, slightly lobed, margins undulate with denticulate proliferations ………*H.dilatata*
1. Fronds divided, large, proliferations from both margins and surface ……*H.venusta.*

Halymenia C. Agardh

REFERENCES: Umamaheswara Rao (1969: 44), V. Krishnamurthy & H. Joshi (1970: 20), Untawale, Dhargalkar & Agadi (1983: 30), Desikachary, V. Krishnamurthy & M. Balakrishnan (1990: 249 - 250). Subba Rao (1992: 182), Silva *et al*. (1996: 196).

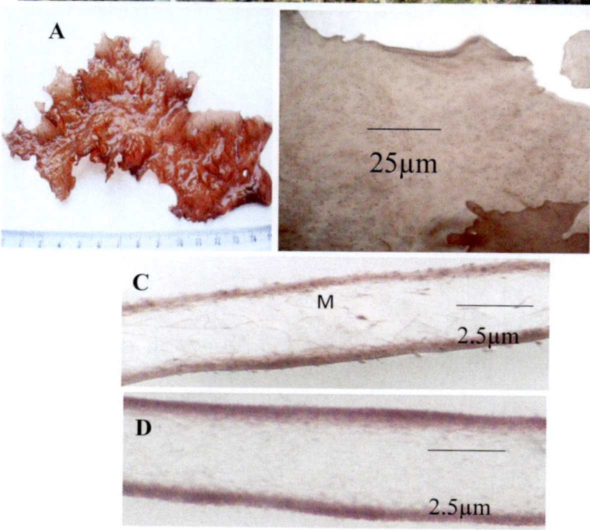

Halymenia C. Agardh

Halymenia dilatata Zanardini
Description: Plant rose to red in colour; attaching on small disc; stipe very short; broad lamina, variable in shape; flabellate to oblong; lobes often imbricate; margin with numerous denticulate proliferations and frilled appearance; medulla large, composed of transverse medullary filament; cortex thin few layered.
Common: On small hard substratum in 15m deep.
Distribution: Manapad, Mulli island, Appa island, Anaipar island, Salli island and Koswari island
A. Habit of the plant
B. Thallus showing marginal proliferations.
C. Transverse section of thallus showing medullary filaments (M)
D. Longitudinal section of thallus

Halymenia C. Agardh

REFERENCES: Sridhar *et al*. (1991: 202), Bhakuni *et al*. (1992: 515), Kaliaperumal *et al*. (1992: 13), Subba Rao *et al*. (1992: 181, 182), Rama Rao *et al*. (1993: 106), Silva *et al*. (1996: 201).

Halymenia C. Agardh

Halymenia venusta **Børgesen**

Description: Plant light rose red in colour, upto 30 cm tall, slightly lubricous, large and attached by small basal discs, frond foliose, variable in shape palmately divided into several lobes; margin and surface of the blades are densely proliferated; proliferations few millimeter to may centimeters in size, repeatedly divided, narrow with sharply acute tips.

Common: On small hard substratum in 15 cm deep.

Distribution: Kottai pattinam, Bharathi nagar, Ervadi, Sadaimuniyan valasai, Mannapad, Vattakottai, Manali island, Mulli island, Valai island, Appa island, Anaipar island and Koswari island.

A. Habit of the plant

B. Plant showing many proliferations

C. Surface view of thallus with reproductive bodies

D. Transverse section of thallus

E. Longitudinal section of thallus

Helminthocladia J. Agardh

***Helminthocladia* J. Agardh**

Plant lightly calcified or sometimes muscilagenous; thallus cylindrical; terminal cells of the cortical cells enlarged in some plants; carpogonial branches lateral, 3-5 celled, first division of the carpogonium vertical, oblique or transverse; gonimoblast filaments producing carpospores or carpo-tetraspores.

Key to the species:

Plants pinnately branched, large pyriform terminal cells

……. *H.clavadosii f. indica*

Helminthocladia J. Agardh

REFERENCES: V. Krishnamurthy & H. Joshi (1970: 17), Gopalakrishnan (1971: table II), Untawale, Dhargalkar & Agadi (1983: 27), M. Dave *et al*. (1987: table 1), Desikachary, V. Krishnamurthy & M. Balakrishnan (1990: 131 – 132, fig. 22, pl. VIII), Silva *et al*. (1996: 121).

Helmin
(Lamouro
forma *indi*
Descriptio
purple in
tall, attach
discs, sin
Thallus n
radially
filaments
terminal ce
filaments la
Rare: On
in shaded p
Distributio
A. Habit of
B. Micro
fronds
C. Thallus
branchlets
D. Tran
showing
cortical cel
E. Longitu

Helminthocladia J. Agardh

Helminthocladia clavadosii (Lamouroux ex Duby) forma _indica_ Desikachary

Description: Plant dark purple in colour, 30-50cm tall, attached by small basal discs, single or branched. Thallus multi axial with radially arranged cortical filaments 4-5 times divided; terminal cells of the cortical filaments large, pyriform

Rare: On rocky substratum in shaded place.

Distribution: Manapad.

A. Habit of the plant

B. Microscopic view of fronds

C. Thallus with many small branchlets

D. Transverse section showing radially arranged cortical cells (C)

E. Longitudinal section.

***Heterosiphonia* Montagne**

Thallus with dorsi-ventral or bilateral organization; creeping axis fixed by means of rhizoids; erect axis cylindrical or slightly flattened, upto 6 cm in height, producing lateral branches, regularly alternating in two rows; the lateral branches are generally multi - seriate at the base ending in uni-seriate bifurcated branchlets.

Key to the species:

Erect fronds branched upto 3rd order, ultimate branches spine like ...*H.muelleri*

Heterosiphonia Montagne

REFERENCES: Untawale, Dhargalkar & Agadi (1983: 34), Chennubhotla, Kaliaperumal, Kalimuthu, (1987: 108 – 109), M. Dave *et al*. (1987: table 1), V. Krishnamurthy & Varadarajan (1992: 137, figs. 1 – 4), Silva *et al.* (1996: 442), Desikachary, V. Krishnamurthy & M. Balakrishnan (1998: 282).

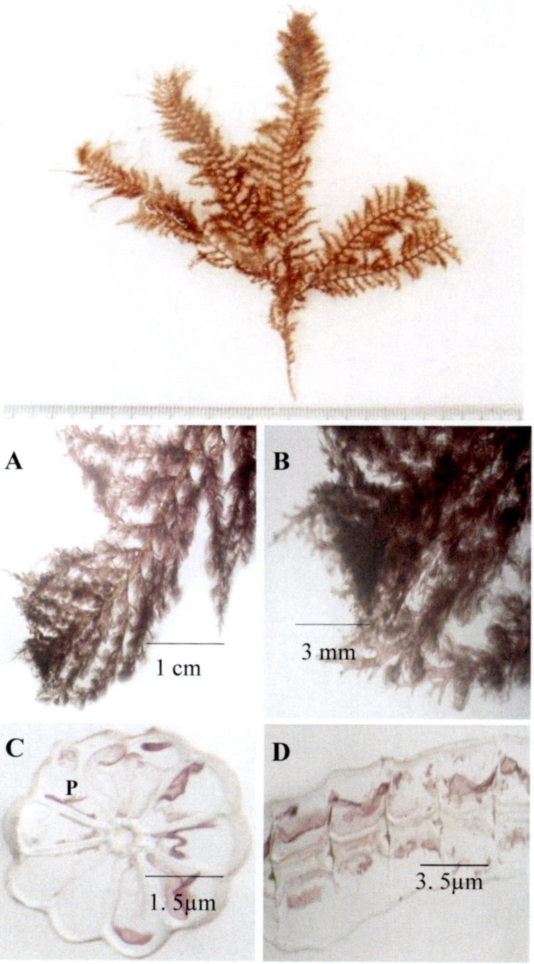

Heterosiphonia Montagne

***Heterosiphonia muelleri* (Sonder) De Toni**
Description: Plant dark red to brownish red in colour ; plants attached by a holdfast; erect axis 10-15cm in height, 2-3mm in thickness; pinnately branched upto third order; heavily corticated towards the base; branches alternate in two rows as both sides of the main axis; ultimate branches spine like and polysiphonous at the base and monosiphonous at distal portion. Axial cells surrounded by ten pericentral cells;
Uncommon: On small coral rocks in 10m deep.
Distribution: Mully island, Valimunai island, Anaipar island and Koswari island
A. Habit of the plant showing pinnate branching
B. Upper portion showing spine like branches
C. Transverse section showing axial and 10 pericentral cells (P)
D. Longitudinal section

Hypnea Lamouroux

Thallus bushy, macroscopic, well branched and mostly cylindrical. An apical cell provides central axis from where lateral filaments are produced; filaments are tightly packed giving the thallus the appearance of a solid cylinder with large colourless cells in the centre and a cortex of gradually smaller and more pigmented, chloroplast rich cells.

Key to the species:-

1. Single axis, freely branched or caespitose …….2
1. Several axis arising from base, tufted, sparingly branched …….7
2. Frond freely branched, branches not clustered or caespitose …….3
2. Front alternately branched, often caespitose …….4
3. Branches denulate at the apex …….*H.nigrescens*
3. Main branches thick, curved with short spine like laterals; branches end in hook like structure …….*H.musciformis*
4. Branches thin, alternate …….5
4. Branches thick …….6
5. Spine like branches horizontally extended…….*H.hamulosa*
6. Plants 5-10 cm, branches with ……….. to acuminate ends ……..*H.pannosa*
7. Branches covered with short spinous branchlets …….8
8. Spinous branches often forked into stellate structure …….*H.valentiae*
8. Spinous branches simple …….*H.flagelliformis*

Hypnea Lamouroux

REFERENCES: V. Krishnamurthy *et al*., (1990: 271, figs. 3, 6), Kaliaperumal *et al.* (1992: 13), Silva *et al*. (1996: 299), Zaidi & Mairh (1997b: 39), Desikachary, V. Krishnamurthy & M. Balakrishnan (1998: 160), Zaidi & Mairh (1999b: 47 - 65).

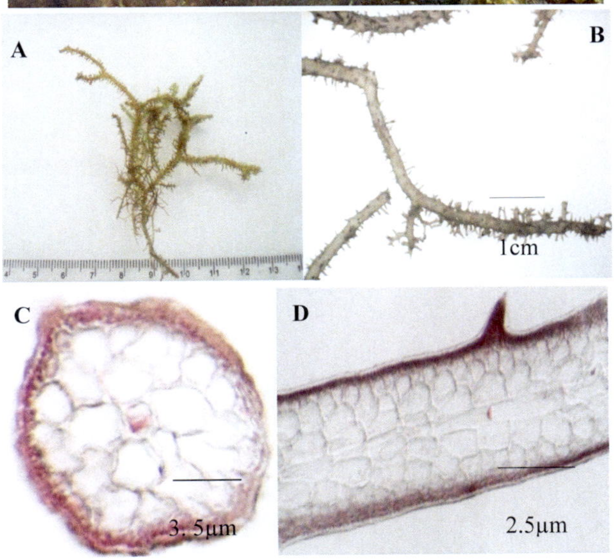

Hypnea Lamouroux

Hypnea flagelliformis **Greville ex J. Agardh**

Description: Thallus greenish to purplish red in colour, 10- 15 cm tall with several fronds arising from a disc, fronds sparsely branched, flagelliform; fronds covered with ramuli from base to apex, ramuli densely crowded, giving the frond a tomentose appearance, simple, subulate.

Uncommon: attached on rocks

Distribution: Pudumadam, Valinokkam, Manapad, Valai island and Koswari island.

A. Habit of the plant

B. Branch covered with ramuli

C. Transverse section of frond

D. Longitudinal section of frond.

Hypnea Lamouroux

REFERENCES: Lamouroux (1800[1797 – 1800]: 169), Silva *et al.* (1996: 299), Desikachary, V. Krishnamurthy & M. Balakrishnan (1998: 157).

Hypnea ham
Lamouroux

Description:
green to bro
Fronds are
branched,
intricately
covered with w
means of spine
very apex; spi
on all sides,
Fertile ramuli i

Rare: on ro
hard substratu

Distribution:
Kanyakumari,

A. Habit of the
B. Plant sho
branches with
C. Transvers
frond.
D. Longitudi
frond.

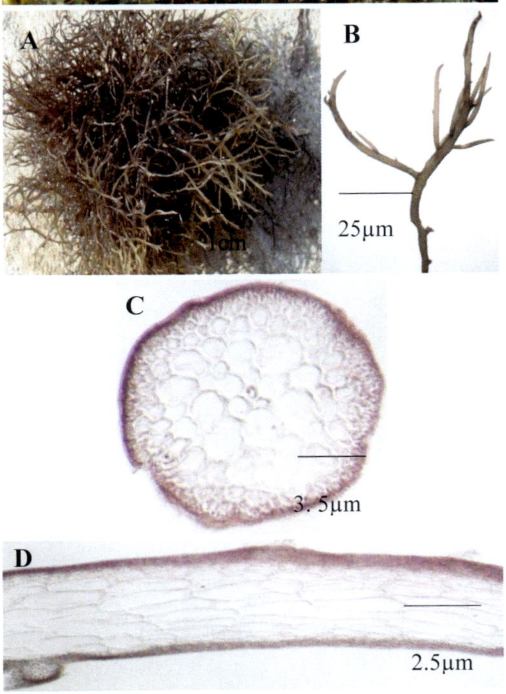

Hypnea Lamouroux

Hypnea hamulosa **(Esper) Lamouroux**

Description: Plant light green to brown in colour. Fronds are alternately branched, branches intricately caespitose, covered with whole length by means of spines except at the very apex; spines coming out on all sides, 2 mm. long. Fertile ramuli inflated.

Rare: on rocks and other hard substratum

Distribution: Kanyakumari, Kovalam

A. Habit of the plant

B. Plant showing alternate branches with spine.

C. Transverse section of frond.

D. Longitudinal section of frond.

Hypnea Lamouroux

REFERENCES: Zaidi & Mairh (1997b: 39), Desikachary, V. Krishnamurthy & M. Balakrishnan (1998: 156), Kaliaperumal *et al.* (1998: 141 - 151), Zaidi & Mairh (1999b: 47 - 65), Sivakumar & Rengasamy (2000: 67 – 74), Prema *et al.* (2000: 25 – 29).

Hypnea
(Wulfen) Lan
Description:
red in colour,
length; tha
attached later
occupying th
sea in deptl
meters; free
branches cove
spine like bra
the branches
and incurred f
Common:
substratum
Distribution:
pattinam, Ka
island, Mully
island, Vali
Anaipar islan
Koswari islan
island.

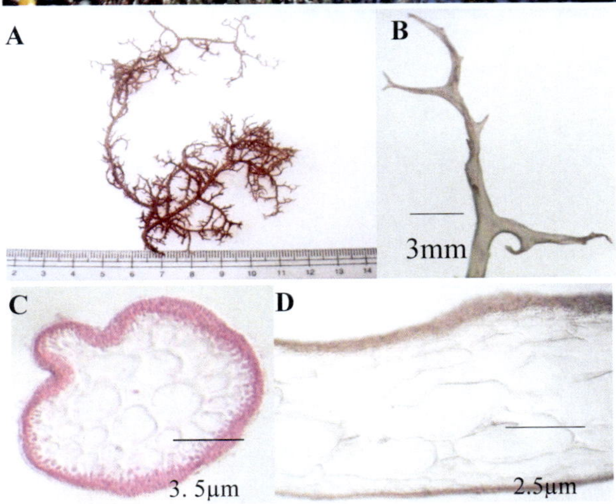

A. Habit of th
B. Branches
structure
C Transverse
frond
D. Longitudin
frond

Hypnea Lamouroux

Hypnea musciformis **(Wulfen) Lamouroux**

Description: Plants pinkish red in colour, upto 30 cm in length; thallus initially attached, later free floating; occupying the floor of the sea in depths upto three meters; freely branched; branches covered with short spine like branches, apices of the branches often thickened and incurred forming hook.

Common: on rocky substratum

Distribution: Arasaneri pattinam, Karankadu, Hare island, Mully island, Valai island, Valimunai island Anaipar island, Van island, Koswari island and Krusadai island.

A. Habit of the plant
B. Branches with hook like structure
C. Transverse section of frond
D. Longitudinal section of frond

Hypnea Lamouroux

REFERENCES: V. Krishnamurthy (1980: 54), Untawale, Dhargalkar & Agadi (1983: 31), Nair *et al.* (1986: 737), Nair *et al.* (1993: 186, 193), Silva *et al.* (1996: 303), Desikachary, V. Krishnamurthy & M. Balakrishnan (1998: 155).

Hypnea ni **ex Agardh**

Descriptio brown in with sever from a ba sparsely br 15cm in hei ramuli fro ramuli de giving the appearance. globose, se the ramuli.

Rare: on s

Distributio

A. Habit of

B. Branch w

C. Transv frond

D. Longitu frond

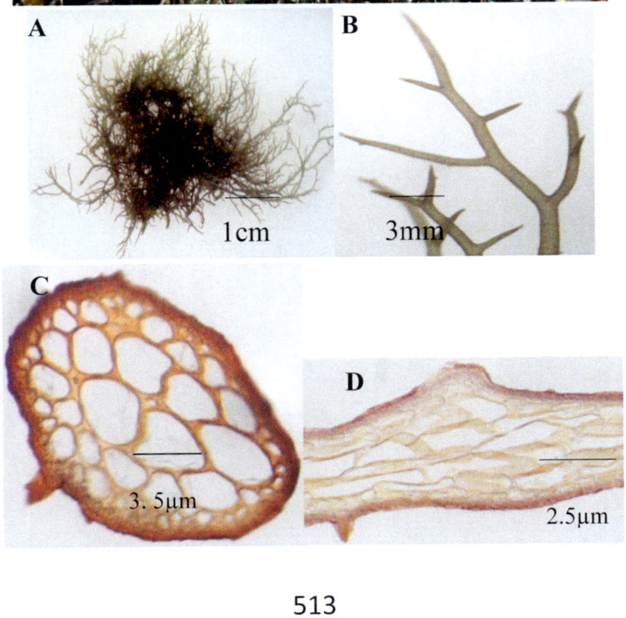

Hypnea Lamouroux

Hypnea nigrescens **Greville ex Agardh**

Description: Plant green to brown in colour. Thallus with several fronds arising from a basal disc; branch sparsely branched, about 10-15cm in height; covered with ramuli from base to apex, ramuli densely crowded, giving the frond a tomentose appearance. Cystocarps globose, sessile at the base of the ramuli.

Rare: on small rocks

Distribution: Nochiurani

A. Habit of the plant

B. Branch with ramuli

C. Transverse section of frond

D. Longitudinal section of frond

Hypnea Lamouroux

REFERENCES: Kalimuthu, Kaliaperumal & Ramalingam (1992: 119 – 126), Thomas & Subbaramaiah (1994: 93), Silva *et al*. (1996: 304), Zaidi & Mairh (1997b: 39), Desikachary, V. Krishnamurthy & M. Balakrishnan (1998: 158), Zaidi & Mairh (1999b: 47 - 65).

Hypnea pan

Description:
pink red in
cm in heig
spongy tuft
broken, intr
with conic
branches s
pyramidly r
median pa
apex. Carpo
of somewhat
upper branch

Common: o
substratum

Distribution
Nalupanai,
Villoondithe
Rameshwara
Kanchiranku
light house,
Ervadi,
valasai and
of Gulf Of M

A. Habit of
B. Plant
branching in
C. Transve
frond
D. Longitu
frond

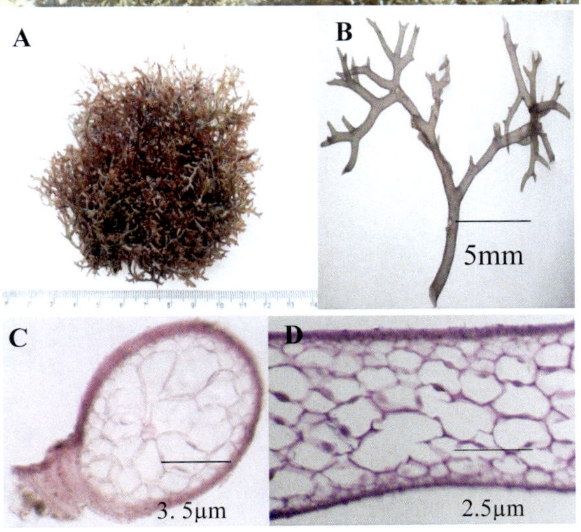

Hypnea Lamouroux

***Hypnea pannosa* J. Agardh**

Description: Plants purple to pink red in colour; 5 to 10 cm in height, growing as spongy tufts, brittle, easily broken, intricately branched with conical tips; fertile branches smooth at base, pyramidly ramose from the median part towards the apex. Carpogonia is one side of somewhat swollen parts of upper branches.

***Common*:** on rocks and coral substratum

Distribution: Munaikadu, Nalupanai, Villoonditheertham, Rameshwaram, Kanchirankudi, Keelakarai light house, Bharathi nagar, Ervadi, Sadaimuniyan valasai and in all 20 islands of Gulf of Mannar.

A. Habit of the plant

B. Plant showing ramose branching in apex

C. Transverse section of frond

D. Longitudinal section of frond

Hypnea Lamouroux

REFERENCES: Nageswara Rao *et al*. (1999: 129 – 138), Zaidi & Mairh (1999b: 47 - 65), Kaliaperumal *et al*. (1998: 141 - 151), Zaidi *et al*. (1999: 66 - 82), Sivakumar & Rengasamy (2000: 67 – 74), Prema *et al*. (2000: 25 – 29).

Hypnea valen
Montagne

Description: P
to pinkish re
Thallus with m
arising from
alternately
branches coesp
covered with s
throughout the
dense toward
Spines given
and held horiz
lower part of
patent towards

Habitat: Comn
corals and
substratum.

Distribution: K
Arasaneri
Karankadu An
Munaikadu, I
house, Nalupa
theertham,
Thonithurai,
nagar Seenia
Nochiurani,
Thalathoppu,
Panjal, Manap
Chinnamuttam,
Kanyakumari,
all 20 islands
Mannar.

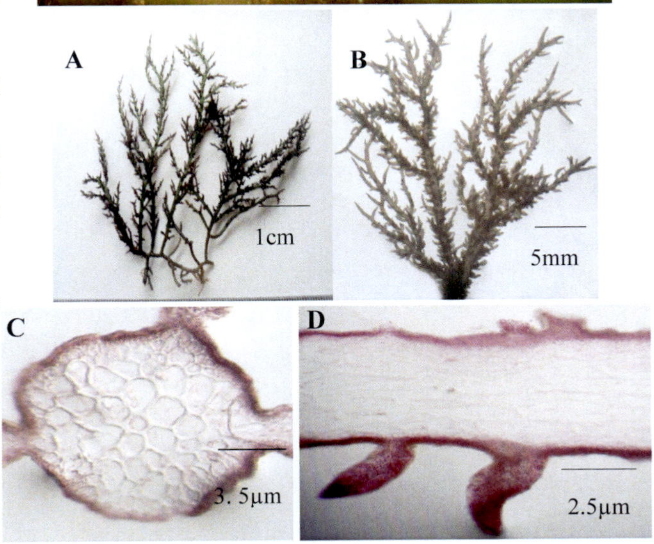

A. Habit of the
B. Branches de
with ramuli
C. Transverse s
D. Longitudina

Hypnea Lamouroux

Hypnea valentiae **(Turner) Montagne**
Description: Plants greenish to pinkish red in colour. Thallus with many erect axes arising from a base; axes alternately branched, branches coespitose, densely covered with spinous ramuli throughout the length, less dense towards the apex. Spines given out all round and held horizontally in the lower part of branches and patent towards the tip.
Habitat: Common on rocks, corals and other hard substratum.
Distribution: Kottai pattinam Arasaneri pattinam Karankadu Amma pattinam Munaikadu, Pamban light house, Nalupanai Villoondi theertham, Ariyankundu, Thonithurai, Velankanni nagar Seeniappa dharga Nochiurani, Pudumadm, Thalathoppu, Valinokkam, Panjal, Manapad, Leepuram, Chinnamuttam, Kanyakumari, Kovalam and all 20 islands of Gulf of Mannar.
A. Habit of the plant
B. Branches densely covered with ramuli
C. Transverse section
D. Longitudinal section

Jania Lamouroux

Jania Lamouroux

Thallus generally pink; forming erect tufts of cylindrical, calcareous segments, with non-calcified articulations, regularly bifurcated. Only one conceptacle per segment, usually at the base of a bifurcation, but also terminal on a branchlet; uniporate conceptacles inflated, with a terminal pore for discharge of spores or male gametes.

Key to the species:

1. Plants small. Less than 2 cm high ……2
2. Plants over 2 cm high ……3
2. Intergenicula 100-150 um wide, 3—5 times as long as broad …*J. capillaceae*
3. Intergenicula 100- 200 um wide, 5- 10 times as long as broad …*J. adhaerens*
4. Plants 2-5 cm in high …….4
4. End segments of thallus rounded … ….*J. iyengarii*
4. End segments of thalli not rounded ….. 5
5. Intergenicula 2-5 times as long as broad the ultimate one is straight ….. *J. rubens*

Jania Lamouroux

REFERENCES: V. Krishnamurthy (1990: figs. 4, 5, 7, 8), Thomas & Subbaramaiah (1991: table 2), Bhakuni et al. (1992: 515), Jagtap (1993: table I), Rama Rao *et al*. (1993: 106), Silva *et al*. (1996: 240), Desikachary, V. Krishnamurthy & M. Balakrishnan (1998: 69).

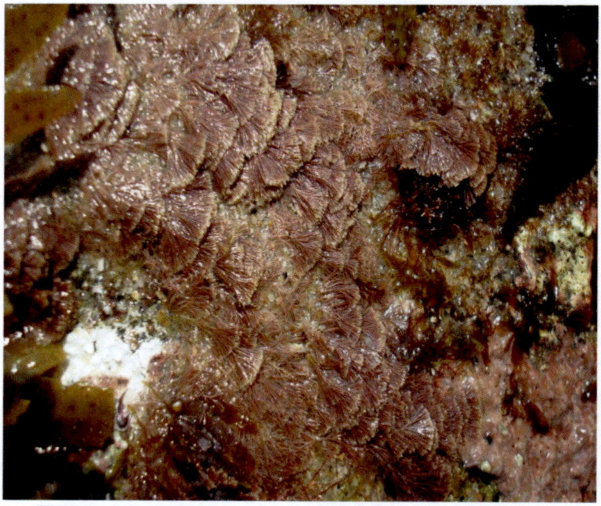

Jania Lamouroux

Jania adhaerens **Lamoueoux**

Description: Plants small, upto 2 cm high; thallus erect but creeping, small upto 2 cm high; basal attachment disc minute; erect fronds cylindrical, tapering slightly toward blunt apices. Segments heavily calcified; medullary cells tiered, parallel; cortical cells roundly rectangular. Joints flexible, uncalcified, occurring at base of branch and at regular intervals between branches; cortex lacking. Holdfast crustose like. Tetrasporangia oval, zonately divided.

Common: on hard surfaces or epiphytic on other marine plants upto 18 m deep.

Distribution: Kottaipattinam, Munaikadu, Rameshwaram, Pamban, Seeniappadharga, Nochiurani, Pudumadam, Kanchirankudi, Valinokkam, Vattakottai, Hare island, Mully island, Valai island and Valimunai island.

A. Habit of the plant

B. Plant showing dichotomous branch

C. & D. Branches showing blunt apices.

Jania Lamouroux

REFERENCES: Harvey (1853: 84 – 85), S. Dixit (1968: 20), Untawale, Dhargalkar & Agadi (1983: 39), Silva *et al*. (1996: 241), Desikachary, V. Krishnamurthy & M. Balakrishnan (1998: 69).

***Jania* Lamouroux**

***Jania capillaceae* Harvey**
Description: Plants small, upto 2 cm high, rose red in colour; branching widely dichotomous. Branches cylindrical, often recurved; apices pointed or occasionally rounded. Segments calcified, 300 – 600 µm long; joints flexible, uncalcified, occurring at base of branches at regular interval. Holdfast disc like. Tetrasporangia conceptacle. Medullary cells parallel, elongated, in one tier.
Uncommon: on rocks and other hard substratum upto 15m depth
Distribution:
Chinnamuttam, Kanyakumari, Kovalam.
A. Habit of the plant
B. Plant showing dichotomous branches

Jania Lamouroux

REFERENCES: E. Ganesan (1966: 46 – 58, figs. 1 – 30, pl. I), V. Krishnamurthy & H. Joshi (1970: 19), Subbaramaiah (1972: 395), Subbaramaiah (1974: 59), Untawale, Dhargalkar & Agadi (1983: 30), Silva *et al*. (1996: 242), Desikachary, V. Krishnamurthy & M. Balakrishnan (1998: 70).

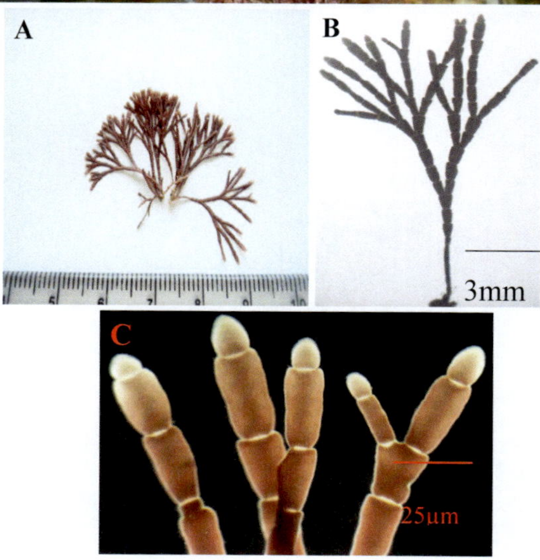

Jania Lamouroux

Jania iyengarii **E. Ganesan**
Description: Plants 2-3 cm high, forming dense tufts; fronds arising from basal crust; branching dichotomous, proliferations rare; branches more profuse in upper portions; apices of ultimate branches rounded. Intergenicula short in the lower portions, progressively longer in upper portions; 3-6 times as long as broad. Male plants sparsely branched. Female plants profusely branched. Cystocarpic conceptacles born on apices of branches.
Uncommon: on hard substratum like rocks and coral stones
Distribution: Ariyankundu, Chinnamuttam, Kanyakumari and Kovalam
A. Habit of the plant showing profuse upper portion
B. Plant showing dichotomous branch
C. Branch with round apex

Jania Lamouroux

REFERENCES: Kalimuthu, Kaliaperumal & Ramalingam (1992: 123), V. Kumar (1993a: 139, 140), V. Kumar (1993b: 47), Silva *et al*. (1996: 244), Vanillakumari & Subba Rangaiah (1997b: 22 – 23), Desikachary, V. Krishnamurthy & M. Balakrishnan (1990: 71).

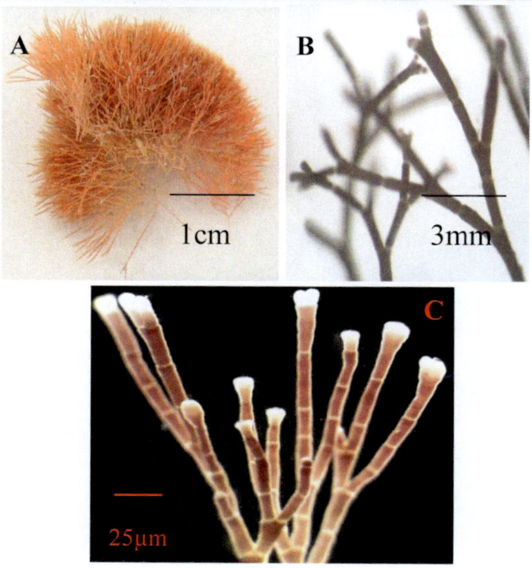

Jania Lamouroux

Jania rubens **(Linnaeus) Lamouroux**

Description: Thallus as tightly packed clumps or cushion like tufts upto 5 cm high; rose red in colour. Branching narrowly dichotomous; apices pointed to occasionally rounded, Segments barrel shaped, 0.6 to 2.5 mm long, calcified; medullary cells elongated; cortex lacking. Joints flexible, uncalcified occurring at regular intervals between branches. Tetrasporangia conceptacles terminal with short antennae and a single large apical pore.

Common: on rocks or dead coral in calm waters upto 30 m depth.

Distribution: Karankadu Thalathoppu, Sallimunthal, Keelakarai, Manapad, Panjal, Chinnamuttam Kanyakumari, Kovalam, Mulli island and Anaipar island.

A. Habit of the plant

B. Branch showing narrowly dichotomous

C. Branch showing rounded apex

***Kappaphycus* Doty**

Plants cylindrical, gelatinous, spines completely absent. Medulla composed of small hyphal cells and large cells in many species. Cystocarps present on thalli, tetrasporangia scattered, zonate.

Key to the species:

Plants cylindrical, branching irregular, axes smooth or with simple spine ….*K.alvarezii*

Kappaphycus Doty

REFERENCES: Doty (1988: 177), Ohno & Critchley (1993: 79), Mairh *et al*. (1997: 41), P.S.N. Rao & Umamaheswara Rao (1999: 93), P.S.N. Rao & Tigga (2000: 14).

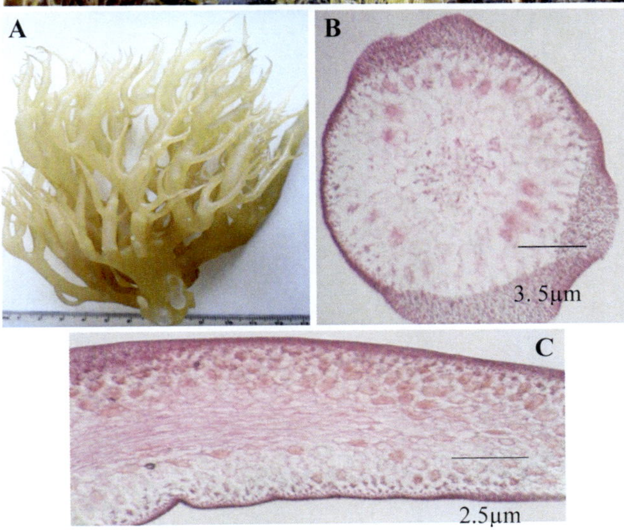

Kappaphycus Doty

Kappaphycus alvarezii (Doty) Doty ex P. Silva

Description: The plant has terete branches that are irregularly divided. Spines absent. The shape of the branch tips and branch length, diameter and density and colour vary with environment. Plants are commercially cultivated.

***Distribution**:* Cultivated

A. Habit of the plant

B. Transverse section

C. Longitudinal section

***Laurencia* Lamouroux**

Plants erect, solitary or in dense turfs. Thallus macroscopic, 2-30cm long, with a discoid or fibrous base and a fleshy texture; mostly cylindrical and sometime compressed; radially branched; short, club shaped branches are characteristic of most species; the apical cell is sunken in a pit and surrounded by trichoblasts; medulla are large, vertically elongated cells, cortex is small, radially elongated cells; polysiphonous structure is often obscured by the development of extensive cortication. Cystocarp with pericarp partially united with adjoining vegetative cells: Tetrasporangia developed on specially defined branches.

Key to the species

1. Surface cells of branches and branchlets radially elongated or palisade like in sectional view 2
1. Surface cells of branches and branchlets not radially elongated or palisade like 3
2. Branches covered with tuberculate branchlets which is present even in young plants ......*L.papillosa*
2 Ultimate branchlets not short ......*L.cruciata*
3 Fronds not compressed 4
3 Fronds compressed 10
4 Lenticular thickening abundant in medullary cells 5
4 Lenticular thickening absent or very few in medullary cells 8
5 Plants 4-6 cm tall, caespitose, attached by rhizoids 6
6 Branching scare .... *L.rigida*
6. Branching more profuse 7
7. Fronds about 1 mm thick, alternately branched ... ...*L. indica*
7 Fronds filiform, oppositely or verticillately branched ....*L. nidifica*
8 Plants small, 1.5 – 3.0 cm tall 9
8 Plants large, surface layer with small cells, medullary cells without Lenticular thickening ...... *L. obtusa*
   a) Plants tufted 2-3 cm tall, rigid, surface cells quadrate in Section, medullary cells small ....*L.obtusa var.rigidula*

Laurencia Lamouroux

9 Branches cylindrical, short ……*L.caraibica*

10 Branches cylindrical, alternate to irregular ……*L.microcladia*

11 Branches compressed, dichotomously, pinnately branched Lower parts of branches smooth, upper parts covered with branchlets ……*L.ceylanica*

12 Branching distichous, pinnate, branchlets 3-5 mm long, narrow below, broa dened above ……*L.pedicularioides*

Laurencia Lamouroux

REFERENCES: Howe (1920: 556), Subbaramaiah (1974: 59), Untawale, Dhargalkar & Agadi (1983: 3, 40), Annapurna & B. Rao (1987: 89 – 92, figs. 1 – 4), Desikachary, V. Krishnamurthy & M. Balakrishnan (1998: 344).

Laurencia

Descriptio

cylindrical

compressec

protruding

not radial

palisade lik

without lei

on their wa

Common:

in high wav

Distributio

Panjal,

Kovalam.

A. Habit of

B. Plant sl

branches

C. Transvei

D. Long

showing m

without thi

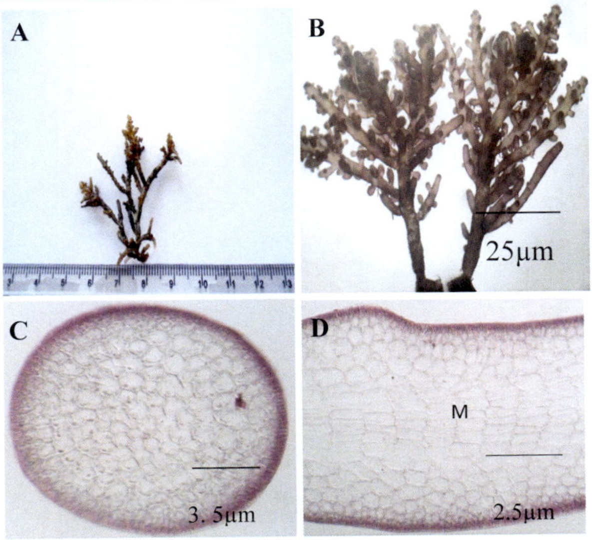

Laurencia Lamouroux

***Laurencia caraibica* P. Silva**

Description: Plants cylindrical or slightly compressed; surface cells not protruding in cross section, not radially elongated or palisade like; medullary cells without lenticular thickening on their walls.

Common: attached on rocks in high wave action regime.

Distribution: Manapad, Panjal, Kanyakumari, Kovalam.

A. Habit of the plant

B. Plant showing cylindrical branches

C. Transverse section

D. Longitudinal section showing medullary cells (M) without thickening

Laurencia Lamouroux

REFERENCES: V. Krishnamurthy & H. Joshi (1970: 26), Subbaramaiah (1974: 59), Untawale, Dhargalkar & Agadi (1983: 37), Kaliaperumal *et al.* (1992: 13), Silva *et al.* (1996: 505), Desikachary, V. Krishnamurthy & M. Balakrishnan (1998: 345).

***Laurencia* Agardh**
Description several ere from a comr 3-6 cm in broad, bran spreading, ri axes from th from the distichous, p compressed, portion, tuburculate l Surface elongated, medullary lenticular thi
Common: o hard surface splash zones
Distribution Panjal, Kovalam.
A. Habit of t
B. Plant sho
C. Transvers
D. Longitud

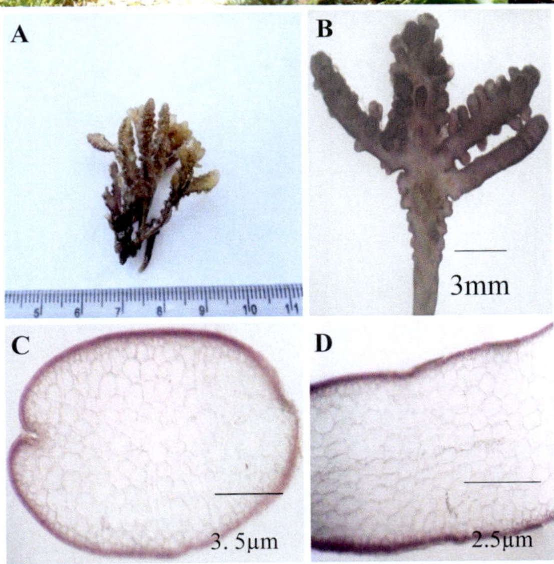

***Laurencia* Lamouroux**

***Laurencia ceylanica* J. Agardh**

Description: Frond with several erect axes arising from a common discoid base; 3-6 cm in height, 2-3 mm broad, branching sometimes spreading, rigid, branching of axes from the base as well as from the upper parts distichous, pinnate; branches compressed, smooth in lower portion, covered with tuburculate branchlets above. Surface cells radially elongated, palisade like, medullary cells without lenticular thickening.

Common: on rocks and other hard surfaces in a heavy tidal splash zones.

Distribution: Manapad, Panjal, Kanyakumari, Kovalam.

A. Habit of the plant
B. Plant showing flat branch
C. Transverse section
D. Longitudinal section

Laurencia Lamouroux

REFERENCES: Jagtap (1987: table I), R. Parekh & Chauhan (1987: table 1), R. Parekh & Chauhan (1990: table 1), Silva *et al*. (1996: 507), Siddhanta *et al*. (1997: 91 – 98), Desikachary, V. Krishnamurthy & M. Balakrishnan (1998: 346).

Laurencia cru

Description: in colour; 4-5 uniformly cyl base to ap branched, distichous (branches an distinctly pa sparse, ste cylindrical arranged, fe swollen and o medulla with thickening.

Common: o coral stones.

Distribution: Chinnamuttam Kanyakumari, Valai island an island.

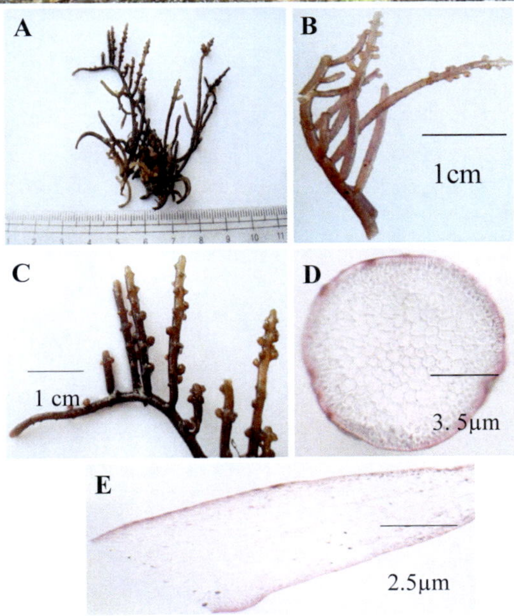

A. Habit of the
B *&C* Pl cylindrical bra
D. Transvers frond
E. Longitudin frond

Laurencia Lamouroux

***Laurencia cruciata* Harvey**

Description: Plant dark red in colour; 4-5 cm in height; uniformly cylindrical from base to apex, pinnately branched, branches distichous or opposite; branches and branchlets distinctly patent; ramuli sparse, sterile ramuli cylindrical cruciately arranged, fertile ramuli swollen and obtuse. Cells of medulla without lenticular thickening.

Common: on rocks and coral stones.

Distribution: Nochiurani, Chinnamuttam, Kanyakumari, Kovalam, Valai island and Upputhanni island.

A. Habit of the plant

B &C Plant showing cylindrical branches

D. Transverse section of frond

E. Longitudinal section of frond

Laurencia Lamouroux

REFERENCES: Balasundaram (1985: 180 ff), V. Krishnamurthy & Balasundaram (1990: 11 ff., fig. 5 table 1), V. Krishnamurthy *et al.* (1990: fig. 3), Desikachary, V. Krishnamurthy & M. Balakrishnan (1998: 349), Silva *et al.* (1996: 512), Desikachary, V. Krishnamurthy & M. Balakrishnan (1998: 348).

Laurencia indi
Description: Pl
height; coespit
about 1 mm
attached by rhi
dense, opposit
Branchlets simp
below and sim
cylindrical to c
cells not radiall
palisade like in
Medullary cells
thickenings on t
Common: on r
stones in subtid
Distribution: N
Villoondi theert
Ariyankundu, I
Thonithurai, M
island, Van isla
Koswari island.
A. Habit of the
B. Plant showin
alternate branch
C. Transverse s
lenticular thick
medullary cells
D. Longitudinal

Laurencia Lamouroux

Laurencia indica **Hauck**

Description: Plants 4-6 cm in height; coespitose, cylindrical about 1 mm in thickness, attached by rhizoids; branches dense, opposite or alternate. Branchlets simple or branched below and simple above, sub cylindrical to clavate. Surface cells not radially elongated or palisade like in cross section. Medullary cells with lenticular thickenings on their walls.

Common: on rocks and coral stones in subtidal region.

Distribution: Nalupannai, Villoondi theertham, Ariyankundu, Pamban, Thonithurai, Manapad, Valai island, Van island and Koswari island.

A. Habit of the plant

B. Plant showing opposite and alternate branch

C. Transverse section showing lenticular thickening in medullary cells (M)

D. Longitudinal section

Laurencia Lamouroux

Laurencia
Kützing
Descriptio
pale green
colour. E
alternate
Branches
cylindrical
generally
compact,
dichotomo
filaments,
in term
Surface c
pigmented.
spherical t
Rare: on
deeper wate
Distributio
A. Habit of
B. Plant sho
branches
C. Transver
frond
D. Longituc
frond

Laurencia Lamouroux

Laurencia microcladia **Kützing**

Description: Thallus bushy, pale green to purple green in colour. Branching dense, alternate to irregular. Branches and branchlets cylindrical. Branchltes generally spiral, rarely compact, apices blunt with dichotomously branched filaments, filaments sunken in terminal depression. Surface cells oval, deeply pigmented. Cystocarp spherical to urn shaped

Rare: on coral stones in deeper waters.

Distribution: Koswari island

A. Habit of the plant

B. Plant showing alternate branches

C. Transverse section of frond

D. Longitudinal section of frond

Laurencia Lamouroux

Laurencia
Agardh
Description:
in height; atta
of rhizoids, f
base, caespit
and filiforn
opposite, rare
irregular; bra
upper parts
numerous,
Surface cells
elongated in
Medullary
lenticular thi
walls.
Common: on
shaded places.
Distribution:
theertham,
dharga,
Manapad,
Chinnamuttam.
A. Habit of the
B. Plant show
branches
C. Transverse
frond
D. Longitudin
frond

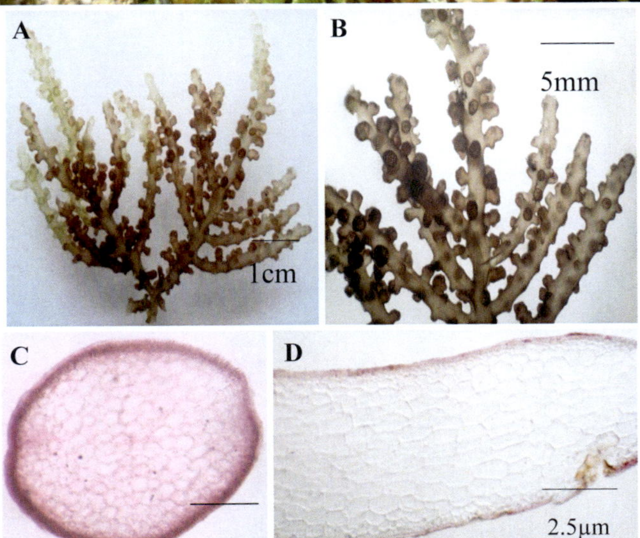

Laurencia Lamouroux

Laurencia nidifica J. Agardh

***Description**:* Plants 5-6 cm in height; attached by means of rhizoids, fronds matted at base, caespitose, cylindrical and filiform. Branching opposite, rarely alternate or irregular; branchlets in the upper parts of branches numerous, cylindrical. Surface cells not radially elongated in cross section. Medullary cells with lenticular thickening in the walls.

Common: on coral stones in shaded places.

Distribution: Villoondi theertham, Seeniappa dharga, Valinokkam, Manapad, Panjal, and Chinnamuttam.

A. Habit of the plant
B. Plant showing alternate branches
C. Transverse section of frond
D. Longitudinal section of frond

Laurencia Lamouroux

REFERENCES: V. Krishnamurthy *et al.* (1990: figs.4,5,7), Padmini Sreenivasa Rao (1991: 37 table 1, 2), Kalimuthu, Kaliperumal & Ramalingam (1992: 124), Desikachary, V. Krishnamurthy & M. Balakrishnan (1998: 350), Silva *et al.* (1996: 515).

Laurencia ob
Lamouroux
Description:
with pink tips,
height; caespit
Branching pi
opposite or o
upper branches
ramuli cylindri
slightly const
Superficial ce
elongated. Me
lenticular
Cystocarps sess
Uncommon: a
stones
Distribution:
Villoondi theert
A. Habit of th
pinnate branche
B. Plant sh
branch.
C. Surface view
D. Transverse s
E. Longitudinal

Laurencia Lamouroux

Laurencia obtusa **(Hudson) Lamouroux**
Description: Plants greenish with pink tips, upto 25 cm in height; caespitose, cylindrical. Branching pinnate, branches opposite or often verticillate; upper branches slightly shorter; ramuli cylindrical to subclavate, slightly constricted at base. Superficial cells not radially elongated. Medullary cells not lenticular thickening. Cystocarps sessile or stalked.
Uncommon: attached on coral stones
Distribution: Nalupanai and Villoondi theertham
A. Habit of the plant showing pinnate branches
B. Plant showing opposite branch.
C. Surface view of thallus
D. Transverse section of frond.
E. Longitudinal section of frond

Laurencia Lamouroux

REFERENCES: Grunow (1874: 45), Borgesen (1938: 231 – 232), Borgesen (1945: 58 – 59, figs. 34 - 35), V. Krishnamurthy *et al.* (1990: figs. 4), Silva *et al.* (1996: 517), Desikachary, V. Krishnamurthy & M. Balakrishnan (1998: 351).

Laurencia obt
Lamouroux
Grunow

Description:
consisting of a
axis; 2-3 cm
from a discoi
Superficial cell
quadrate, n
elongated or
Medullary cell
in diamete
lenticular thick

Uncommon:
rocks

Distribution: N

A. Habit of the
B. Plant showi
branches
C. Transverse s
frond
D. Longitudina
showing witho
thickening in m

Laurencia Lamouroux

Laurencia obtusa **(Hudson) Lamouroux var. *rigidula* Grunow**
Description: Plants consisting of a tuft or erect axis; 2-3 cm high, arising from a discoid base, rigid. Superficial cells more or less quadrate, not radially elongated or palisade like. Medullary cells up to 50 μm in diameter, without lenticular thickening.
Uncommon: attached on rocks
Distribution: Nalupanai.
A. Habit of the plant
B. Plant showing opposite branches
C. Transverse section of frond
D. Longitudinal section showing without lenticular thickening in medulla (M)

Laurencia Lamouroux

REFERENCES: Kamat *et al*. (1992: table 1), Jagtap (1993: table I), Silva *et al*. (1996: 517), Kaliaperumal, Ezhilvalavan & Ramalingam (1997: 38), Siddhanta *et al*. (1997: 91 – 98), Desikachary, V. Krishnamurthy & M. Balakrishnan (1998: 353).

Laurencia Lamouroux

Laurencia papillosa **(C. Agardh) Greville**

Description: Plant dark red to brownish red in colour; 8-15 cm in height; attached to the rocks and coral stones by discoid holdfast; erect axis irregularly branched; branches covered with short wart like tuberculate branchlets, superficial cells palisade like; medullary cells thick wall but no lenticular thickening on the walls. Cystocarp ovoid or spherical.

Common: on hard substratum like rocks and coral stones.

Distribution: Arasaneri pattinam, Karankadu, Pamban light house, Seeniappa dharga, Nochiurani, Bharathi nagar, Valinokkam, Manapad, Idinthakarai, Leepuram, Chinnamuttam Vattakottai, Kanyakunari, Kovalam and all 20 islands of Gulf of Mannar.

A. Habit of the plant

B. Branches covered with small branchlets

C. Transverse section

D. Longitudinal section

Laurencia Lamouroux

REFERENCES: Siddhanta *et al.* (1995: 329), Desikachary, V. Krishnamurthy & M. Balakrishnan (1998: 349,355), Silva *et al.* (1996: 519), Siddhanta *et al.* (1997: 91 – 98), Pasricha & Vijayaraghavan (1998: 19 – 22).

Laurencia pe
Børgesen

Description:
in colour; ε
attached by
compressed,
breadth and
branching pi
branches irre
ultimate bran
narrow belo
apical portior
with uniforr
Sexual plants

Rare: on roc
substratum.

Distribution:

A. Habit of th
B. Branch shc
secondary bra
C. Transverse

Laurencia Lamouroux

Laurencia pedicularioides **Børgesen**

Description: Plants rosy red in colour; 8-15 cm in tall, attached by disc; fronds compressed, 3-4 mm in breadth and 1 mm thickness; branching pinnate, secondary branches irregularly produced, ultimate branchlets 3 mm long, narrow below, broadened at apical portion. Medullary cells with uniformly thick walls. Sexual plants not known.

Rare: on rocks and other hard substratum.

Distribution: Manapad.

A. Habit of the plant

B. Branch showing small secondary branches

C. Transverse section of frond

Laurencia Lamouroux

REFERENCES: J. Agardh (1876: 651 – 652), De Toni (1903: 789), Subbaramaiah (1974: 59), Untawale, Dhargalkar & Agadi (1983: 37), Silva *et al*. (1996: 520), Desikachary, V. Krishnamurthy & M. Balakrishnan (1998: 357).

Laurencia
Descripti
6cm in
cylindrical
compresse
branched.
quadrate
sometimes
than broa
usually n
thickening
only rarely
Uncomm
coral ston
deep.
Distributi
island, M
island
island.
A. Habit o
B. Plant
branches
C. Trans
frond
D. Longi
frond

Laurencia Lamouroux

***Laurencia rigida* J. Agardh**

Description: Plants upto 6cm in height; caespitose, cylindrical or slightly compressed, sparsely branched. Superficial cells quadrate in cross section, sometimes slightly longer than broad. Medullary cells usually not with lenticular thickening on their walls, only rarely seen.

Uncommon: on rocks and coral stones upto 10 meter deep.

Distribution: Krusadai island, Mully island, Valai island and Upputhanni island.

A. Habit of the plant

B. Plant showing sparse branches

C. Transverse section of frond

D. Longitudinal section of frond

***Leveillea* Decaisne**

Thallus macroscopic, mostly prostrate, with dorsi-ventral organization; attached to the substratum by tufts of rhizoids; main axis without cortication and with 7 pericentral cells, producing alternate leaflet branches. Young fragments produce three pericentral which may increase 5 to 6 in older parts. Cystocarp conical, on the abaxial side of the determinate branches. Tetrasporangia in a single series in a bladeless indeterminate branch.

Key to the species:

Plants dorsi-ventral, main axis with 7 pericentral cells with incurved tips …..*L.jungermannioides*

Leveillea Decaisne

REFERENCES: Jagtap (1987: table I), V. Krishnamurthy & Balasundaram (1990: 8), V. Krishnamurthy *et al*. (1990: figs. 4, 57, 8), Kaliaperumal, Rajagopalan & Chennubhotla (1992: 44), Silva *et al*. (1996: 524), Desikachary, V. Krishnamurthy & M. Balakrishnan (1998: 325).

***Leveillea* Decaisne**

***Leveillea jungermannioides* (Hearing & G. Martens) Harvey**
Description: Thalllus is creeping, dorsiventral axes, upto 5 cm in length. Plants dark red to brownish red in colour. Attached to the substratum by rhizoids protruding from the ventral surface, producing series of leaf like substances along the dorsal surfaces. Sometimes branches arising from the basal segments of determinate branches. Apices of main axis incurved.
Uncommon: on coral stones and other small stones.
Distribution: Nalupanai, Pullivasal and Valimunai island
A. Branch producing leaf like structures
B. Plant showing basal branches

Liagora Lamouroux

Plants well branched with many bifurcations, calcification light or absent; thallus more or less slippery when alive. A compact medulla of colourless filaments produce radially disposed, lateral filaments with chloroplasts.

Key for the species

1. Plants dichotomously branched 2
1. Plants pinnately branched, not dichotomously divided 3
2. Lateral proliferations very few or absent .... L.indica
3. Lateral branches very few, short, rarely divided, main axis with longitudinal furrow or slightly compressed, plants heavily calcified ....L. albicans
3 Lateral branches many, muscilagenous or lightly calcified 4
4 Fertilized carpogonium transversely divided into Row of 3 cells ....L. mannarensis
4. Fertilized carpogonium not divided into row of 3 cells 5
5. Male and female plants similar in branching ....L. orientalis

Liagora Lamouroux

REFERENCES: G. Rao & Umamaheswara Rao (1986: 103), Chennubhotla, Kaliaperumal & Kalimuthu (1987: 104 – 105), Desikachary, V. Krishnamurthy & M. Balakrishnan (1990: 142 – 144, fig. 29A, B, 31, pls. XII, XV.B), Umamaheswara Rao, (1990: table. 3), Bhakuni *et al.* (1992: 515), Kalimuthu, Kaliaperumal & Ramalingam (1992: 123).

Liagora
Lamouro

Descri
pink ir
10 cm
cylindr
branche
are v
undivic
divided
calcifie
slightly
of b
curved.

Common
high tidal

Distributi
Nalupanai
Thonithur
Manapad,
Koswari i:

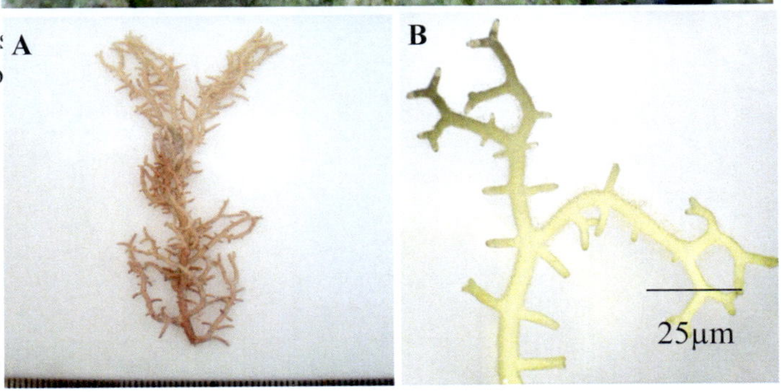

A. Habit o
B. Plant
branches

Liagora Lamouroux

Liagora albicans **Lamouroux**

Description: Plant light pink in colour; more than 10 cm in height, thallus cylindrical, pinnately branched, lateral branches are very few, short, undivided or rarely divided. Plants heavily calcified. Main axis slightly compressed. Apex of branches slightly curved.

Common: on rocks with high tidal influence.

Distribution: Munaikadu, Nalupanai, Pamban, Thonithurai, Pudumadam, Manapad, Van island and Koswari island.

A. Habit of the plant

B. Plant showing pinnate branches

Liagora Lamouroux

REFERENCES: V. Krishnamurthy & Sundararajan (1985: 53 – 55, 57, 59, figs.47 – 65, pl. I: figs. 3, 4), Desikachary, V. Krishnamurthy & M. Balakrishnan (1990: 139 – 140, fig. 24). Silva *et al.* (1996: 125).

Liagora
Krishnamu
Sundarara
Descriptio
cm in hei
substratum
calcificatio
region and
region.
dichotomou
proliferatio
Medulla
medullary
assimilatory
and dichoto
Common:
and rocks.
Distributio
Villoondi
Ariyankund
Thonithurai
island
Koswari isl
A. Habit of

Liagora Lamouroux

Liagora indica **V. Krishnamurthy & Sundararajan**

Description: Plant erect, 4-5 cm in height, attached to substratum by basal disc; calcification is more in lower region and lesser in upper region. Main axis dichotomously branched, proliferations totally absent. Medulla multiaxial, medullary filaments parallel; assimilatory filaments radial and dichotomously branched.

Common: on coral stones and rocks.

Distribution: Nalupanai, Villoondi theertham, Ariyankundu, Pamban, Thonithurai, Manapad, Valai island, Van island and Koswari island.

A. Habit of the plant

Liagora Lamouroux

REFERENCES: V. Krishnamurthy & Sundararajan (1985: 53 – 56, 59, 59 – 62, figs.6 – 95, pl. I: figs. 5, 6), Desikachary, V. Krishnamurthy & M. Balakrishnan (1990: 144 – 145, fig. 32). Silva *et al*. (1996: 126).

***Liagora** m* **Krishnamurtl Sundararajan**

Description: cm high; att disc, giving o axis 3 mm branched, common, more Medullary fila assimilatory fi produced, branched with

Common :on hard substratur

Distribution: Villoonditheer Thalathoppu, Keelakarai, M Koswari isl

A. Habit of th pinnate branch

B. Branch proliferations

C. Another vi the plant

D. Close view proliferations.

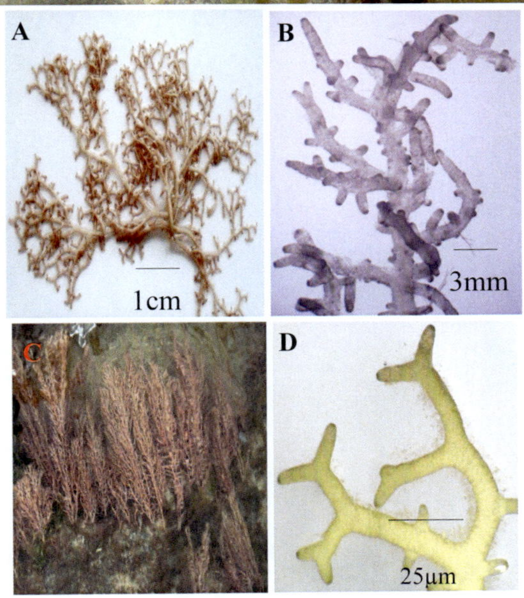

Liagora Lamouroux

Liagora mannarensis V. Krishnamurthy & Sundararajan
Description: Plant erect, 12 cm high; attached by basal disc, giving out 3 main axis; axis 3 mm thick, pinnately branched, proliferations common, more on male plants. Medullary filaments elongated; assimilatory filaments radially produced, dichotomously branched with several forking.
Common: on rocks and other hard substratum.
Distribution: Villoonditheertham, Thalathoppu, Sullimunthal, Keelakarai, Manapad, and Koswari island.
A. Habit of the plant showing pinnate branches
B. Branch showing many proliferations
C. Another view of habitat of the plant
D. Close view of branch with proliferations.

Liagora Lamouroux

REFERENCES: Weber – van Bosse (1913: 111), Weber – van Bosse (1914: 275), S. Dixit (1966: 14), V. Krishnamurthy & Sundararajan (1987a: 73, 75, 77, figs. 2, 15 28), Desikachary, V. Krishnamurthy & M. Balakrishnan (1990: 147 – 148, fig. 29H) Silva *et al*. (1996: 126).

Liagora orie
Description:
35 cm in h
lightly calc
prominent lc
3 mm thicl
base, lateral
length. Assi
repeatedly
branched; l
filaments el
slightly broa
Habitat: Ra
Distribution
A. Habit of t
B. Plant sho
branch
C. Plant sho
branching.

Liagora Lamouroux

***Liagora orientalis* J. Agardh**

Description: Plants large, upto 35 cm in height; cylindrical, lightly calcified, with very prominent long central axis; 2-3 mm thick, branched form base, lateral branches in varied length. Assimilatory filaments repeatedly dichotomously branched; lower assimilatory filaments elongated and upper slightly broader.

Habitat: Rare: on rocks

Distribution: Manapad.

A. Habit of the plant

B. Plant showing cylindrical branch

C. Plant showing repeated branching.

Lophocladia (J. Agardh) Schmitz

***Lophocladia* (J. Agardh) Schmitz**

Plants cylindrical, upto 10 cm tall; segments have central axial cell and four pericentral cells; spermatangia in clusters, cylindrical on trichoblast. Cystocarps flask shaped with long neck; tetrasporangia are in contorted strichidia.

Key to the species:

Branches exogenous, trichoblast short and broad, neck short. ……*L.lallemandii*

Lophocladia (J. Agardh) Schmitz

REFERENCES: V. Krishnamurthy & H. Joshi (1970: 26), Umamaheswara Rao (1970b: 368), K. Srinivasan (1973: 1, pl. I – M), Dave *et al.* (1987: table 1), Silva *et al.* (1996: 525), Desikachary, V. Krishnamurthy & M. Balakrishnan (1998: 311).

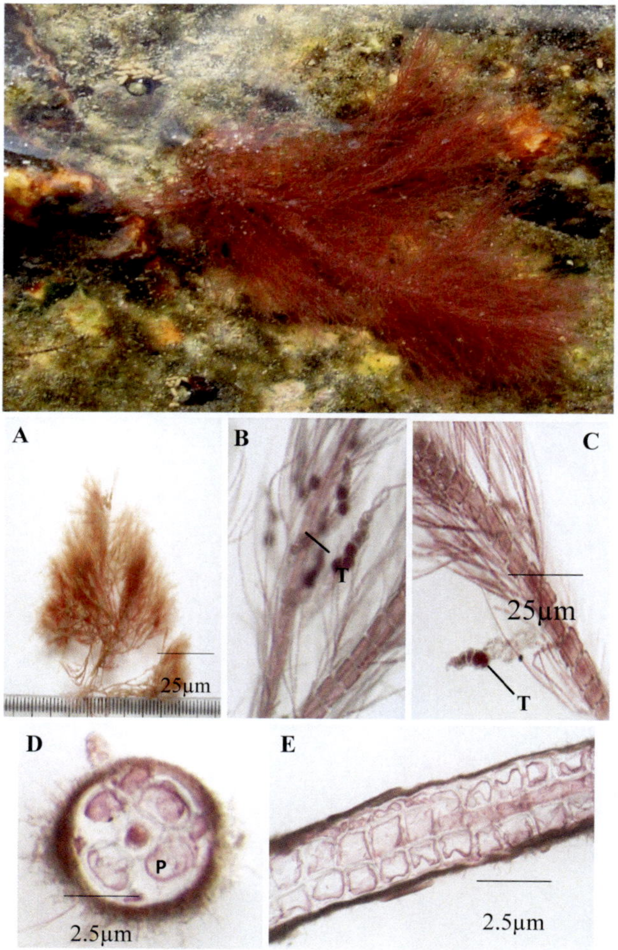

Lophocladia (J. Agardh) Schmitz

Lophocladia lallemandii **(Montagne) Schmitz**

Description: Thallus bushy, 10 cm high; rose pink in colour; branching widely dichotomous. Branches 0.5 – 2.0 mm diameter, tapering distally, covered with fine hair like filaments of four pericentral cells. Filaments 60 µm diameter, deciduous, dichotomously branched. Holdfast inconspicuous, disc like or attached by rhizoids from decumbent branchlets. Tetrasporangial strichidia at first dichotomy of filament, tetrasporangia spherical.

Uncommon: attached to rocks or pebbles, to 20 m. deep.

Distribution: Shingle Island and Mully Island.

A. Habit of the plant

B. & C. Branches showing tetrasporangial strichidia (T)

D. Transverse section showing 4 pericentral cells. (P)

E. Longitudinal section

Martensia **Hearing**

Thallus foliaceous, membranous, with a cylindrical stalk that gives rise to expanded leafy blades, Branches dichotomously or irregularly lobed; 1 cell layer thick; periodic growth phases give the thallus a curious grid like appearance forming a sort of net. Spermatangia in rounded sori occurring on lamellae; cystocarps globular, projecting from the margin of the lamellae of the net.

Key for the species:

1. Frond with flabellate blades …… 2
1. Frond with cuneate blades ……*M. sp.*
2. Lamellae in mature portion of the thallus, 30 - 70um apart. …… *M. fragilis*
2. Lamellae in mature portion of the thallus 85- 175 um apart ……..*M. indica*

Martensia Hearing

REFERENCES: Huisman & Walkar (1990: 430), Miller (1990: 419), V. Krishnamurthy & Varadarajan (1991b: 67, figs. 27 – 32), Silva *et al*. (1996: 460), Desikachary, V. Krishnamurthy & M. Balakrishnan (1998: 259).

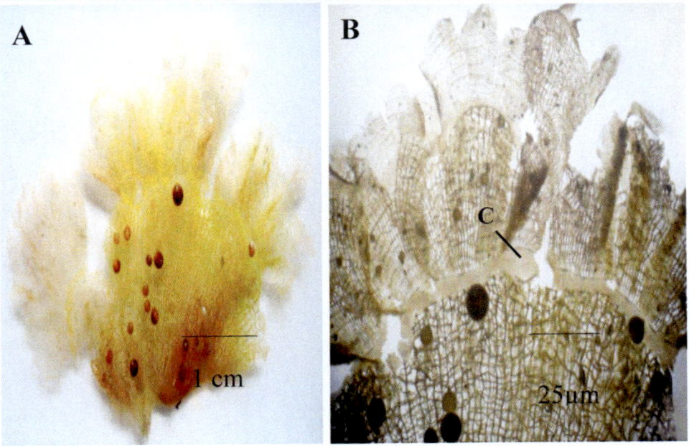

Martensia Hearing

Martensia fragilis **Harvey**
Description: Plants 2-5 cm tall; pale red in colour; attached to substratum by rhizoids; fronds very thin `and fluffy, consisting of flabellate blades arising from stalk of compact tissue. Lamellae interconnected by transversely extended laterals, giving rise to net like portion of the frond. Lamellae 30- 70 um apart. Spermatangia in sori on the lamellae in the net region of the thallus. Cystocarps globose, produced on the lamellae.
Uncommon: attached on rocks or coral stones.
Distribution: Pamban pass, Thonithurai.
A. Habit of the plant
B. Branch showing cystocarps (C) on lamellae

Martensia Hearing

REFERENCES: V. Krishnamurthy & Thomas (1977: 49 – 50, 103, figs.12 – 18, 25, 26), Untawale, Dhargalkar & Agadi (1983: 36), Silva *et al.* (1996: 461), Desikachary, V. Krishnamurthy & M. Balakrishnan (1998: 273).

Martensia Krishnamu & Thomas

Description: green in c height; att expanding to arranged distally by radiating la tangential fi Marginal ce diameter and 85- 175 um 2 mm in dia stalked. Tetr lamellae.

Uncommon and coral st deep.

Distribution: Sadaimuniya Krusadai isl Valimunai Upputhanni i

A. Habit of t

B. Thallus sh

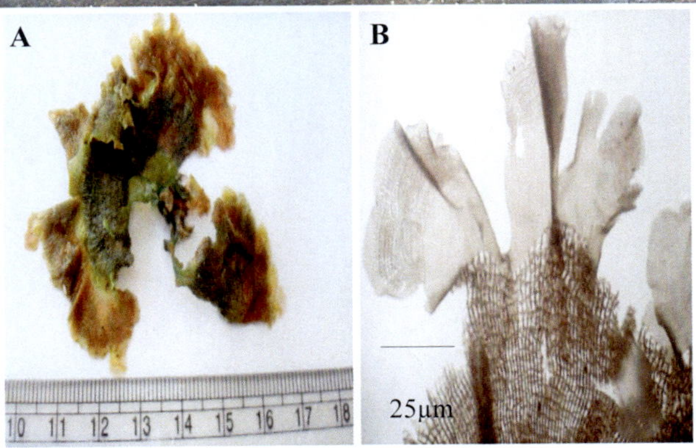

Martensia Hearing

Martensia indica **V. Krishnamurthy & Thomas**

Description: Plant dark grey green in colour; 3-8 cm in height; attached by stalk expanding to a blade of solidly arranged cells, continued distally by a network of radiating lamellae joined by tangential filaments of cells. Marginal cells 15-40 um in diameter and those of lamellae 85- 175 um apart. Cystocarp 1-2 mm in diameter, globose and stalked. Tetrasporangia on the lamellae.

Uncommon: attached on rocks and coral stones to 10 meter deep.

Distribution: Pamban, Sadaimuniyan valasai, Krusadai island, Mully island, Valimunai island and Upputhanni island.

A. Habit of the plant

B. Thallus showing lamellae.

Martensia Hearing

A

Marter

Descrip
colour;
rhizoids
basal,
frond;
consisti
mm in
alternat
lamella
Cystoca
tetraspo
lamella
Habitat
rocks
Distribi
island.
A. Habi
B. Thall

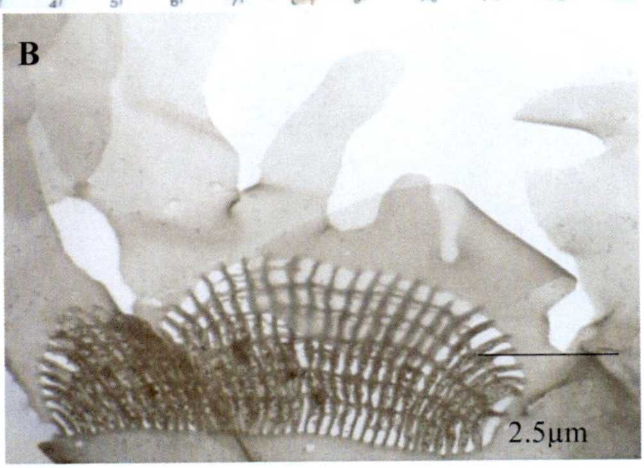

Martensia Hearing

***Martensia* sp.**

Description: Plants yellow in colour; attached to rocks by rhizoids arising from the basal, solid region of the frond; 5-7 cm in height; consisting of blades 4- 15 mm in width; blades with alternate net like region; lamellae 50-65µm apart. Cystocarp globular, sessile; tetrasporangia born on lamellae.

Habitat: Rare: attached on rocks

Distribution: Krusadai island.

A. Habit of the plant

B. Thallus showing lamellae

Neurymenia J. Agardh

Neurymenia J. Agardh

Thallus, bushy, macroscopic upto 20cm long, with a cylindrical stem at the base and ribbon like branches, upto 2cm broad and 200 µm thick with prominent midrib that produce other branch from midrib; the midribs present lateral, opposite that extend to the margin of the blades in small tooth like projections; the tips of the blades are in rolled. Fertile branches arising from the base of the second order branches of the axis. Cystocarps oblong, shortly stalked. Tetrasporangia in elongate, flattened strichidia.

Key to the species:

Plants with alternate lateral veins and inrolled apices *N. fraxinifolia*

Neurymenia J. Agardh

REFERENCES: Nair *et al*. (1986: 738), Chennubhotla, Kaliaperumal & Kalimuthu, (1987: 110 – 111), V. Krishnamurthy *et al*. (1990: 271), Rama Rao *et al*. (1993: 106), Silva *et al*. (1996: 529), Desikachary, V. Krishnamurthy & M. Balakrishnan (1998: 327).

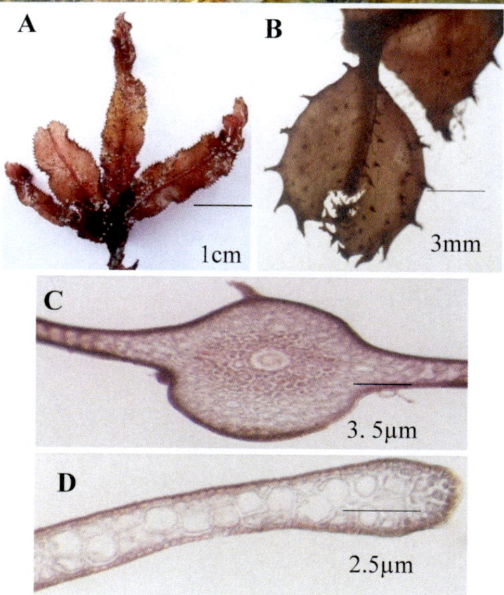

Neurymenia J. Agardh

***Neurymenia fraxinifolia* (Mertens ex Turner) J. Agardh**

Description: Plants bushy, upto 20 cm in height; attached by discoid holdfast. Main axis forms long, cylindrical, frequently and irregularly branched stipe, continuing as the midrib of elongate blade with prominent midrib and clear alternate lateral veins; axis with inrolled apex. Cross section of the blade showing 2 layers of medullary cells surrounded by a layer of small celled cortex. Segments producing 5 pericentral cells, one is dorsal and other is lateral and lateral cells produce wing like extension. Cystocarps oblong with short stout stalk and thick pericarp.

Rare: attached on rocks and coral stones upto 10 m deep.

Distribution: Manapad, Kanyakumari, Kovalam and Appa island

A. Habit of the plant showing prominent mid rib

B. Branches showing many proliferations

C. Transverse section

D. Longitudinal section

***Nitophyllum* Greville**

Thallus delicate, foliose, dichotomously, or sub-dichotomously or irregularly lobed; blade without midrib and veins; branch tips lacking a distinct apical cell; with just one layer of cells in the upper part and with a margin of small cells. Cystocarp sessile, scattered on the thallus; tetrasporangia in clear rounded sori arranged in a sub-marginal series.

Key to the species:

Frond dichotomously branched, cystocarp sessile …..*N.punctatum*

Nitophyllum Greville

REFERENCES: S. Dixit (1966: 19), V. Krishnamurthy & H. Joshi (1970: 24), Untawale, Reddy & Ambiye (1989: table 1), V. Krishnamurthy & Varadarajan (1991b: 63 – 65, figs. 9 -19), Silva *et al*. (1996: 463), Desikachary, V. Krishnamurthy & M. Balakrishnan (1998: 270).

Nitophyllu
(Stackhou
Descriptio
colour; 3-
attached tc
minute di
fronds
dichotomou
margin be
single laye
and 2-3 la
Spermatang
position. C
scattered
region of th
Common:
Distributio
Shingle
island,
Valimunai
island,
Upputhann
A. Habitat
B. Surface
C. Longitu
thallus.

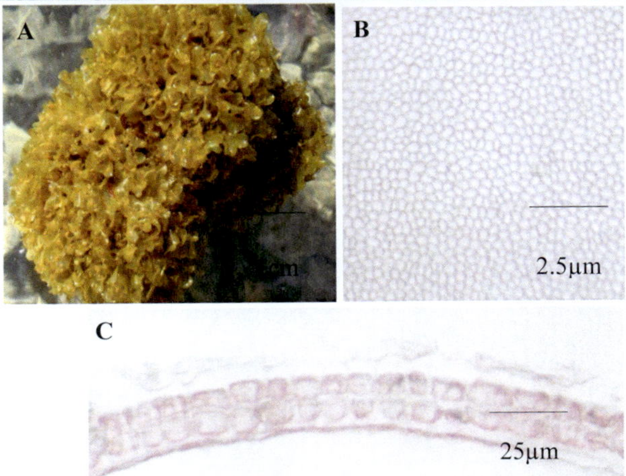

Nitophyllum Greville

***Nitophyllum punctatum* (Stackhouse) Greville**
Description: Plant yellow in colour; 3-15 cm in height; attached to substratum by a minute disc like holdfast; fronds irregularly, dichotomously lobed, upper margin being entire; frond single layer in upper portion and 2-3 layers at the base. Spermatangia in submarginal position. Cystocarps sessile, scattered over the upper region of the thallus.
Common: attached on rocks.
Distribution: Rameshwaram, Shingle island, Krusadai island, Manali island, Valimunai island, Anaipar island, Salli island and Upputhanni island.
A. Habitat of the plant
B. Surface view of thallus
C. Longitudinal section of the thallus.

***Peyssonnelia* Decaisne.**

Plants much calcified, crustose, attached to the substratum by simple or branched rhizoids; hypothelium with radiating filaments; perithelium multilayered with vertical row of filaments. Tetrasporagia intermingled with paraphyses in nemathecia.

Key to the species:

Perithelium with vertically ascending rows ……*P.obscura*

Peyssonnelia Decaisne

REFERENCES: Subbaramaiah (1971: 21), Untawale, Dhargalkar & Agadi (1983: 29), Untawale, Reddy & Ambiye (1989: table 1), Desikachary, V. Krishnamurthy & M. Balakrishnan (1990: 222 - 223), Sindhu & Panikkar (1995b: 147 – 150), Silva *et al*. (1996: 213).

***Peyssonnelia* Decaisne**

***Peyssonnelia obscura* Weber-van Bosse**

Description: Plants dark red in colour; crustose, orbicular, attached by unicellular rhizoids arising from the lower surface; margins free; thallus composed of single row of hypothelial cells and vertical rows of perithelial filaments with 4-6 celled.

Common: attached to rocks and other hard substratum.

Distribution: Kanyakumari, Kovalam, Hare island, Mully island, Appa island, Anaipar island, Salli island, Upputhanni island, Van island and Koswari island.

A. Habit of the plant

B. Transverse section of thallus

C. Longitudinal section of thallus showing hypothelial cells (H) and perithelial cells (P)

***Polysiphonia* Greville**

Thallus macroscopic, a few millimeters to several centimeters long, filamentous, radially organized with a discoid base or creeping filaments attached by unicellular rhizoids; all branches have a polysiphonous structure with or without cortication. Cystocarps globose, carpospores terminal or gonimoblast filaments, tetrasporangia in the upper part of the fertile branchlets.

Key for the species:

1. Spernatangial cluster with a terminal sterile cell

…….*P.tuticorinensis*

1. Spermatangial cluster without sterile cell at the apex

…….. *P.subtilissima*

Polysiphonia Greville

REFERENCES: Sreenivasa Rao (1965b: 301), Sreenivasa Rao (1967: 171), V. Krishnamurthy & H. Joshi (1970: 26), Untawale, Dhargalkar & Agadi (1983: 36), Silva *et al.* (1996: 538), Desikachary, V. Krishnamurthy & M. Balakrishnan (1998: 295).

Polysiphon **Tseng**

Descriptio purple; tuf height; p attached rhizoids; sparsely a dichotomou with uni lower pa woven tog Segments throughout long as b almost equ and again long near t ovate.

Uncommo intertidal a

Distributio Manapad, Chinnamut Valai isla island.

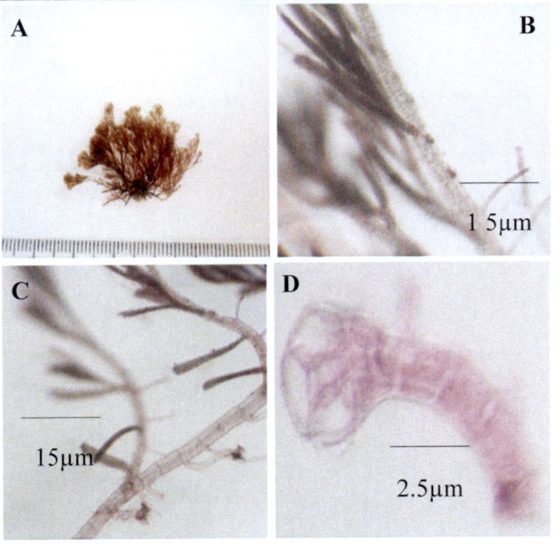

A. Habit of
B & C I branches
D. Trans frond

Polysiphonia Greville

Polysiphonia subtilissima **Tseng**

Description: Plants brownish purple; tufts about 2 cm in height; primary filaments attached by short, stout rhizoids; erect filaments sparsely and distantly sub dichotomous, occasionally with unilateral branches; lower parts of filaments woven together by rhizoids. Segments shorter than broad throughout; generally half as long as broad at the base, almost equal in mid portions and again about one third as long near the tips. Cystocarps ovate.

Uncommon: on rocks in intertidal and subtidal regions

Distribution: Valinokkam, Manapad, Kuduthalai, Chinnamuttam, Mully island, Valai island and Anaipar island.

A. Habit of the plant

B. & C. Different views of branches

D. Transverse section of frond

Polysiphonia Greville

REFERENCES: V. Krishnamurthy *et al.* (1990: figs. 5), Maya & Nair (1992: 26), P. Subba Rao *et al.* (1992: 181), Anilkumar & Panikkar (1994: 53: figs. 1 – 4), Silva *et al.* (1996: 547), Desikachary, V. Krishnamurthy & M. Balakrishnan (1998: 301).

***Polysiphoni*
Børgesen

***Description*.** filaments in height; u branches all filaments; lo filaments gradually u as long as pericentral well develop on the firs trichoblast, Cystocarp el

Uncommon pebbles.

***Distribution*:** Kovalam anc

A. Habit of tl
B Plant shov
C. Branch wi
D. Transvers thallus.
E. Longitudi thallus.

Polysiphonia Greville

Polysiphonia tuticorinensis **Børgesen**
Description: Thallus dense of filaments in a tuft; 12 cm in height; upwardly erect branches all along the main filaments; lower parts of main filaments broad, tapering gradually upwards, segments as long as broad, with four pericentral cells; trichoblast well developed. Spermatangia on the first forking of a trichoblast, elongate, conical. Cystocarp elongate.
Uncommon: on rocks and pebbles.
Distribution: Kanyakumari, Kovalam and Koswari island.
A. Habit of the plant
B Plant showing branches
C. Branch with cystocarp (C)
D. Transverse section of thallus.
E. Longitudinal section of thallus.

***Portieria* Zanardini**

Thallus erect, flattened, branching dense and alternate in a single plane, sometimes with an inconspicuous basal midrib. Attached to the substratum by a discoid holdfast. Structurally the thallus is uniaxial, with a clear central row of cells surrounded by loose globular cells that decrease in diameter towards cortex. When collected the plants emit a strong smell that attracts small ants. Tetrasporangia zonate and cushion like cystocarp.

Key for the species:

1. Plants 3-3.5 cm in high, tips spinulose ……*P.spinulosa*

1. Plants over 10 cm high, tips incurved, sparse branches ………. *P.hornemannii*

Portieria Zanardini

REFERENCES: Maya & Nair (1992: 32), Subba Rao *et al*. (1992: 181), Rama Rao *et al*. (1993: 106), Silva *et al*. (1996: 215 – 216), Parekh & Mairh (1997: 40), Parekh & Mairh (1999: 99 - 105).

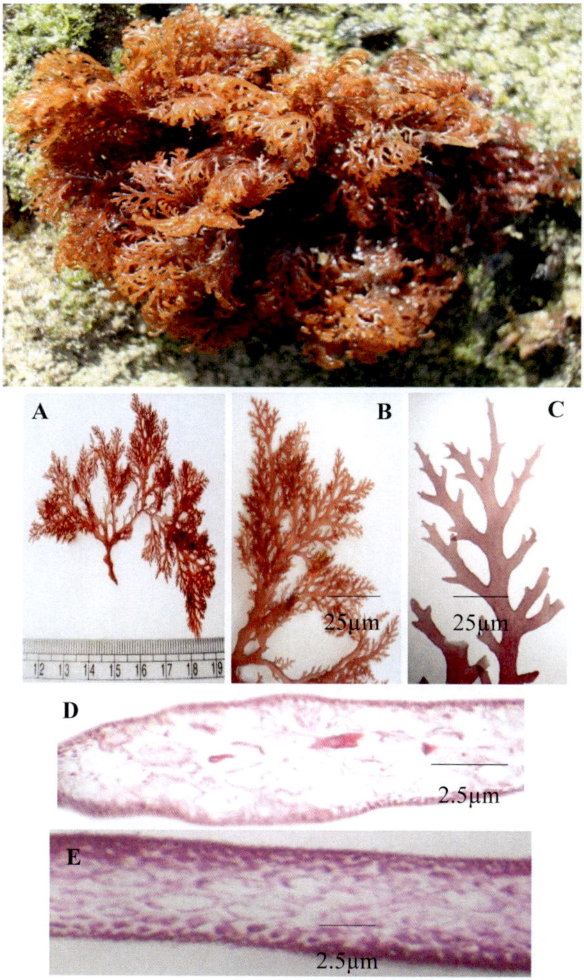

Portieria Zanardini

***Portieria hornemannii* (Lyngbye) P. Silva**

Description: Plants erect, saxicolous; upto 15cm high; bright red in colour; somewhat cartilaginous in consistency; the erect fronds compressed, terete in lower portion, distinctly flattened in upper portion; branching divaricate, pinnate; main branches 2-5mm wide, pinnules narrower, tip of pinnules markedly incurved; margin entire or rarely dentate. Cortex compact; inner cortical cells large, outer cortical cells progressively smaller, gland cells abundant in the outer cortex.

Common: on rocks and coral stones and pebbles

Distribution: Pamban light house, Nalupanai, Villoonditheertham, Ariyankundu, Pamban, Thonithurai Keelakarai light house, Keelakarai, Ervadi, Sadaimuniyan valasai, Kanyakumari and Kovalam.

A. Habit of the plant

B. Plant showing pinnate branches

C. Branch showing narrow tip

D. Transverse section

E. Longitudinal section

Portieria Zanardini

REFERENCES: Kutzing (1968: 11, pl. 32: figs. a – c), Menez & Moe (1987: 129), Silva *et al.* (1996: 217).

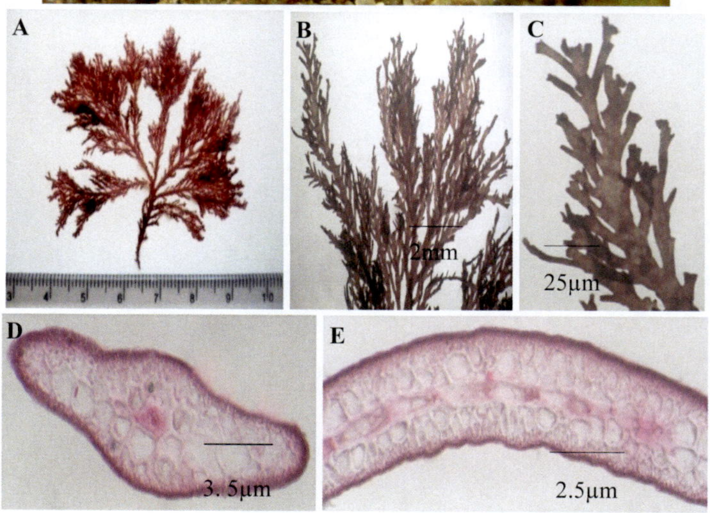

Portieria Zanardini

***Portieria spinulosa* (kützing) P. Silva**

Description: Plants erect, saxicolous; 3-3.5 cm high; gelatinous to cartilaginous in consistency; erect fronds linear, somewhat densely branched; branching bipinnate to tripinnate; branches and pinnules delicate, linear and attenuated with pronouncedly spinulose tips, commonly proliferous also.

Rare: on rocks in intertidal to subtidal regions

Distribution: Kanyakumari and Kovalam.

A. Habit of the plant

B. Plant showing bipinnate branches

C. Branch with many proliferations.

D. Transverse section of frond

E. Longitudinal section of frond

Pterocladia J. Agardh

***Pterocladia* J.Agardh.**

Plants having creeping as well as erect fronds. Most of the characters are similar to Gelidium. Tetrasporangia in ovoid to circular sori formed at main branches; cystocarp unilocular with ostiole on one side of the frond.

Key for the species:

1. Plants less than 2.0 cm tall, tetrasporangia in slightly curved transverse row …..*P. kanyakumariensis*
1. Plants more than 2.0 cm tall, tetrasporangia arranged in sori ……..2
2. Erect fronds 3-6 cm tall, compressed or flattened, irregularly branched ….. *P. heteroplatos*
2. Erect fronds flattened, profusely branched … *P. caloglossoides*

Pterocladia J. Agardh

Pterocladia
(Howe) San
Description
around 2-4 c
cylindrical
flattened
produced fr
bases; reser
pusillum,
cystocarp
Branching p
Uncommon:
rocks in
region.
Distributio
Mahabalipur
Chinnamutta
Kanyakumar
Mully island
A. Habit of t
B. Plant sh
branches
C. Transve
frond
D. Longitu
frond

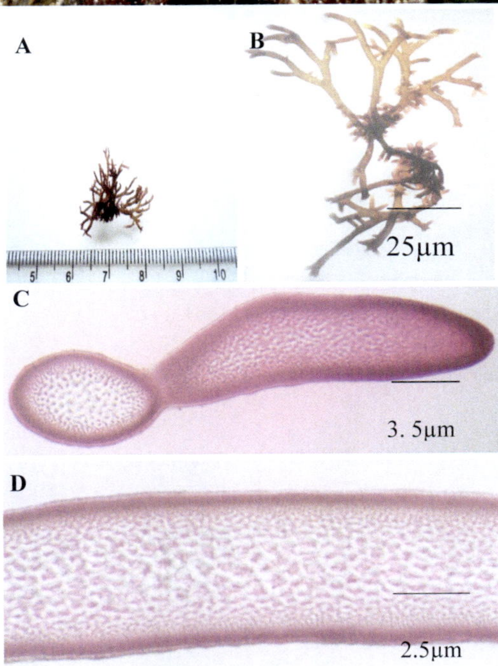

Pterocladia J. Agardh

Pterocladia caloglossoides **(Howe) Santelices**

Description: A small plant around 2-4 cm in height; with cylindrical and partially flattened erect fronds produced from stoloniferous bases; resembling *Gelidium pusillum*, except for cystocarp structure. Branching partially bipinnate.

Uncommon: attached on rocks in lower intertidal region.

Distribution: Muttukadu, Mahabalipuram, Valinokkam, Chinnamuttam, Kanyakumari Kovalam and Mully island.

A. Habit of the plant
B. Plant showing cylindrical branches
C. Transverse section of frond
D. Longitudinal section of frond

Pterocladia J. Agardh

REFERENCES: Umamaheswara Rao & Kaliaperumal (1987: 193 – 198, fig. 1, II, X, XIV), Kaliaperumal (1990: 87 ff., figs. 1 – 4), Umamaheswara Rao (1990: 324, tables 2, 5, 8), Kalimuthu, Kaliaperumal & Ramalingam (1992: 123), Sindhu & Panikkar (1992: 125, figs. 21 – 29), Silva *et al*. (1996: 145).

Pterocladia
(Børgesen) U
Rao & Kalia
Description:
tall; forming
low mats on
attached by
arising from
shoots;
alternately
compressed,
flattened
branches alte
narrow, com
broadened ab
cortex of 2
pigmented ce
colourless
tetrasporangi
sterile marg
unilocular.
Uncommon:
rocks.
Distribution:
Pudumadam
island.
A. Habit of th
B. Transverse
frond.
C. Longitudir
frond.

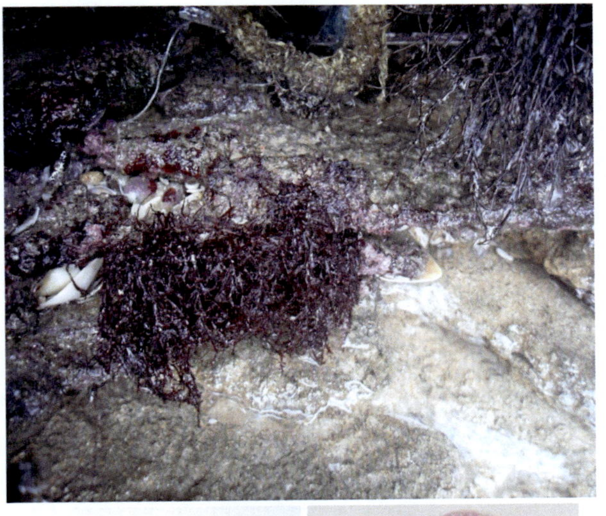

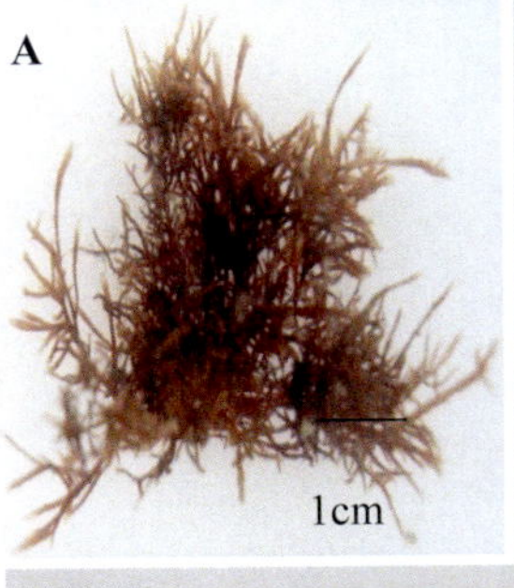

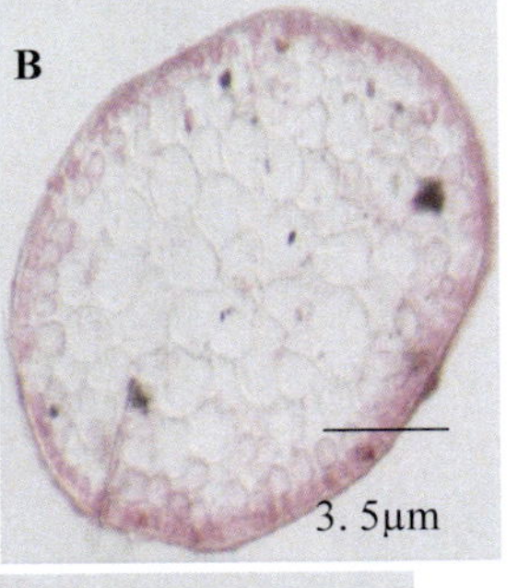

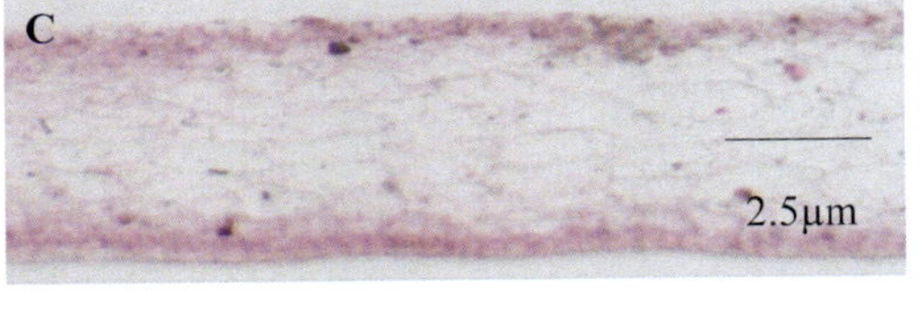

Pterocladia J. Agardh

Pterocladia heteroplatos (Børgesen) Umamaheswara Rao & Kaliaperumal

Description: Plants 3-6cm tall; forming dense tufts or low mats on rocky substrata; attached by discoid holdfast arising from the prostrate shoots; erect shoots alternately cylindrical and compressed, sometimes flattened throughout; branches alternate, opposite, narrow, compressed at base, broadened above; thallus with cortex of 2-3 layers of pigmented cells; medulla of colourless cells; tetrasporangial sori with sterile margins; cystocarps unilocular.

Uncommon: Attached on rocks.

Distribution: Nochiurani, Pudumadam and Mully island.

A. Habit of the plant

B. Transverse section of frond.

C. Longitudinal section of frond.

Pterocladia J. Agardh

REFERENCES: Sindhu & Panikkar (1995a: 23 –29).

***Pterocladia kanyakumarien* Panikkar**

Description: T very small, atta 1.5- 2.0 cm; cor to several ere irregular to pin The erect axes cylindrical cr attached by numerous hepte are oval in cr the base and fl The cortex of composed of pigmented cell cortical layer i Inner cortical than outer. M are comparati than inner c Tetrasporangial cysocarps are the pinnately fertile branches

Uncommon: rocks and hard the infra-littoral

Distribution: Kanyakumari, K Mully island.

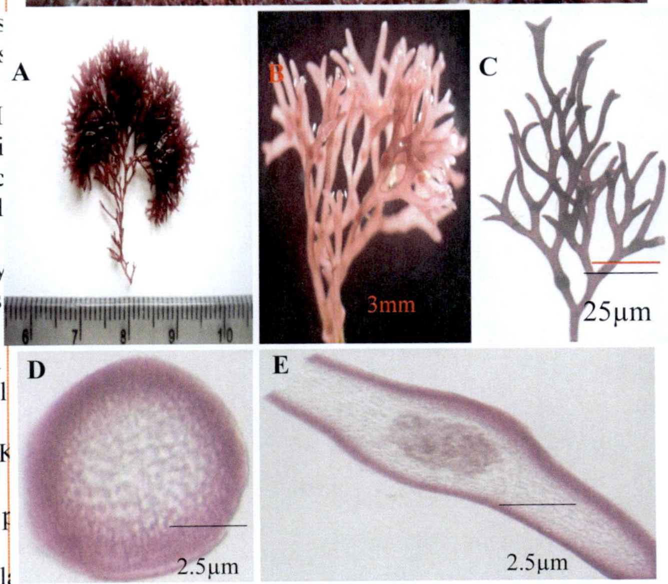

A. Habit of the p

B & C Pl irregular branch

D Transverse se

*E.*Longitudinal

Pterocladia J. Agardh

***Pterocladia kanyakumariensis* Sindhu & Panikkar**

***Description**:* The plants are very small, attain a height of 1.5- 2.0 cm; consisting of one to several erect axes with irregular to pinnate branches. The erect axes arising from cylindrical creeping axis attached by means of numerous heptera. Erect axes are oval in cross section at the base and flattened above. The cortex of erect axis is composed of 4-5 layers of pigmented cells. The external cortical layer is conspicuous. Inner cortical cells bigger than outer. Medullary cells are comparatively smaller than inner cortical cells. Tetrasporangial sori and cysocarps are developed on the pinnately compound fertile branches.

Uncommon: attached on rocks and hard substratum in the infra-littoral fringe

Distribution: Manapad, Kanyakumari, Kovalam and Mully island.

A. Habit of the plant

B. *&* *C*. Plant showing irregular branches

D. Transverse section

E. Longitudinal section

***Rhodymenia* Greville**

Plants erect, 2-40 cm tall, attached by discoid or rhizomatous holdfast, with flat dichotomously or irregularly branched fronds; margin sometime proliferous; thallus with medulla large cells and cortex of small cells; spermatangia in small sori; cystocarp prominent, scattered on the thallus or restricted to younger parts of the thallus. Tetrasporangia scattered in the cortex.

Key to the species:

Thallus dichotomously branched with entire margin … *R.sonderi*

Rhodymenia Greville

REFERENCES: Chennubhotla, Kaliaperumal & Kalimuthu (1987: 108 – 109), M. Dave *et al.* (1987: table 1), Untawale, Reddy & Deshmukhe (1989: table 1, 2), R. Parekh & Chauhan (1990: table 1), Silva *et al.* (1996: 370), Desikachary, V. Krishnamurthy & M. Balakrishnan (1998: 167).

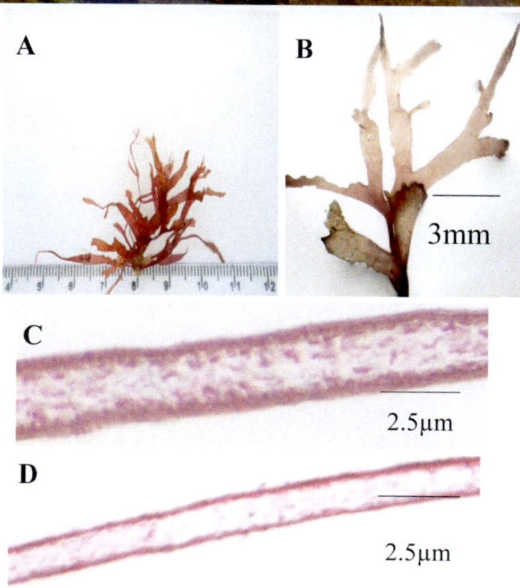

Rhodymenia Greville

Rhodymenia sonderi **P. Silva**
Description: Plants dark red to rose purple in colour; thallus erect, dichotomously divided, upto 15 cm tall; basal disc distinct; erect blades stipitate, lower potions cuneate, upper portion linear; margin of the blade entire, rarely with proliferations; fronds slightly thickened at the margins, more or less concave at one side. Cystocarps hemispherical, protruding and slightly constricted at base. Tetrasporangia shield shaped.
Uncommon: attached to rocks and small stones upto 10 m deep.
Distribution: Manapad, Kanyakumari, Kovalam and Mully island.
A. Habit of the plant
B. Plant showing dichotomous branches
C. Transverse section of the frond.
D. Longitudinal section of the frond.

Sarconema Zanardini

***Sarconema* Zanardini**

The macroscopic thallus is erect, upto 20 cm tall, cylindrical and well branched with tapering tips. Structurally muti-axial with a cortex of small cells, a sub-cortex of larger cells and a filamentous medulla. Tetrasporangia transversely divided. Cystocarps scattered.

Key to the species:

Plants filamentous, dichotomously branched with tapering tips ... *S.filiforme*

Sarconema Zanardini

REFERENCES: Panikkar, Ampili & Chauhan (1992: 61), Jagtap (1993: table 1), Nair *et al*. (1993: 186, 193), Kaliaperumal *et al.* (1995: 183 - 193), Silva *et al.* (1996: 336) Desikachary, V. Krishnamurthy & M. Balakrishnan (1998: 1351).

Sarconema
(Sonder) K
Description
red in colo
cm in heig
the substrat
holdfast. Th
1-3 mm in
and dichoto
narrowing
the tip; the
thallus for
tufts. In s
packed
rectangular
parenchyma
composed
polygonal c
composed c
thick wal
filaments.
Common:
and pebbles
Distribution
Arasaneripa
Ammapattin
Kuduthalai,
Kanyakuma
Anaipar is
island and V
A. Habit of t
B. Branch sh
C. Transvers
D. Longitud

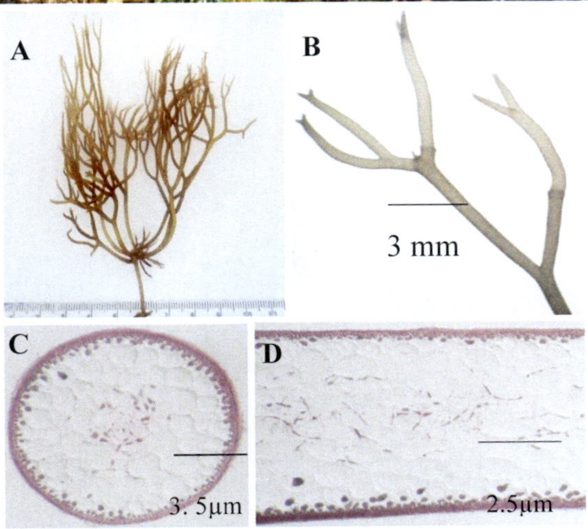

Sarconema Zanardini

***Sarconema filiforme* (Sonder) Kylin**

Description: Plant purplish red in colour; erect, upto 15 cm in height and attached to the substratum by a branched holdfast. Thallus solid, terete, 1-3 mm in diameter, regularly and dichotomously branched, narrowing gradually towards the tip; the repeatedly forked thallus forming dense, broad tufts. In section, a densely packed peripheral cells rectangular in shape; a parenchymatous tissue composed of round to polygonal cells; central cells composed of densely packed thick walled longitudinal filaments.

Common: attached to rocks and pebbles upto 5 meter deep

Distribution: Kottai pattinam, Arasaneripattinam, Ammapattinam, Nalupanai, Kuduthalai, Panjal, Leepuram Kanyakumari, Kovalam, Anaipar island, Upputhanni island and Van island.

A. Habit of the plant

B. Branch showing pointed tip

C. Transverse section

D. Longitudinal section

***Scinaia* Bivona-Bernardi**

Thallus soft, dichotomously branched, cylindrical and sometimes compressed; plant dark red to pink red in colour. Plants attached by a discoid hold fast. Cortex form from the lateral branches of medullary filaments. Epidermal cells formed by colourless utricles. Cystocarps sunk in the cortex.

Key to the species:

Palisade like utricle with flattened outer walls ……*S.bengalica*

Scinaia Bivona-Bernardi

REFERENCES: P. Subba Rao *et al*. (1985: 33), V. Krishnamurthy & Sundararajan (1987: 113, 115, figs. 14 – 25, 57), Desikachary, V. Krishnamurthy & M. Balakrishnan (1990: 162, figs. 34, D – F, pl. XIV), B. Kaliaperumal *et al*. (1992: 13), Subba Rao *et al*. (1992: 181), Silva *et al*. (1996: 113).

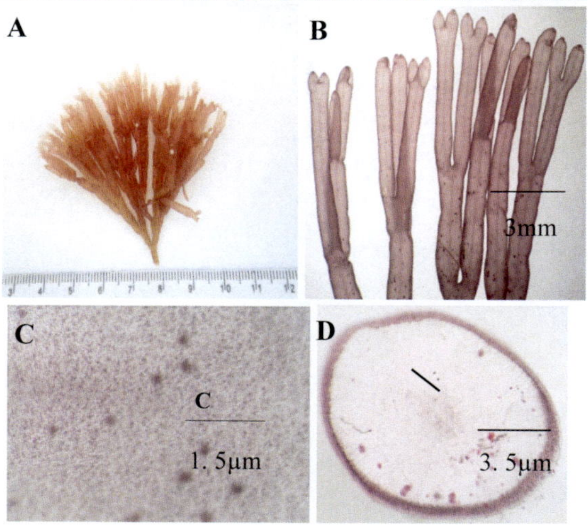

Scinaia Bivona-Bernardi

Scinaia bengalica **Børgesen**
Description: Plants dark red in colour; erect, upto 12 cm tall; attached by small discs; fronds flattened or complanate, dichotomously branched with round or acute apices; 3-5 mm broad, constrictions absent; thallus multiaxial; utricles interspersed with a few coloured cells; cystocarps scattered, immersed in the thallus, globular to pyriform in shape.
Uncommon: attached on rocks and pebbles upto 5 meter depth.
Distribution: Manapad, Van island and Koswari island.
A. Habit of the plant
B. Plant showing dichotomous branches
C. surface view of thallus with cystocarps (C)
D. Transverse section of frond.

Solieria J. Agardh

Solieria J. Agardh

The macroscopic thallus is erect, upto 30cm tall, cylindrical and irregularly branched, upto 5mm broad, fleshy, branches constricted below, subacute above. Structurally multi axial with a cortex of small cells, a sub cortex of larger, loose cells and a filamentous medulla. Cystocarps and tetrasporangia scattered all over the thallus.

Key to the species

1. Plants upto 50 cm tall, upto 5cm in width, branching irregular and infrequent ……*S. robusta*
1. Plants upto 15 cm tall, 2-4 cm width, branching profuse on all sides …….*S. filiformis*

Solieria J. Agardh

Solieria filij
P. W. Gabri
Description
slippery, fir
height; rose
colour;
dichotomous
above. Mair
0.5- 2.0 mm
base, taperin
apices sh
Medullary
arranged. C
thick; inner
outer corte
Surface cel
elongated.
spherical, er
branches
swelling.

Common: or
stones and peb
waters.

Distribution: I
pattinam, Aras
Ammapattinan
Ponnagaram,
Manapad.

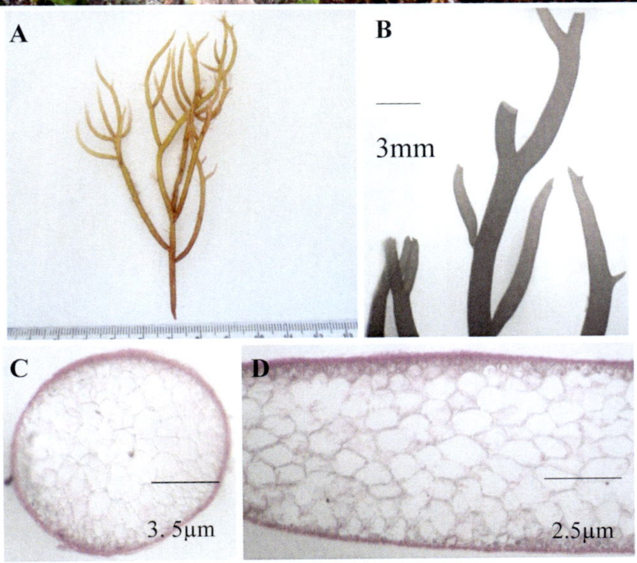

A. Habit of the
alternate branc
B. Branch witl
C. Transverse
frond
D. Longitudin
frond

Solieria J. Agardh

Solieria filiformis **(Kützing) P. W. Gabrielson**

Description: Thallus fleshy, slippery, firm, 5-24 cm in height; rose red to pink in colour; branching dichotomous below, alternate above. Main axis cylindrical, 0.5- 2.0 mm diam; pinched at base, tapering towards apices; apices sharply pointed. Medullary filaments loosely arranged. Cortex 4-6 cells thick; inner cortex 2-4 layers; outer cortex 1-2 layers. Surface cells spherical to elongated. Cystocarps spherical, embedded in larger branches causing surface swelling.

Common: on rocks, small stones and pebbles in subtidal waters.

Distribution: Kottai pattinam, Arasaneripattinam, Ammapattinam, Ponnagaram, Kodikkarai, Manapad.

A. Habit of the plant with alternate branches

B. Branch with pointed apices

C. Transverse section of frond

D. Longitudinal section of frond

Solieria J. Agardh

REFERENCES: Nair *et al.* (1993: 186, 193), Kaliaperumal *et al.* (1995: 183 – 193), Silva *et al.* (1996: 338), Desikachary, V. Krishnamurthy & M. Balakrishnan (1998: 142), Nasheer *et al.* (1998: 119 – 124), Kaliaperumal *et al.* (1998: 141 – 151).

Solieria _r_
Kylin
Descriptio
to brownis
attached to
fibrous l
branched
fleshy, te
compressec
mm in (
erect later
are basall
sub-acute
branching
side, somet
arising
damaged b
branch bo
cortical ce
zonate, sc
cortex.
Common:
and hard
intertidal re
depth.
Distributio
Panjal,
Kanyakum
Koswari isl
A. Habit of
B. Plant
branching
C. Branch
D. Trans
frond

Solieria J. Agardh

Solieria robusta **(Greville) Kylin**

Description: Plant grey-red to brownish red in colour; attached to the substratum by fibrous holdfast; thallus branched consisting of fleshy, terete or slightly compressed branches; 1-5 mm in diameter; bearing erect lateral branches which are basally constricted and sub-acute at their tips; branching irregular from all side, sometimes 4-5 branches arising profusely from damaged branch. Carpogoinal branch borne on an inner cortical cell. Tetrasporangia zonate, scattered in the outer cortex.

Common*:* attached on rocks and hard substratum in intertidal region up to 3 meter depth.

Distribution: Manapad, Panjal, Chinnamuttam, Kanyakumari, Kovalam and Koswari island.

A. Habit of the plant
B. Plant showing irregular branching
C. Branch with sub acute tips
D. Transverse section of frond

***Spermothamnion* Areschoug**

Thallus is filamentous having both prostrate and erect filaments. It attached to rhizoids with disc; erect filaments dichotomously branched. Cystocarps naked or sometimes covered by filaments. Tetrasporangia are single or in cluster.

Key to the species:

Cells of prostrate filaments 40 μm broad, 3-4 times longer than broad …*S. speluncarum*

Spermothamnion Areschoug

REFERENCES: Howe (1920: 578), Umamaheswara Rao & Sreeramulu (1970: 41, fig. 59), S. Dixit (1980: 75), Untawale, Dhargalkar & Agadi (1983: 35), Silva *et al.* (1996: 425), Desikachary, V. Krishnamurthy & M. Balakrishnan (1998: 239).

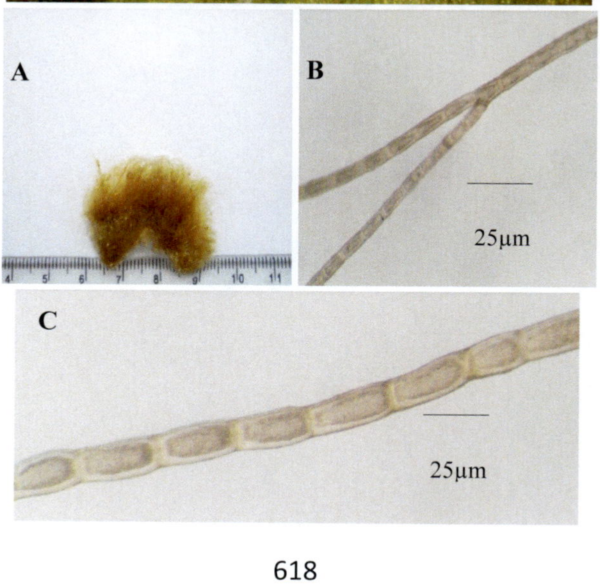

Spermothamnion Areschoug

Spermothamnion speluncarum (Collins & Harvey) Howe

Description: Thallus tufted, bushy, filamentous; purple- pink in colour. Prostrate filaments fixed to the substratum by longer or shorter rhizoids. Branching dichotomous below, soon becoming opposite, uncorticated throughout. Filaments 50-100µm diam; distally tapering; cells of the erect filaments cylindrical, 2- 2.5 times as long as they broad; rarely long, branching irregular, issuing all directions.

Rare: In the crevices of rocks in the infra littoral fringe.

Distribution: Manapad.

A. Habit of the plant

B. & *C.* Branches showing cylindrical filaments

Spyridia Harvey

Spyridia Harvey

Plants erect, bushy, profusely branched, branches alternate; several centimeters long, main axes and primary branches corticated throughout by means of vertically elongated cells with chloroplast; covered by uni-seriate branchlets with a narrow band of small cortical cells at the point of contact between the filaments resembling ceramium thallus. Short spines on the uni-seriate laterals is a characteristic of most species.

Key to the species:

1. Branching radial ……2
1. Branches distichous or pinnate …… 3
2. Branches sparse, fusiform ……. *S. fusiformis*
3. Branchling ending in mucronate tip; occasional single hooked spine on uppermost node ………*S. hypnoides*

Spyridia Harvey

REFERENCES: Umamaheswara Rao (1987: 25), Kamat *et al.* (1991: 56), Kamat *et al.* (1992: table 1), Anilkumar & Panikkar (1993: 21, figs. 7 – 9), Silva *et al.* (1996: 428), Desikachary, V. Krishnamurthy & M. Balakrishnan (1998: 218).

Spyridia
Børgesen
Descripti
pink in
height; co
monolifor
alternately
about
gradually
the a
branched,
short, fu
sparse, fi
about 1m
broad, cor
ring of cel
Rare: atta
coral sto
depth.
Distributi
Leepuram
A. Habit o
irregular b
B&C
attenuated
D. Tra
showing ri
E. Long
frond

Spyridia Harvey

***Spyridia fusiformis* Børgesen**

Description: Thallus whitish pink in colour; 20 cm in height; corticated, terete, sub monoliform, irregularly and alternately branched; axes about 1 mm in width, gradually attenuated towards the apex, irregularly branched, branches sparse, short, fusiform. Branchlets sparse, filiform, deciduous, about 1mm long and 25 µm broad, corticated at base by a ring of cells.

Rare: attached on rocks and coral stones upto 2 meters depth.

Distribution: Nalupanai and Leepuram.

A. Habit of the plant showing irregular branches

B. & *C.* Branches with attenuated tip

D. Transverse section showing ring of cells

E. Longitudinal section of frond

Spyridia Harvey

REFERENCES: Bory de Saint – Vincent (1968: 281 - 282), Silva *et al.* (1996: 428), Desikachary, V. Krishnamurthy & M. Balakrishnan (1998: 219).

Spyridia hy
Saint-Vinc
Descriptio
in colour;
cm in
ramified,
branchlets
distichously
feather like
frond; the
branches
corticated. I
mm long br
and tapering
ending with
with 2 or 3
Uncommon
coral stone
depth.
Distributio
Kanyakuma
Shingle isla
A. Habit o
dense branc
B. Brach
branchlets.
C. Feather l
D. Transver
E. Longitud

Spyridia Harvey

Spyridia hypnoides **(Bory de Saint-Vincent) Papenfuss**
Description: Plants dark red in colour; Thallus terete, 20 cm in height; densely ramified, branches and branchlets arranged distichously and giving a feather like appearance to the frond; the main axis and branches completely corticated. Branchlets about 1 mm long broadest at the base and tapering gradually above ending with mucronate tip or with 2 or 3 curved spine cells.
Uncommon: on rocks and coral stones upto 3 meters depth.
Distribution: Kuduthalai, Kanyakumari, Kovalam and Shingle island.
A. Habit of the plant with dense branchlets
B. Brach covered with branchlets.
C. Feather like branch.
D. Transverse section
E. Longitudinal section

***Tenaciphyllum* Børgesen**

Plant dorsi-ventral, fleshy, 0.5 to 1.5 cm broad, 350-850 μm thick. Thallus irregularly and dichotomously branched showing mat like appearance. Medulla are large cells, cortical cells are pigmented.

Key to the species:

Plants dorsi-ventral, irregularly and dichotomously branched ….*T.lobatum*

Tenaciphyllum Børgesen

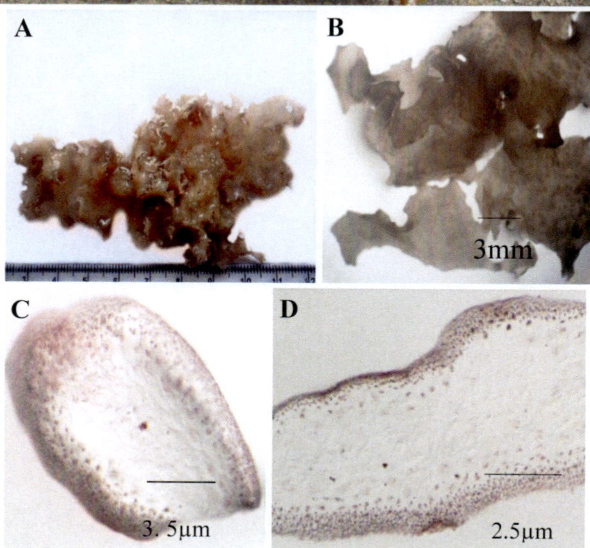

Tenaciphyllum Børgesen

Tenaciphyllum lobatum **Børgesen**

Description: A robust species; more than 10 cm; flattened, lobed thalli, upto 3 mm thick, more or less prostrated and firmly attached to the substratum by several clumps of rhizoides. The cortical region is formed by many rows of densely packed small cells.

Rare: on coral stones and big stones upto 10 meter depth.

Distribution: Sullimunthal, Kanchirankudi and Keelakarai.

A. Habit of the plant

B. Plant showing prostrated branches

C. Transverse section of thallus

D. Longitudinal section of thallus

***Tolypiocladia* Schmitz**

Plants 10-15cm tall, somewhat spongy, irregularly branched with determinate and indeterminate branches, determinate branches sparsely divided with spinous apical cells and trichoblast at the tips of the forks, spirally arranged, filaments with a central axial cell and 4 pericentral cells; spermatangia ovate or cylindrical clusters on long stalks, cystocarps scattered, carposporangia terminal on gonimoblast filaments, tetrasporangia on branches of limited growth.

Key to the species

1. Plants upto 15 cm tall, lax with free branches, apical cell elongated with blunt tip. …………*T.glomerulata*
1. Plants upto 20 cm tall, side branches are partially fuse ……….. *T.calodictyon*

Tolpiocladia Schmitz

Tolypiocla
(Harvey
Silva

Descriptio
reddish b
forming a
2-3 cm
bifurcately
completely
short bran
wide angle
pointed ap
contact
branchlets
rhizoids p
tips.

Rare: on
pebbles.

Distributio
Valimunai
Upputhann

A. Habit of
B. Main
bifurcate bı
C. Branch
branchlets

***Tolpiocladia* Schmitz**

***Tolypiocladia calodictyon* (Harvey ex Kützing) P. Silva**
Description: Thallus light reddish brown in colour; forming a small spongy tufts; 2-3 cm height; main axis bifurcately branched and completely covered with short branchlets; branches at wide angles and ending in a pointed apical cell; when in contact with another branchlets tend to fuse by rhizoids produced at their tips.
Rare: on coral stones and pebbles.
Distribution: Thonithurai, Valimunai island and Upputhanni island.
A. Habit of the plant
B. Main axis showing bifurcate branching
C. Branch covered by short branchlets

Tolpiocladia Schmitz

REFERENCES: Chauhan & Mairh (1978: 35), Untawale, Dhargalkar & Agadi (1983: 38), Chennubhotla, Kaliaperumal & Kalimuthu (1987: 110 – 111), Kalimuthu, Kaliaperumal & Ramalingam (1992: 124), Silva et al. (1996: 555), Desikachary, V. Krishnamurthy & M. Balakrishnan (1998: 305).

Tolypiocladi
(C. Agardh)
Description
cm in height
thread like
Indeterminat
intervals of
more segme
branches al
segment;
elongated
Cystocarp su
Rare: on ex
coral stones.
Distribution
A. Habit of t
B, C & D l
with small b

Tolpiocladia Schmitz

***Tolypiocladia glomerulata* (C. Agardh) Schmitz**

Description: Plants upto 15 cm in height; freely branched, thread like and spongy. Indeterminate branches at intervals of two to seven and more segments; determinate branches almost from every segment; apical cell elongated with blunt tip. Cystocarp subglobose.

Rare: on exposed rocks and coral stones.

Distribution: Koswari island.

A. Habit of the plant

B., C. & D. Branches covered with small branchlets

***Tricleocarpa* Huisman & Borowitzka**

Plants cylindrical and dichotomously branched. Thallus calcified. Assimilatory filaments rarely developed. Cortex 4 layered; cystocarps globular or oval with pericarp.

Key to the species:

Plants dichotomously branched, cortex four layered …..*T.fragilis*

Tricleocarpa Huisman & Borowitzka

REFERENCES: Bhakuni *et al*. (1992: 514), Jagtap (1992: 57), Kalimuthu, Kaliaperumal & Ramalingam (1992: 123), Kamat *et al*. (1992: table 1), Jagtap (1993: table I), Kerkar (1994: table 1).

Tricleocarpa Huisman & Borowitzka

Tricleocarpa fragilis (Linnaeus) Huisman & Towsend

Description: Plants Olive red in colour; 5-6 cm tall; plants calcified, cylindrical, dichotomously branched with articulation in basal portion; branches 1-2 mm in diameter, growing points situated in a deep narrow depression in apices. Cortex 4 layered; medullary filaments loosely interwoven. Calcification strong in younger plants confined to cortex in younger plants and extending to the peripheral portion of the medulla in the older plants.

Rare: attached on small dead coral piece and rocks

Distribution: Nalupanai Kanyakumari.Kovalam.

A. Habit of the plant

B. Plant showing dichotomous branch.

C. Transverse section showing loosely arranged medulla (M)

***Vanvoorstia* Harvey**

The thallus is delicate with overlapping net like lobes, attached by a disc like holdfast. New blades produced from the mid rib portion of older blades fuse with each other and form a net of flattened segments. Cystocarps developing from the midrib region. Tetrasporangia on short blades, four in each segment.

Key to the species:

Fronds net like, blades corticated ....*V.coccinea*

Vanvoorstia Harvey

Vanvoorsti
Harvey ex
Descriptio
in colour;
lobed fron
hard substı
hold fast; 1
long blad(
surface on
are produ
with the ad
in the ventı
corticated (
cylindrical
Tetrasporaı
short blade:
Rare: atta(
coral stone:
Distributio.
Mully islan
A. Habit of
B. Net like
C. Surface
with tetrasp

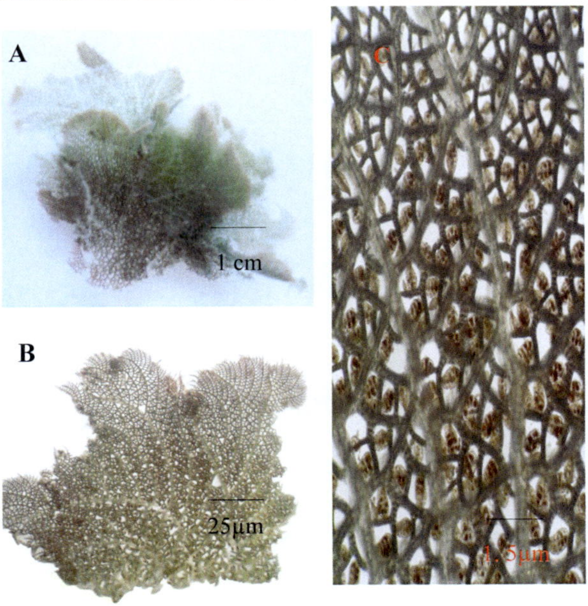

Vanvoorstia Harvey

Vanvoorstia coccinea Harvey ex J. Agardh

Description: Plants light red in colour; having net like lobed frond attached to the hard substratum by rhizoidal hold fast; frond consisting of long blades in the dorsal surface on which short blades are produced, which fuse with the adjoining long blade in the ventral surface. Blades corticated early and assuming cylindrical shape. Tetrasporangia formed in short blades.
Rare: attached on rocks and coral stones to 10 meter deep.
Distribution: Valai island, Mully island, Koswari island.
A. Habit of the plant
B. Net like lobe
C. Surface view of thallus with tetrasporangia

Wrangelia C. Agardh

Wrangelia C. Agardh

Thallus erect, filamentous, few centimeters long, with a central axis of larger cells; with or without rhizoidal cortication; the cells of the main branches are densely covered by short and well branched whorled branchlets. Involucre filaments surround the reproductive structures.

Key to the species:

1. Plant upto 20 cm tall, rhizoidal filaments compact …….*W.penicillata*
1. Plants upto 7 cm tall, determinate branch obtuse …….*W.balakrishnanii*

Wrangelia C. Agardh

REFERENCES: B. Subramanian (1985b: 192 – 193, figs. 8 – 15, 36 - 39), Silva *et al*. (1996: 345), Desikachary, V. Krishnamurthy & M. Balakrishnan (1998: 187).

Wrangelia C. Agardh

Wrangelia balakrishnanii **B. Subramanian**

Description: Plant small, epiphytic on other algae. Branches alternate, distichous; determinate branches ending in cells with rounded apex. Axial cells with four determinate branches near the top; Spermatangial head spherical, stalked.

Rare: attached on rocks and coral stones.

Distribution: Anaipar island.

A. Habit of the branch

B. & C. Branches with crowded branchlets

Wrangelia C. Agardh

REFERENCES: C. Agardh (1828: 138), Borgesen (1945: 18), Agadi & Untawale (1978: 67, table 2), Untawale, Dhargalkar & Agadi (1983: 35), Silva *et al.* (1996: 432), Desikachary, V. Krishnamurthy & M. Balakrishnan (1998: 223).

Wrangelia C. Agardh

Wrangelia penicillata **(C. Agardh) C. Agardh**
Description: Plants upto 20 cm in height. Main branches elongate; determinate branches start from upper cells. Indeterminate branches are regularly alternate: determinate branches sub dichotomous, thin, soft. Basal cell of determinate branches giving rise to rhizoidal filaments. Spermatangial cells spherical.
Rare: attached on rocks in subtidal region
Distribution: Manapad.
A. Habit of the plant
B. & C. Plant showing alternate branches
D. Transverse section of frond

GLOSSARY

abaxial: away from the axis

abruptly- pinnate: Where a pinnate leaf is without an odd leaflet at its extremity

acropetal: in the direction from the base toward the apex

acute: with a sharp angle; ending in a point

adjoining: filaments joining together, weakly or tightly

adventitious branching: supplementary to the normal branching pattern

anastomosing: intermittently joining and separating to form the network

annular: ring-like

antennae: one pair of mobile appendages

antheridium (pl. antheridia): a structure that produces motile male gametes (sperm)

anticlinal (ly): perpendicular to the surface or periphery of a structure

apex (apices): tip, summit

apical: at or near the summit

appendages: a part superadded to another; thus the leaves are appendages to the stem

appendices: subsidiary part or projecting part superadded to another part with distinct function or appearance

arched: roots produced on plant stems in a position above the normal position of roots that serve to brace the plant during and following periods of prolonged inundation in water

arcuate: bent or curved like a bow

articulated: jointed; composed of stiff parts attached to each other by flexible parts

articulation: a constriction between two successive segments; also called geniculum (pl. genicula) and found in a few groups of calcified red and green alga; the articulations are never calcified

asexual: reproduction without the fusion of gametes; involves no transfer of genetic material between individuals

assimilator: in *Caulerpa* used as the upright frond

attenuated: where the breadth is gradually diminished towards either extremity; as the bases of some leaves

axis (axes): main stem or major branch

axial: central filament or core of filaments

base: attached to the base of any organ or part

basal: that extremity at which an organ is attached to its support, and by which the nourishing vessels enter it

bifid: forked; cut in two for about half its length

bifur: dividing into two branches of about equal length.

biflagellate: having two flagella

bipinnate the leaflets on the secondary petioles of a doubly compound leaf, are arranged in a pinnate manner

bisexual: having both sexes

bizonal: relating to two administrative zones

blade: a relatively broad, thin part of the thallus; leaf-like portion of an alga (sometimes also called the lamina)

branch: main side structure on the axes (with limited growth)

branchlet: smaller side structure on the branches (with limited growth)

brittle: hard but liable to break easily

bulbous holdfast: rhizoids getting densely intricate and holding large amounts of sand, resulting in (sub-) cylindrical structures submerged in the sand

bushy: densely branched, forming small bushes

caducous: falling off easily

caespitose: forming a dense, short turf; matted

calcareous: with obvious calcification

carpogonial branch: the short, specialized branch bearing the female reproductive cell (carpogonium) in red algae

carpogonium (pl. carpogonia): an oogonium, or female reproductive cell, in red algae; consists of a swollen basal portion and an elongated terminal projection (the trichogyne) receiving the male gamete

carposporangium (pl. carposporangia): in the red algae, the sporangium which produces diploid carpospores

carpospore: diploid spore formed in carposporangia, by the carposporophyte

carposporophyte: the diploid generation developing on the female gametophyte after fertilization of a carpogonium and producing diploid carpospores

carpets: bed like the appearance of algae, occur in the rocks of the coastal area

carpo-tetraspores: Carposporangia after meiotic division gives rise to carpotetraspores

cartilaginous: firm, tough but flexible

cavity: a hollow space present in the internal structure of algae

cell: basic (microscopic) unit of plant structure, generally consisting of compartments in a viscous fluid surrounded by a wall

cellulose: the common, structural carbohydrate of plant cell walls, composed of beta 1,4 glucosides commonly in the form of microfibrils

central cavity: the main hollow center of the stem

chloroplast: a cell organelle containing chlorophyll

chromatophore: a small, discoid, colorless substance which is devoid of pyrenoids located around the nucleus

claviform: club-shaped appearance

clavate: cylindrical structure or thallus broader at the free extremity; club-shaped

clinging: to adhere closely

clumps: the formation of a dense cluster of flagellated male gametes attaches itself by the longer flagellum to the body of female gametes, occurring immediately before fusion of the gametes

clusters: synonym for raceme

coenocytic: possessing a cell or a filament without septa between the numerous nuclei

compressed: slightly flattened (oval in transverse section)

concave: applied to any surface with a curvilinear depression, or hollow formed without angles

conceptacle: a hollow structure or a cavity enclosing reproductive structures

concentric: thin structure arranged at one circle inside another

contorted: entangled, wreathed, sinuous, with numerous bends

convolute: twisted and rolled up longitudinally

coralline (alga): calcified red alga

cortex: part of the parenchyma cells of the leaf, which composes the outermost portion of the mesophyllum.

cortical: adhering or belonging to the bark

cortication: small cells or rhizoids covering a branch, most often on the basal portion of a thallus; found primarily in the order Ceremiales of the red algae

creeping (repens): lying flat upon the ground, and emitting roots under the surface

cruciate (division): with transverse walls perpendicular to each other; in the same plane, resulting in a cross-like appearance

crustose: forming a crust

cryptostomata: a flask-shaped cavity opening to the surface and containing hairs. (eg. in Fucales, Phaeophyceae)

cuneate: wedge-shaped; broad at one end, tapering by nearly straight lines to the tip

cup-shaped: concave, hemispherical, and tapering below like a drinking cup

curled: when a foliaceous organ is irregularly folded and crimped

cushion: a swollen part of the stem or branches immediately below the leaves

cylindrical: circular in transverse section

cymose: the type of inflorescence that ends in a flower

cystocarp: the complex structure resulting from fertilization in red algae, composed of the internal gonimoblast and the enveloping sterile involucral branches or the pericarp

cytoplasm: the contents of a cell within the plasma membrane

deciduous: falling off easily; not permanent

decumbent: lying flat and loose on the substratum, with tip curving upward

dentate: toothed, with rather large, sharp teeth directed outwards

denticulatus: furnished with small teeth

denudate: stripped

determinate: lateral or branch with a limited growth, fixed in length

detrimental: an environmental or physiological condition that causes mortality

devoid: entirely lacking

diagnosis: description of a new species in Latin; the characters of one species, distinguishing it from other species of the same genus

diaphragm: A thin partition of the cell across a hollow space

dichotomous: forked into two similar parts as a result of the equal division of the apex

dichotomous branching: branching into two

digitate: branching like the fingers on a hand, with numerous branches radiating from the same point

dimorphism: thallus or structures existing in two distinct forms

dioecious: unisexual; male and female reproductive structures produced on separate individuals

diploid: with two homologous sets of chromosomes in each nucleus (2n)

discoid: having the form of a disc, being flat and circular

distal: away from the place of attachment; towards the apex

distichous: branches or branchlets arranged alternately in two opposite vertical rows

distromatic: having two cell layers

disc (disk): a plate or ring of structures derived from the receptacle, and occurring between whorls of floral parts

divaricate: branching at wide angles, widely divergent

dorsal: facing away from the axis or away from the substrate.

dorsiventral: thallus with a distinct morphology on the upper and lower side

ellipsoidal: a 3-dimensional shape, elliptical in all sections through the long-axis

encrusting: forming a crust

endemic: a species only occurring in a single region

entangled: irregularly or confusedly interlaced; as in the case of some branches; or in some fibres of the root; or of the hairy pubescence of some plants

entire: with a smooth margin

epidermal cells: the outermost layer of cells, forming a protective layer on the surface of thallus or plant

epidemic cells: The cells which are affected in large numbers of individuals at the same time

epiphyte: attached to the surface of another algae or plant; not parasitic

epizoic: growing on animals (barnacles, shells of gastropods etc.)

erect: upright

extremities: terminal point

falcate: branch system curved like a sickle

fascicle: a bundle

fasciculate: arranged in small clusters or bundles

fastigiate: numerous branches spreading from a compactly clustered point of origin

felt: a densely intricated mass of thin filaments

fertile: being or containing a reproductive structure

fibrous: consisting of structures resembling fibers (holdfast)

filament: an elongated thread- or hair-like structure; in the cyanoprokaryota, it is used to describe a trichome with a sheath

filamentous: thin, elegant, supple threadlike structure composed of a single or a few rows of cells

filiform: thread-like

fimbria: slender hair-like process (plural: fimbriae)

fimbriate: fringed

flabellate: fan-shaped

flabellum: fan-shaped part of the thallus

flabelliform: having the form of fan shaped or flabellate

flaccid: lacking vigor

flagellum: a whip

flagelliform: flexible, narrow, and tapering, like the thong of a whip

foliaceous: resembling the more usual character of leaves in being- thin and membranous

foliose: leaf-like, flattened like a leaf

forcipate: markedly incurved, like a forceps or pincers

forked: separating into two distinct branches, more or less apart

fragmentation: some branches break off from the mother plant, stay alive, attach to the substratum and go on growing to new plants

frond: erect (upright), mostly compressed part of an alga

fan-shaped: funnel-shaped, synonyms for Flabelliform

fused: joined together

fusiform: spindle-shaped, thicker centrally and tapering to both ends

furcated: forked, branched; when divided into two, not strictly dichotomous

gametangium (pl. gametangia): structure from or in which the gametes (sexual reproductive cells) are formed

gamete: a sexual reproductive cell having the haploid number of chromosomes, e.g. a sperm or an egg cell

gametophyte (gametophytic): the gamete-producing phase of a plant

gamma diversity: or regional diversity, is the diversity of a landscape, or all sites combined

ganglionic cell: a darkly staining cell in certain red algae (e.g. *Halymenia*, *Cryptonemia*) characterized by a central swelling and long and slender arms

gelatinous: with large amounts of jelly, gluey

geniculum (-ula): the uncalcified joint(s) between segments (intergenicula) of coralline algae

germination: a phenomena by which seeds begin to grow when they are placed under the conditions requisite to excite the vital energies of the dormant embryo

geographical: derived from the physical features of an area.

glandular cells: a colorless, usually refractive cell, that specialized to secret a specific substance

gland cell: a small cell with highly refractive content in red algae

globose: spherical, rounded

globular: nearly spherical

globule: the spherical body filled with elastic filaments

gonimoblast: diploid structure, developed after fertilization of the carpogonium, composed of branched filaments producing carposporangia

granules: a small grain, many of which are contained in each grain of pollen

gregarious: growing in groups; clustered

hair: an elongated cell or narrow, branched or unbranched, multicellular filament

haploid: with only one set of chromosomes in each nucleus (n)

hapteron (pl. haptera): a structure for attachment of an erect branch on another branch or any other substrate

hapters: branched, multicellular attachment structures

hemispherical: one half of a sphere

heteromorphic: with a different morphology, often used with regard to gametophytic and sporophytic phases in a life cycle

hirsute: covered with stiff, long, straight hairs

holdfast: basal attachment structure

holocarpy: when the entire cytoplasmic content of a thallus is being transformed in gametes

holotype: the single specimen on which an author based the description of a new taxon

hook: a stiff hair, bristle, or prickle, curved back at the point

hue: tinge, color

hyaline: colorless, transparent

hydrophyte: plant growing (partly) submerged in water

hyphal cells: an elongated cell or filament, distinct from other filaments in the thallus. Some authors use the term rhizine

hypothallium: small sheath-like petiole seated below a peculiar leaf-like form of branch

imbricated: closely overlapping, like root tiles

incrustated: where an outer envelope is firmly attached to the part it covers; as when a pericarp invests the seed so closely that it seems to form a portion of it

incurved: bend towards the main axis

incurred: sustained

indeterminate: lateral branch with a (theoretically) unlimited growth

inflated: swollen

intercalary: between the basis and the apex

intercellular: lying between cells

interdichotomy: part of the thallus between two dichotomies

intergeniculum (-ula): the calcified segments of articulated coralline algae; parts between the uncalcified joints

intercalated: the cells or structures positioned in between pre-existing cells or structures

intermediary: situated or coming between

intermingled: merged or combined together

internodes: The part of a plant stem between the joints or nodes

internodium: part of a stem or rhizome between 2 nodes

intertwined: twist or twine together

intertidal zone: the coastal zone between mean high and mean low tide levels

interwoven: to mix or blend together

intricately: having a lot of small parts that are arranged in a complicated way

inrolled: rolled inwards

involucre (involucrum): radially arranged and generally incurved filaments surrounding reproductive structures in red algae

Involucral: surrounding a structure or organ

inwardly: towards the center/to curve internally

iridescence: glowing or shining; reflecting an interplay or rainbow-like colors as when seen from different angles

isodiametric: with (approximately) equal dimensions in two or more directions

isomorphic: with the same (or similar) morphology; often used with regard to gametophytic and sporophytic phases in a life cycle

isotype: duplicate specimen of the holotype

juvenile: young specimen

lacerate: irregularly divided by deep incisions

lacinae /lacunae: Cavities within a structure

lagoons: a shallow lake of salt or brackish water separated from the sea

lamina: The flat blade or flattened, the upper part of stalked algae

lamellated/ lamellae: composed of thin layers or plates of tissues

lanceolate: Lance – shaped, long and narrow, tapering to both ends, especially to the apex

lateral (adj.): on the side of

lateral (subst.): a side axis or side branch

leaf: in algae sometimes used to describe an expanded portion of the thallus that is analogous, but not homologous, with the leaf of a vascular plant

leathery: tough, but still flexible

lectotype: a specimen designated from the original material as the nomenclatural type if no holotype was indicated at the time of publication, or if it is missing, or if it is found to belong to more than one taxon

lenticular: looking like a lens, double convex

ligulate: strap-shaped, relatively broad when compared to its length

linear: narrow, with parallel sides and several times longer than broad, like a grass leaf

littoral: intertidal region with rocks

lobed: with rounded (fan-shaped) parts or margins

loculi: a chamber of a reproductive organ

lubricous: smooth and slippery

lumen: a central cavity in a cell or a thallus

macroalgae: algae visible with the naked eye, as opposed to microalgae for which a microscope is needed for their observation

macroscopic: large enough to be seen with the unaided eye

margins/ marginal: pertaining at the edge of the structure

mat: a densely interwoven, horizontal growth form

medulla: inner tissue, the central region of the thallus, internal to the cortex; the pith

median: In the middle

meiosis: nuclear division by which the number of sets of chromosomes is reduced from two (2n) to one (n)

membranous: (membranaceous), forming a thin blade or membrane

meristematic/ meristem: A tissue that initiates growth

mesh: the opening or space enclosed by the threads of a net between knot

midrib: a central, longitudinal thicker part of a (mostly strap-like) blade

mid littoral: to the middle portion of the intertidal zone

moniliform: resembling a string of beads

monoecious: bisexual; male and female reproductive structures produced on a single individual

monosiphonous: composed of a single tubular structure

monospecific: composed of a single species

monochromatic: composed of a single layer of cells

morphology: form

mucilaginous gel: having a slimy or slippery texture

mucronate: abruptly ending in a short, stiff point

multinucleate: with several nuclei per cell or thallus compartment

multiaxial: having an axis with several apical cells that give rise to nearly parallel filaments; found in many red algae

multilayered: consisting of several layers of cells

multiseriate: thallus formed by more than two rows of cells

multizonal: with more than one zones

muricate: rough, provided with short and hard tubercles

naked flower: flower without a perianth

nemathecium (pl. nemathecia): an elevation on the thallus surface containing reproductive structures

node: the place of the axes where laterals or branches arise

nomenclature: giving a name to an organism; this happens according to rules which have been internationally recognized. For Algae the International Code of Botanical Nomenclature applies

notched: with marginal indentations (angular cuts)

nuclei: plural form of the nucleus, directing growth, metabolism, and reproduction

oblique: neither perpendicular nor parallel; slanting

oblong: rectangular, but in leaves usually rounded at the corners

obtuse: blunt or slightly rounded at the end

obtusely angled: being greater than a right angle

oogonium (pl. oogonia): the female gametangium, containing one or several immobile gametes (oospheres)

opposite: a type of branching in which there are two structures (branchlets) at the same level

ovate/oval: egg-shaped in two-dimensional outline, broadest near the base or point of attachment

overlapping: extend over to cover partly

ovoid: egg-shaped in outline

paddle-like: a short pole with a wide flat part at one end or both ends

palisade: a row of elongate structures at right angles to the surface

palmate: divided in the manner of an outspread hand

paraphyses: sterile filament between or around reproductive cells

parenchymatous: In red algae, boxy, plant-like cells that have complex, multicellular structure

pectinate: with closely packed side branchlets on one side, like the teeth of a comb

peltate: umbrella-shaped, horizontal disc with a perpendicular stalk attached at the center

percurrent: extending through the entire length

perforated: presenting holes

periaxial cell: a cell cut off from an axial cell but shorter and orientated obliquely or at right angles to it (e.g. *Ceramium*)

pericarp: a sterile envelope surrounding the gonimoblast in a cystocarp

pericentral cell: a cell cut off from an axial cell and remaining similar in size and orientation to it (e.g. *Polysiphonia, Dasya*)

phorophyte: a plant that carries epiphytes

phycobilins: water-soluble, accessory pigments in the red algae, mainly phycoerythrin (red) and phycocyanin (blue)

pillar: a slender, free-standing, vertical support

pigmented: colored

piliferous zone: bearing hairs or setae

pinna(e): a compressed to a complanate side branch

pinnate: feather-like, with a main axis and branchlets on two opposite rows in one plane

pinnule: a lateral (branchlet) of a pinnate branch

pitted: provided with small depressions

plastid: cell organelle in which photosynthesis takes place

plumose: like a plume, feather-like

pluricellular: composed of several cells

plurilocular: many-celled, each cell containing a single spore

pneumatophore: a (mangrove) root growing vertically and upwardly

polychotomous: dividing in several (sub)equal parts from a single point

polygonal: a plane geometric figure with numerous sides

polyhedral: An irregular shaped stage in the life cycle of *Hydrodictylon*

polyphyletic: descended from several, unrelated ancestors

polysiphonous: axes composed of a central axis surrounded by a series (at least 3) pericentral cells (eventually covered by a cortex), visible on a transverse section. In surface view, those algae without a cortex, seem to be composed of several, closely packed filaments ('siphons')

polystichous: In certain Phaeophyceae, have the thallus truly parenchymatous

pores: In Corallinales (Rhodophyta), the opening above a reproductive structure.

primitive plant body: with a relatively simple structure

prismatic: like a prism

procumbent: horizontally spread over the substratum but only attached at the basis

prokaryotic: organism lacking a nuclear membrane around the chromosomes

proliferation: a branchlet (or bladelet) formed as irregularly offshoots, smaller than side branches; frequently formed after grazing or erosion of the original branch (or blade)

profuse/ly: produced in large amounts

projections: a part that juts out

prostrated: parallel to the ground or lying along the substratum

protuberances: Bulging outward or an outward bulge

protruding: something that sticks out from a surface

propagule: branchlet with a special morphology, detaching from the mother plant and each of them producing a new juvenile

prostrate: lying down on and tightly adhering to the substratum

prothallus: the upper layer of tissue of a crustose alga

protrusion: protuberance; bulging part

proximal: nearest to the point of attachment

pseudolateral: a lateral branch derived from a displaced apex

pseudoparenchymatous: Tissue with the appearance of parenchyma but delivered from the close aggregation of uniseriate filaments, usually into the medulla of inflated or subspherical cells

pyramidal: shaped like a pyramid

pyrenoid: an intracellular refractive cytoplasmic body associated with plants in many algae (e.g. *Ulva*), associated in starch formation in green algae

pyriform: pear-shaped, with the broadest part toward the top end

quadrangular: with four corners

quadriflagellate: with four flagellates

quadrate: Rectangular or four-sided

racemose: the receptacle is borne on short stalks lying along a common axis

rachis: the main axis of the erect frond (assimilator); mainly used in *Caulerpa*

radial: radiating or developing uniformly from a central point or an axis

ramuli: last order branchlets

ramification: the pattern of branching of an organism

ramose: richly branched

receptacle: swollen part of a branch bearing reproductive structures

recurved: bent away from the main axis frequently bent downwards to the substratum

reniform: kidney-shaped

repent: creeping along the substrate

reproduction: the formation of new individuals by sexual or non-sexual means

reproductive: concerned in reproduction/ in sexual increase

resupinate: thallus horizontally spreading from a vertical wall

reticulate: in the form of a two- or three dimensional net-like structure

reticulum: a two- or three dimensional network

rhizine: an unbranched filament with small diameter and very thick cell wall, resulting in a very small central opening (lumen); strongly refractive on transverse section

rhizoid: a single- or few-celled attachment structure of an alga

rhizoidal: The cell present in the basal region of filament helps in anchoring

rhizomatous: Having rhizomes as its form of food storage or mode of spread

rhizome: a creeping (mostly cylindrical) structure from which erect branches arise

rhizophore: an aerial root of a mangrove tree, starting from the basal part of the stem or a branch, bending downwardly and reaching the bottom

robust: strongly or stoutly built

rosette: (bladelets) arranged around a central point

rufous: brownish-red

rugose: wrinkled or ridged

sac: shaped like a pouch or bag

sand-dwelling: attached in the sandy substratum

saxicolous: Growing on rocks

segment: in calcified algae: the large calcified part between the non-calcified joints; in (some) red algae: one central cell and surrounding pericentral cells (if present) and cortex (if present)

semi-rosette: the partially circular arrangement of leaves

septate: with transverse walls

serrate: saw-edged, with sharp teeth pointing forwards

sessile: not stalked; attached directly to the axis or a substratum

sexual: a process involving the sex in the multiplication of the plants

sinuous: irregularly curving in and out in one plane (a filament or a blade margin)

sinuate: wavy, with deep wave-like indentations in a horizontal plane along the margin

siphon: aseptate, tubular structure

siphonous: with a thallus formed of multinucleate, tubular unit cells (Siphons)

slippery: surface which is difficult to move on or to hold.

sorus (sori): a well-demarcated surface patch containing a group or cluster of reproductive structures

solitary: single or occurs alone.

sparingly: in a restricted or infrequent manner.

spare/ sparsely: Scattered.

spatially: relating to the position, area, and size.

species: a basic unit of classification; a set of organisms with similar characteristics and which can breed with each other

speckled: with darker (lighter) dots on a lighter (darker) background

spermatangium (-gia): the cell producing the male reproductive cell (spermatium) in red algae

spermatium (-a): non-motile male reproductive cell in red algae

spindle-shaped: thicker centrally and tapering to both ends (= fusiform)

spines: a stiff, sharp-pointed projection on a cell or tissue

spinose: bearing spines; provided with stiff, sharp-pointed projections

spinulose: bearing small spines

spiny: beset with spines

spiralized: coiled

sporangium (pl.sporangia): a cell producing one or several spores

spore: asexual cell able to germinate into a new plant

sporophyte: the spore-producing phase of a plant parts

spur: a horn-like extension to an organ

stalks: basal part of the bladder like thallus narrowed into a short structure

starch: a storage polysaccharide composed of alpha 1, 4 and 1,6 linked glucose residues

stellate: star-shaped

stenohaline: tolerant only for a narrow range of salinity

stenoionic: tolerant only for a narrow range of ion concentration

stenotherm: tolerant only for a narrow range of temperature

stipe: stalk; part between the holdfast and the frond

stipitate: provided with a stalk

stolon: a creeping (mostly cylindrical) structure from which erect branches arise

stoloniferous: bearing stolon-like structures

strichidium (-dia): a specialized, generally inflated branch producing tetrasporangia

strips: to remove or to wrest away

strap-shaped: ribbon-like or girdle-like

stupose: having a tuft of matted filaments

subopposite: nearly opposite but not exactly opposite

subacute: somewhat acute

subclavate: somewhat club-shaped

subcompressed: somewhat flattened

subcortical: beneath the cortex or cortical layer

subcortex: the inner part of the cortex, often differentiated from the outer cortex as, for example in Chordariaceae (Phaeophyceae)

subcuneate: somewhat wedge-shaped or somewhat with the broad end apical

subcylindrical: Nearly cylindrical

subflabellate: Nearly fan-shaped

subglobose: not quite spherical

sublittoral: the region above the eulittoral (or above high tide level) occupied by organisms with marine affinities, the spray zone

submembranous: partly or nearly membranous

suborbicular: nearly circular

subquadrate: almost rectangular/ four-sided

substantial: of considerable size

subspherical: almost spherical

subtrichotomous: somewhat branching into three divisions

subtidal: coastal area under the low water level, down to the circalittoral = lower limit of development of seagrasses

subtidal fringe: between mean and spring low tide levels; a narrow zone with some typical organisms

sub-terete: almost terete.

substratum: the material (such as rock) on which the alga is growing (substrate)

subulate: narrow and tapering gradually to a fine point

submarginal series: nearly to or slightly inside of margin

succession: the more or less predictable changes in the composition of communities following a natural or human disturbance

succulent: full of juice or sap

superficial: on the surface of the substratum

supralittoral: the coastal zone above high tide level

supralittoral fringe: the coastal zone between mean and spring high tide levels

sunken: immersed

swarmers: Flagellated reproductive cell whether gamete or spore

swept: to carry with a long, swinging or dragging motion

swollen: enlarged and solid

tangential filaments: parallel to tangents

tapering: gradually narrowing toward a point

taxonomy: the principles and methods for the classification of living organisms

tenacular cell: specialized cell achieving attachment with adjacent cells or filaments

tenaculum/tenacula: minute, lobed, hapteroid cells.

tendril: a slender climbing organ formed by the modification of part of a plant.

terete: circular in cross-section, tubular in three dimensions

tertiary: referring to a third stage

tetrahedral: cells contiguously placed as at the tips of a four-sided pyramid; - crystal:pyramid-shaped crystal

tetrasporangium (-gia): cell in which four spores are formed by meiosis

tetraspore: haploid spore, formed by meiosis (therefore mostly grouped by four) on the diploid tetrasporophyte

tetrahedrally: having four triangular faces

tetrasporangia: In brown algae, nonmotile spores(tetraspores) produced within specialized cells known as tetrasporangia

tetrasporophyte: diploid plant in red algae originating from the diploid carpospore, in which meiosis will take place resulting in the haploid tetraspores

tetrasporic phase: The diploid phase in the life cycle of the red algae.

tetrachotomous: four forked/ branching into four divisions.

thallus: the relatively undifferentiated multicellular plant body of a non-vascular plant (e.g. an alga)

tier: one of several layers or levels.

tips: a pointed or somewhat sharply rounded end

tomentose: thickly and evenly covered with hairs

torn: when marginal incisions are deep and irregular

trabacula: (a small beam), A connection, like a crossbar, uniting contiguous teeth in some Mosses

trabecular spine: a spine-like projection of cell wall material in the cell

transverse: In a plane perpendicular to the axis, or longitudinal direction

transformation: In a parasexual process in prokaryotes, a transfer of genetic information utilizing naked extracellular DNA

triangular: a solid part whose transverse section

approaches a triangle with acute angles

tribuliform: inverted triangular in surface view, with rounded upper angles and a small bulge in the middle of the upper side

trichoblast: a colorless, usually (dichotomously) branched, hair-like appendage, produced near branch apices in Rhodophyta

trichotomous: dividing into three (similar) parts or branching apically and equally in groups of three

tridichotomous: the main axis forking into three branches, which gives rise to two branches

trifurcate: with three forks or branches

trilobate: presenting three lobes

tripinnate: where the leaflets are arranged in a pinnate manner

triquetrous: (three-cornered). With three faces and edges

tristromatic: composed of three cell-layers (visible on cross-section)

truncate: cut off rather abruptly at the tip

tubular: a cylindrical figure and hollow

turbinate: as an inverted cone

turgidus: swollen, thick, and as if swollen, but not inflated with air.

tufts/tufted: clustered or clumped

turf algae: short, mat-forming, densely intricated seaweeds

type locality: the place where the type specimen of a taxon has been collected

type specimen: specimen after which the original description of an organism is based (remark: this does not mean that this specimen is 'typical')

undulate: being wavy, regularly curving up and down in a three dimensional way (margin of a blade)

uncorticated: lacking the cortex

uniaxial: A thallus axis made up of single filament

unilateral: along a single side of an axis

unilocular: one-chambered, having one loculus or chamber

uninucleate: with a single nucleus per cell

uniporate: with a single pore

uniseriate: composed of a single linear row of cells; a filament

unisexual: of one sex; bearing only male or only female reproductive organs

unistratose: composed of a single layer

urceolate: shaped like an urn (or pitcher), with a swollen middle and narrowing top

utricle: swollen end of siphons forming a palisade-like surface layer in *Codium* and various Udoteaceae (e.g. *Halimeda*)

vacuole: a large cavity within the cell filled with cell sap and surrounded by a plasma membrane (tonoplast).

vein: series of larger cells (mostly longer as compared to the neighboring ones) forming longitudinal rows within a smaller-celled tissue

vegetative: not associated with reproductive cells or sexual reproductive capacity

ventral: at the lower side

verticil: whorl

verticillate: in whorls; whorled

vesiculate: with vesicle

vesicular: bladder-like

warty: a surface covered with small round protuberances, especially in fruit, leaves, twigs, and bark

whorl: structures arranged in a circle or vertical around the axis

xanthophylls: brown accessory pigments in the brown algae

zonate: with the cross walls parallel to each other; in a tetrasporangium resulting in the four tetraspores

zoospores: a motile reproductive cell, not involved in sexual reproduction, resulting from mitosis (mitospore) or meiosis (meiospore)

zygote: diploid cell, formed after fertilization of an egg cell

BIBILIOGRAPHY

Agadi, V. V. and A. G. Untawale. 1978. "Marine algal flora of Goa coast." *Seaweed Research and Utilization*. 3: 56-70. 3 figs. 2 tables.

Agardh, C. A. 1820a. "Icones algarum ineditae." Fasc. 1. Lumdae [Lund]. [4] pp. pls. I-X.

Agardh, C. A. 1822-1823. "Species algarum." Vol. 1, Part 2. Lumdae [Lund]. [4] pp. [I-VIII +] 169-398 (1822), 399-531 (1823).

Agardh, C. A. 1824. "Systema algarum." Lumdae [Lund]. XXXVIII+ 312 pp.

Agardh, C. A. 1828. "Systema algarum." Vol. 2, Sect. 1. Gryphiae [Greifswald]. LXXVI + 189 pp.

Agardh, J. G. 1846. "Anadema, ett nytt slagte bland Algerna. Ofversigt af Kongl." [Svenska] vetenskaps–Akademiens Forhandlingar 3: 103-104.

Agardh, J. G. 1848a. "Species genera et ordines algarum." Volumen primum: algas fucoideas Complectens. Lundae] Lund]. VIII + 363 pp.

Agardh, J. G. 1848b. "Anadema, ett nytt slagte bland Algerna." Kongl. Svenska vetenskaps–Akademiens Handlingar 1846: 1-16, pl. XVII.

Agardh, J. G. 1876. "Species genera et ordines algarum." Voluumen tertium: de Florideis curae posteriors. Part1. Lipsiae [Leipzig]. VII + 724 pp.

Agardh, J. G. 1889. "Species Sargassorum Australiae descriptae et dispositae. Kongl. Svenska Vetenskaps-Akademiens Handlingar." [ser. 4], 23 (3). 133 pp. XXXI pls.

Ambiye, V. and A. G. Untawale. 1991. "*Galaxaura rugosa* (Ellis et Sol.) Lamour. From India." *Seaweed Research and Utilization* 13: 97-100. 5 figs.

Anbazhagan, C. V., S. Govindan and R. Rengasamy. 2001. "Potential of *Gracilaria verrucosa* and *Ulva fasciata* in domestic sewage treatment." *In* National Symposium on Marine Algal research in India: Prospect and Retrospect, 2001. Programme and Abstracts, Bharathidasan University, Tiruchirappalli. P. 23.

Anilkumar, C., P. Sindhu and M. V. N. Panikkar. 1993. "Species of the genus *Spyridia* Harvey (Ceramiales, Rhodophyta) from Kerala." *Seaweed Research and Utilization* 16: 17-22, 9 figs.

Anilkumar, C., P. Sindhu and M. V. N. Panikkar. 1994. "Species of *Polysiphonia* Greville (Ceramiales, Rhodophyta) from the Kerala coast. India." *Bionature* 14: 53-53. 16 figs.

Annapurna, B. and B. G. S. Rao. 1987. "Cytological studies in *Laurencia nana* and its significance in the evolutionary system of the red algae." *Indian Journal of Botany* 10: 89-92. 4 figs.

Asolkar, R. N. S. Wahidulla, C. G. Naik, S. Y. Kamat and V. P. Kamat. 1997. "Steroids from marine green alga *Cladophora* species. *In* V. P. Kamboj eds. Chemistry and Biology of Herbal Medicine." *Publ. Society of Biosciences* Muzaffarnagar. Pp. 27-31.

Awasthi, A. K. 1989. "A contribution of Andaman and Nicobar Islands and its significance." *In* A. C. Shukla ed. *Advances in applied phycology* – II. New Delhi: Today & Tomorrow's. Pp. 77-81. 4 tables.

Babu, B. and R. Rengasamy. 2001. "Histochemical and ultrastructural studies on the alginophyte *Cystoseira trinoides*." *In* National symposium on Marine Algal Research in India: Prospect and Retrospect, 2001. Programme and Abstracts, Bharathidasan University, Thiruchirappalli.P.30.

Balakrishnan S, and K. Krishnamurthy. 1993. "A morphological and histochemical study on the hairs of *Padina boergesenii* (phaeophyta)." *Seaweed Research and Utilization* 16: 227-236. 15 figs.

Balakrishnan S. and K. V. Krishnamurthy. 1998. "Occurrence of calcium oxalate crystal from a brown alga *Padina boergesenii* Allender & Kraft." *Seaweed Research and Utilization* 20 (1&2): 115-117.

Balakrishnan, S., M. Ravichandran and N. Kaliaperumal. 1992. "Studies on the distribution and standing crop of algae at Muthupet Estuary, Tami Nadu." *Seaweed Research and Utilization* 15: 63-68. 2 figs. 1 table.

Balasundaram A. 1985. "Seasonal succession of algae in the mid – algal zone on the Thiruchendur shore." *In* V. Krishnamurthy & A. G. Untawale eds. Marine plants, Madras: *Seaweed Research and Utilization Association* Pp. 179-190. 3 figs.

Bhakuni, D. S. B. N. Dhawan, H. S. Garg, A. K. Goel, B. N. Mehrotra, R. C. Srimal and M. N. Srivastava. 1992. "Bioactivity of marine organisms: part VI - Screening of some marine flora from Indian coasts." *Indian Journal of Experimental Biology* 30: 512-517. 2 tables.

Bheemasankar Rao, Ch, G. Trimurtulu, D. Venkata Rao, S. C. Bobzin, D. M. Kushlan, and D. J. Faulkner. 1991. "Diterpenes from the brown alga *Dictyota divaricata* of the Indian Ocean." *Phytochemistry* 30: 1971-1976.

Biswas, K. 1949a. "Common fresh & brackish water algal flora of India and Burma with a catalogue of Indian Myxophyceaea (blue-green algae), Chlorophyceae (green algae), Xanthophyceae (yellow -green algae) and Bacillariophyceae (diatoms)." Part 1." Records of the *Botanical Survey of India* 15 (1). [V +] iv + 105 pp. X pls.

Biswas, K. 1949b. "Common fresh & brackish water algal flora of India and Burma." Part II." Records of the *Botanical Survey of India* 15 (2). [V +] iii + 169pp.

Borgesen, F. 1914. "The marine algae of the Danish West Indies. Part 2. Phaeophyceae." *Dansk Botanisk Arkiv* 2 (2). 66 [68] pp. 44 figs.

Borgesen, F. 1934. "Some marine algae from the northern part of the Arabian Sea with remarks on their geographical distribution." Kongelige Danske Videnskabernes Selskab,
Biologiske Meddeleser 11 (6). 72 pp. 8 figs. II pls.

Borgesen, F. 1935. "A list of marine algae from Bombay." Kongelige Danske Videnskabernes Selskab, *Biologiske Meddeleser* 12 (2). 64 pp. 25 figs. X pls.

Borgesen, F. 1936. "Some marine algae from Ceylon. Ceylon Journal of Science, Botany 12: 57-96. 12 figs.

Borgesen, F. 1937. "Contribution to a South Indian marine algal flora – II." *Journal of the Indian Botanical Survey* 16: 311-357. 20 figs.

Borgesen, F. 1938. "Contribution to a South Indian marine algal flora – III." *Journal of the Indian Botanical Survey* 17: 205-242. 13 figs. pls. VII, VIII.

Borgesen, F. 1940. "Some marine algae from Mauritius. I. Chlorophyceae." Kongelige Danske Videnskabernes Selskab, *Biologiske Meddeleser* 15 (4). Pp. 26 figs. III pls.

Borgesen, F. 1943. "Some marine algae from Mauritius. III. Rhodophyceae. Part 2. *Gelidiales*, *Cryptonemiales*, *Gigartinales*" Kongelige Danske Videnskabernes Selskab, *Biologiske Meddeleser* 19(1): 85 Pp. 42 figs. I pl.

Borgesen, F. 1945. "Some marine algae from Mauritius. III. Rhodophyceae. Part 4. *Ceramiales*" Kongelige Danske Videnskabernes Selskab, *Biologiske Meddeleser* 19(10): 68 Pp. 35 figs.

Borgesen, F. 1946. "Some marine algae from Mauritius. An additional list of species to Part I Chlorophyceae." Kongelige Danske Videnskabernes Selskab, *Biologiske Meddeleser* 20 (6): 64 Pp. 35 figs.

Borgesen, F. 1948. "Some marine algae from Mauritius. Additional lists to the Chlorophyceae and Phaeophyceae." Kongelige Danske Videnskabernes Selskab, *Biologiske Meddeleser* 20 (12): 55 Pp. 24 figs. II pls. + map.

Borgesen, F. 1953. "Some marine algae from Mauritius. Additional to the parts previously published, V." Kongelige Danske Videnskabernes Selskab, *Biologiske Meddeleser* 21 (9): 62 Pp. 20 figs. III pls.

Bornet, E. 1888. "Algues du voyage au golfe de Tadjoura recueillies par M. L. Faurot." *Journal De Botanique [Morot]* 2: 17–20.

Bory de Saint-Vincent 1834. "Hydrophytes, Hydrophytae. *In* C.Belanger, Voyage aux Indes-Orientales. Pendant les annees 1825, 1826, 1827, 1828 et 1829. Botanique. Crytogamie. Pairs." Pp. 159-178, pls XV, XVI.

Chacko, P. I. S. Mahadevan and R. Ganesan. 1955. "A guide to the field study of the fauna and flora of Krusadai Island, Gulf of Mannar. Contribution from the Marine Biological Station, Krusadai Island, Gulf of Mannar," No 3. Madras: Government Press, 16 pp., 1 fig.

Chaugule, B. B. and H. D. Deodhar. 1985. "Notes on Konkan algae. III. *Champia compressa* Harvey." *Seaweed Research and Utilization* 7: 63-64, 3 figs.

Chaugule, B. B, A. G. Quadri and S. Goswamy. 1989. "Siphonaceous green algae of the Konkan coast," *Seaweed Research and Utilization* 11: 107-115, including III pls. [=58 figs.]

Chauhan V. D. 1978. "Report on the survey of marine algae resources of Maharashtra coast." *Salt and Research industry* 14 (1): 1-10. map, 4 tables.

Chauhan, V. D. and O. P. Mairh. 1978. "Report on survey of economic seaweeds resources of Saurashtra coast, India." *Salt Research and Industry* 14 (2): 21-41. 5 figs. [map], 11 tables.

Chennubhotla, V. S. K., N. Kaliaperumal and S.Kalimuthu. 1978. "Seasonal changes in growth, fruiting cycle and oospore output in *Turbinaria conoides* (J. Agardh) kutzing." *Botanica Marina* 21: 67–69, 2 figs. 1 table.

Chennubhotla, V. S. K., S. Mathew and I. Joseph. 1990. "A note on the occurrence of *Porphyra kanyakumariencis* (Bangiales: Rhodophyta) along the Kerala coast." *Seaweed Research and Utilization* 13: 1-4. 1 fig. 1 table.

Chiang, Y. M. 1970. "Morphological studies of red algae of the family Cryptonemiaceae." University of California Publications in Botany 58. vi + 95 pp., including 10 pls. 34 figs.

Chowdary, Y. B. K. and S. Singh. 1984. "Observations on the cytology of *Neomeris annulata* Dickie." *Phykos* 23: 171-177, figs. A-F. 1-10a.

Collins, F. S. and A. B. Hervey. 1917. "The algae of Bermuda" Proceedings of the *American Academy of Arts and Sciences* 53: 1-195, including VI pls.

Dave, M. J, R. G. Parekh, B. K. Ramavat, Y. A. Doshi and V. D.Chauhan. 1987. "Protein contents of red seaweeds from Gujarat coast." *Seaweed Research and Utilization.* 10: 17-20. 1 table.

Dave, M. J. and V. D. Chauhan. 1985. "Protein contents of brown seaweeds from Gujarat coast, west coast of India." *Indian Journal of Marine Science* 14: 46-47. 1 table.

Dave, M. J. and E. J. Lewis. 1979. "Free and combined amino acid composition in species of *Bryopsis* from Saurashtra coast." *Journal of Marine Biological Association of India* 18: 172-175. 2 tables.

Dave, M. J. and R. G. Parekh. 1975. "Protein content of green seaweeds from Saurashtra coast." *Salt Research and Industry* 11 (2): 41-44. 1table.

Dave, M. J. and R. G. Parekh. 1994. "Amino acid composition of marine green alga *Caulerpha scalpelliformis* (Chlorophyta)." *Indian Journal of Marine Science* 23: 59-60.

Dave, M. J. and R. G. Parekh. 1995. "Amino acid of some marine green algae of Okha coast. *In* National Symposium on Algal Potential and its Exploitation." Programme and Abstracts, American College, Madurai, India. P.16.

Dave, M. J., R. G. Parekh and V. D. Chauhan. 1990. "Amino acid of three species of *Codium* from Saurashtra coast." *Seaweed Research and Utilization* 12: 41-45. 2 tables.

Dawson, E. Y. 1954. "Notes on tropical pacific marine algae." Bulletin of the Southern California. *Academy of Sciences* 53: 1-7. 4figs.

Dawson, E. Y. 1963. "Marine red algae of Pacific Mexico, part 6. Rhodymeniales." Nova Hedwigia 5: 437-476. pls. 77-95.

Decaisne, J. 1842. "Essais sur une classification des polypiers calciferes de lamouroux." Annales des sciences naturelles, Botanique, ser. 2, 17: 297–380, pls. 14-17 (1842a): ibid. 18: 96- 128 (1842b). [Second part, appearing in volume 18, bears subtitle, "Memoire sur les corallines ou polypers calciferes".].

Desikachary, T. V., V. Krishnamurthy and M. S. Balakrishnan. 1990. "Rhodophyta. II. Taxonomy Part II A. Madras, India: Madras Science Foundation. [vi +] 279 pp. + 51 figs., XLII pls. [Bangiophycidae by Krishnamurthy, Balakrishnan & Desikachary; Palmariales to Gelidiales by Desikachary, Balakrishnan & Krishnamurthy; Cryptonemiales and Hildenbrandiales by Balakrishnan, Krishnamurthy & Desikachary].

Desikachary, T. V., V. Krishnamurthy and M. S. Balakrishnan. 1998. "Rhodophyta. II. Taxonomy continued. Part II B. Chennai, India:" *Madras Science Foundation*. [vi +] 359 pp. + 89 figs., XXXI pls.

De Toni, G. B. 1895. "Sylloge algarum." Vol. III. Fucoideae patavii [padova]. XVI + 638 pp.

De Toni, G. B. 1903. "Sylloge algarum." Vol. IV. Florideae. Sectio III. patavii [padova]. pp. 775, 775 bis, 776, 777 bis, 777–1522 + 1523–1525 [Index].

De Toni, G. B. 1905. "Sylloge algarum." Vol. IV. Florideae. Sectio III. patavii [padova]. pp. 1523– 1973.

Dhargalkar, V. K. 1985. "Studies on photosynthesis and respiration in some marine macroalgae of the Goa coast." *In* V. Krishnamurthy & A. G. Untawale (eds.), Marine plants. Madras; *Seaweed Research and Utilisation* Pp.79–88. 4 figs. 2 tables.

Dixit, S. C. 1940. "Algae of the Malwan harbour." *Journal of the University of Bombay* 8(5): 82-98. 5 figs.

Dixit, S. C. 1966. "Species list of Indian marine algae determined by Borgesen." *Journal of the University of Bombay* 32(3&5, Sect. B): 3-23.

Dixit, S. C. 1968. "Species list of Indian marine algae - II." *Journal of the University of Bombay* 36 (3&5, Sect. B): 9–24.

Dixit, S. C. 1970. "Species list of Indian marine algae - III." *Journal of the University of Bombay* 39 (Sc.): 99-130.

Dixit, S. C. 1980. "Species list of Indian marine algae - IV." *Journal of the University of Bombay* 48/49 (Sc.): 56-80.

Doty, M. S. 1988. "Prodromous ad systematic Eucheumatoideorum : a tribe of commercial seaweeds related to *Eucheuma* (Solieriaceae, Gigartinales)." *In* I. A. Abbott ed. Taxonomy of economic seaweed with reference to some pacific and Caribbean species. La Jolla: California Sea grant College program [Report T-CSGCP-018]. Vol. 2. Pp. 159-207. 37 figs. 1 table.

Durairatnam, M. 1962. "Some marine algae from Ceylon – 1." *Fisheries Research Station, Ceylon* Bulletin 15: 1-16. including 2 pls. 6 figs.

Foersskal, P. 1775. "Flora aegyptiaco-arabica." Post mortem auctoris edidit Carsten Niebuhr. Havniae [Copenhagen]. 32 + CXXVI + 219 [-220] pp., frontispiece [map].

Gajaria, S. C. and R. J. Patel. 1985. "Study of marine Cladophorales of Gujarat." Genus *Cladophora* Kuetz. *In* V. Krishnamurthy & A. G. Untawale eds. marine plants. Madras:" *Seaweed Research and Utilisation Association.* Pp. 51–58. 8 figs.

Gajaria, S. C. & R. J. Patel. 1988. "Contribution to our knowledge of *Cladophorales.*" IV. Genus *Chaetomorpha* Kuetz. *Seaweed Research and Utilisation* 11. 1–8. 29 figs. 1 table.

Ganesan, E. K. 1968. "Studies on the morphology and reproduction of the articulated corallines – III. *Amphiroa* Lamouroux emend." Weber van Bosse. *Phykos* 6: 7–28. 39 figs. I pl.

Ganesan, M, K. Eswaran and O. P. Mairh. 1997. "Effects of light intensity, salinity and source of nitrogen on the growth and biochemical composition of *Ulva fasciata* Delile. (Chlorophyta). *In* National Symposium on Recent trends in Algal Research.1997." Programme and Abstracts, Andhra University, Visakhapatnam. P. 17.

Ganesan, M., K. Eswaran, P. V. Subba Rao and C. R. K. Reddy. 2000. "Seasonal Variation in biomass and alginic acid content of *Sargassum wightii* and *Turbinaria conoides* from Mandapam coast, South East coast of India." *In* National Symposium on Seaweeds of India: Biodiversity and Biotechnology. 2000. Programme and Abstracts, Bharathidasan University, Thiruchirappalli. P. 26.

Gepp, A. and E. S. Gepp. 1908. "Marine algae (Chlorophyceae and Phaeophyceae) and marine phanerogams of the "Sealark" Expedition, collected by J. Stanley Gardiner, M. A., F. R. S., F. L. S. Transactions of the Linnean Society of London." Second Series, Botany 7: 163–188. pls. 22–24.

Gepp, A. and E. S. Gepp. 1909. "Marine algae (Chlorophyceae and Phaeophyceae) and marine phanerogams of the "Sealark" Expedition, collected by J. Stanley Gardiner, M. A., F. R. S., F. L. S. Transactions of the Linnean Society of London." Second Series, Zoology 12: 373–398, pls. 47–49. [Reprinting of Gepp & Gepp 1908].

Gepp, A. and E. S. Gepp. 1911. "The Codiaceae of the Siboga Expedition including a monograph of *Flabellaceae* and *Udoteaceae*." Expedition monographie 62, Leiden. 150 pp., XXII pls.

Gmelin, J. F. 1792. "Caroli a Linne…Systema naturae per regna tria naturae" Editio decimal tertia. Vol. 2, part 2. Lipsiae [Leipzig]. Pp. 885–1661.

Gopalakrishnan, P. 1970. "Some marine algae from the Gulf of Kutch." *Phykos* 8: 61–67.

Gopalakrishnan, P. 1971. "Some observations on the shore ecology of the Okha coast. *Journal of the Marine Biological Association of India* 12: 15–34. 3 figs. 1a, b.

Gopinathan, C. P. and R. C. Panigrahy. 1983. "Mariculture potential of Andaman and Nicobar island–an indicative survey:" Seaweed resources. *Bulletin of the Central Marine Fisheries Research Institute* [Cochin, India] 34: 47-51, II pls.

Greville, R. K. 1848 – 1849. "Algae orientales:- Descriptions of new species belonging to the genus *Sargassum*." Annals and Magazine of Natural History, ser. 2, 2: 203-206, pl. IV; 274-277, pl. V; 431-434. pl. XIII (1848); Ibid. 3: 106-109. pl. IV; 216-219. pl. IX; 254-257. Pl. X; 503-506. pl. XI (1849).

Greville, R. K. 1850. "Algae orientales: - Descriptions of new species belonging to the genus *Sargassum*." Transactions of the Botancial Society [Edinburgh]. 3: 85 -102, pls. VI-XI; 229-232, pl. XV. [Reprinting of Greville, 1848–1849].

Grunow, A. 1867. "Algen der Fidschi-Tonga-und Samoa-Inseln, gesammelt von Dr. E. Graeffe." *Journal des Museums Godeffroy* [Hamburg] 3: 23–50.

Grunow, A. 1874. "Algae. In E. Fenzl. ed." Reise der osterreichischen Fregatte Novara um die Erde in den Jahren 1857, 1858, 1859. Botanischer Theil. Erster Band. Sporenpflanzen. Wien. Pp. [1]–104, pls I, Ia, II -XI.

Haneefa koya, C. N. A. K. V. Nasser and Gulshad Mohammed. 1998. "Productivity of the coral reef alga *Halimeda gracilis* Harv. Ex J. Ag. At Minicoy Island, Lakshadweep." *In* National Symposium on *Seaweeds and their Utilisation* 1998. Programme and Abstracts, Madurai Kamaraj University, Madurai. P. 10.

Harvey, W. H. 1849. "A manual of the British marine algae." London. Lii + 252pp. 27 pls.

Harvey, W. H. 1853. "Nereis boreali-americana. Part II. Rhodospermeae." Smithsonian Contributions to Knowledge 5 (5). 258 pp. pls. XIII–XXXVI.

Harvey, W. H. 1858. "Nereis boreali-americana. Part II. Chlorospermeae." Smithsonian Contributions to Knowledge 10 (2). 140 pp., pls. XIII–XXXVII - L.

Hauck, F. 1886–1889. Ueber einige von J. M. Hildebrandt im Rothen Meere und Indischen "Ocean gesammelte Algen." Hedwigia. 25: 165-168, 217-221, [1] fig. (1886); ibid. 26: 8-21, 41-45 (1887); ibid. 27: 86-93 (1888); ibid. 28: 188-190 (1889).

Hieroonymus, G. 1985. "Cholophyceae." *In* A. Engler (ed.), Die Pflanzenwelt Ost-Afrikas. Theil C. Berlin." Pp. 21-24.

Holmes, E. M. 1896. "New marine algae." *Journal of Botany* 34: 349–351.

Hooker, W. J. 1844. "Icones plantarum." Ser. 2. fasc. 7. London. Pp. [i]–VIII, pls. DCI–DCC (with text).

Howe, M. A. 1920. "Algae. In N. L. Briton and C. F. Millspaugh, The Bahama flora." New York. Pp. 553–618.

Huisman, J. M. and D. I. Walker. 1990. "A catalogue of the marine plants of Rottnest Island, Western Australia, with notes on their distribution and biogeography." Kingia 1: 349-459. 2 figs. 1 table.

Jagtap, 1985a. "Studies on littoral flora of Andaman Islands." *In* V. Krishnamurthy & A. G. Untawale eds. Marine plants. Madras:

Seaweed Research and Utilization Association. Pp. 43–50. 4 figs. 1 table.

Jagtap, 1987. "Distribution of algae, seagrasses & coral communities from Lakshadweep Islands, eastern Arabian Sea." *Indian Journal of Marine Sciences* 16; 256–260. 3 fig. 3 tables.

Jagtap, 1992. "Marine flora of Nicobar group of Islands in Andaman Sea." *Journal of Marine Sciences* 21: 56–58. 1 fig. 1 table.

Jagtap, 1993. "Studies on littoral and sublittoral macrophytes around the Mauritius coast." *Atoll Research Bulletin* 382. 10 Pp. + 3 figs. 1 table.

Jayasankar, R. and J. R. Ramalingam. 1993. "Photosynthetic pigment of marine algae from Mandapam coast." *Seaweed Research and Utilization* 16: 41–43. 1 table.

Jayasankar. R, J. R. Ramalingam and N. Kaliaperumal. 1990. "Biochemical composition of some green algae from Mandapam coast." *Seaweed Research and Utilization* 12: 37–40. 1 table.

Kaladharan, P, C. P. Gopinathan, K. B. Bindhu, K. S. Smitha and R. Gireesh. 2000. "Isolation and culture of protoplasts from agar yielding seaweed *Gracilaria edulis* (Gmelin) Silva." *Seaweed Research and Utilization* 22 (1&2): 61–65.

Kaliaperumal, N. 1990. "Effects of environmental factors on diurnal periodicity of tetraspore output in some red algae of Visakhapatnam coast." *Seaweed Research and Utilization* 12: 87–95. 4 figs.

Kaliaperumal, N., V. S. K. Chennubhotla, S. Kalimuthu and J. R. Ramalingam, 1986. "Growth phenology & spore shedding in *Gracilaria arcuata* var. *Arcuata* (Zanardini) Umamaheswara Rao & *G. corticata* var. *cylindrica* (J. Agardh) Umamaheswara Rao (Rhodophyta)." *Indian Journal of Marine Sciences* 15: 107-110. 3 figs. 2 tables.

Kaliaperumal, N and S. Kalimuthu. 1993. "Need for conservation of economically important seaweeds of Tamil Nadu coast and time table for their commercial exploitation. *Marine Fisheries Information service. Technical and Extension Series* 119: 5–12. 14 figs, 2 tables.

Kaliaperumal, N., S. Kalimuthu, K. Muniyandi, J. R. Ramalingam, V. S. K. Chennubhotla, K. Rama Rao, P. V. Subba Rao, P. C. Thomas, S. H. Zaidi and K. Subbaramaiah. 1992. "Seaweed resources of Tamil Nadu coast: sector, III. Valinokkam – Kilakkarai." *Seaweed Research and Utilization* 15: 11–14. 1 fig. 1 table.

Kaliaperumal, N., S. Kalimuthu and J. R. Ramalingam. 1988. "Seasonal variation in growth, alginic acid mannitol contents in *Cystoseira trinoidis* (Forsskal) C. Agardh and *Hormophysa triquetra* (Linnaeus) Kuetzing." *Indian Journal of Botany* 11: 17-21. 2 figs.

Kaliaperumal, N., S. Kalimuthu and J. R. Ramalingam. 1990. "Studies on phycocollid contents from seaweeds of Tamil Nadu coast." *Seaweed Research and Utilization* 12: 115-119. 2 tables.

Kaliaperumal, N, S. Kalimuthu and J. R. Ramalingam. 1992. "Studies on agar content of *Gracilaria corticata* var. *corticata* and *G. corticata* var. *cylindrica*." *Seaweed Research and Utilization* 15: 191-195. 1 fig. 2 tables.

Kaliaperumal, N., S. Kalimuthu and J. R. Ramalingam. 1996. "Effect of repeated harvesting on the growth of *Sargassum* spp. and *Turbinaria conoides* occurring in Mandapam area." *Seaweed Research and Utilization* 15: 191-195. 1 fig. 2 tables.

Kaliaperumal, N., M. S. Rajagopalan and V. S. K. Chennubhotla. 1992. "Field cultivation of *Gracilaria edulis* (Gmelin) Silva in the lagoon of Minicoy (Lakshadweep)." *Seaweed Research and Utilization* 14: 133–107. 1 table, map.

Kaliaperumal, N., V. S. K. Chennubhotla, S. Kalimuthu, J. R. Ramalingam and S. Krishna Pillai. 1995. "Distribution of seaweeds off Kattapadu-Tiruchendur coast, Tamil Nadu." *Seaweed Research and Utilization* 17: 183-193.

Kaliaperumal, N., S. Kalimuthu, K. Muniyandi, J. R. Ramalingam, S. Krishna Pillai, V. S. K. Chennubhotla and M. S. Rajagopalan. 1996. "Distribution of Marine Algae and seagrass off Valinokkam-Kilakkarai Tamil Nadu coast." *Seaweed Research and Utilization* 18: 73-77.

Kaliaperumal, N, V. S. K. Chennubhotla, S. Kalimuthu, J. R. Ramalingam, S. Krishna Pillai, K. Muniyandi., K. Rama Rao, P. V. Subba Rao, P. C. Thomas, S. H. Zaidi and K. Subbaramaiah. 1998. "Distribution of Marine Algae and seagrass off Valinokkam-Kilakkarai Tamil Nadu coast." *Seaweed Research and Utilization* 20: 141-151.

Kaliaperumal, N. and Soniya Sukumaran. 1999. Spore shedding at different environmental factors in *Sargassum wightii* (Greville) J. Agardh and *Turbinaria conoides* (J. Agardh)

Kutzing." *In* National Symposium on Recent Trends in Marine Algal Research in India.1999. Programme and Abstracts, Annamalai University, Parangipettai. P. 15.

Kaliaperumal, N., R. Ezhilvalavan and J. R. Ramalingam. 1997. "Study on salinity tolerance and acclimatization of some commercially important seaweeds." *In* National Symposium on Recent Trends in Algal Research.1997." Programme and Abstracts, Andhra University, Visakhapatnam. P. 38.

Kalimuthu, S., N. Kaliaperumal and J. R. Ramalingam. 1992. "Distribution and seasonal changes of Marine algal flora from seven localities around Mandapam." *Seaweed Research and Utilization* 15: 119-126. 1 fig. 2 table.

Kamat, S. Y., S. Wahidulla, C. G. Naik, L. D'Souza, V. Jayasree, V. Ambiye, D. S. Bhakuni, A. K. Goel, H. S. Garg and R. C. Srimal. 1991. "Bioactivity of marine organism. Part III: screening of marine algae of the Indian coast for biological activity." *Mahasagar–Bulletin of the National Institute of Oceanography [India].* 24: 53–57. 1 table.

Kamat, S. Y., S. Wahidulla, L. D'Souza, C. G. Naik, V. Ambiye, D. S. Bhakuni, A. K. Goel, H. S. Garg and R. C. Srimal. 1992. "Bioactivity of marine organism. Part VI: Antiviral evaluation of marine algae of the Indian coast." *Botanica Marina* 35: 161-164. 1 table.

Kannupandi, T., A. Shanmugam, V. Tamil Mani & S. Anand. 1998. "Prospects and problem of *Gracilaria edulis* and *G. verrucosa* culture in Vellar estuary." *In* Symposium on *Seaweeds and their Utilisation.* 1998. Programme and Abstracts, Madurai Kamaraj University, Madurai. P. 20.

Kerkar, V. 1994. "Mineralogical studies on calcareous algae." *Current Science* 66: 381-382. 1 fig. 1 table.

Kesava Rao, Ch. S. Chinneraj, S. N. Inamdar and A. G. Untawale. 1991. "Arsenic content in certain marine brown algae and mangroves from Goa coast." *Indian journal of Marine Sciences* 20: 283-285. 2 tables.

Kesava Rao, Ch. and V. K. Indusekhar. 1989. "Distribution of certain cations and anions in seaweeds and seawater of Saurashtra coast and their geochemical significance." *Indian Journal of Marine Sciences* 18: 37-42. 1 fig. 3 tables.

Kraft, G. T. and M. J. Wynne. 1979. "An earlier name for the Atlantic North American and red algae *Neoagardhiella baileyi* (Solieriaceae, Gigartinales)." *Phycologia* 18: 325-329. 7 figs.

Krauss, F. 1846. "Pflanzen des Cap-und Natal-Landes, gesammelt und zusammengestellt von Dr. Ferdinand Krauss (Schluss) Flora" 29: 209-219.

Krishnamurthy, K. and L. Kannan. 1982. "A comparison of photosynthetic productivity of phytoplankton and seaweeds in a tropical estuary." *Seaweed Research and Utilization* 5: 33-35.

Krishnamurthy, V. 1989. "Criteria for the classification of *Gracilaria greville*." *Seaweed Research and Utilization* 11: 73-81. 10 figs.

Krishnamurthy, V. 1991. "*Gracilaria* resources of India with particular reference to Tamil Nadu coast." *Seaweed Research and Utilization* 14: 1-7. VI tables.

Krishnamurthy, V. and A. Balasundaram. 1990. "Vertical distribution of marine algae at Tiruchendur, South India." *Seaweed Research and Utilization* 12: 1-22. 5 figs. in text and on suppl.

Krishnamurthy, V. A. Balasundaram, M. Baluswami and K. Vardarajan. 1990. "Vertical distribution of marine algae on some South Indian shores." In V. N. Raja Rao ed. *perspectives in phycology*. New Delhi: Today and Tomorrow's. Pp. 267-281. 20 figs.

Krishnamurthy, V. and H. V. Joshi. 1969. "The Species of Ulva L. From Indian waters." *Botanical Journal of the Linnean Society*. [London]. 62: 123-130. 19 figs. 1 pl. 1table.

Krishnamurthy, V. and H. V. Joshi. 1970. "A check-list of Indian marine algae." Bhavnagar: Central Salt and Marine Chemicals Research Institute. 36 [37=Errata] pp.

Krishnamurthy, V. and N. Rajendran. 1987a. "Gonimonema, a structural feature of *Gracilaria.*" *Seaweed Research and Utilization* 10: 9-15. 17 figs. 1 table.

Krishnamurthy, V. and N. Rajendran. 1987b. "New species of *Gracilaria Greville* from Tamil Nadu." *Seaweed Research and Utilization* 10: 69-70.

Krishnamurthy, V., P. V. Raju and P. C. Thomas. 1980. "On augmenting seaweed resources of India." *Journal of the Marine Biological Association of India.* 17: 181-185. 2 tables.

Krishnamurthy, V. and M. Sundararajan. 1985. "Four new taxa of *Nemaliales* from India." *Seaweed Research and Utilization* 8: 43-63. 95 figs. I pl.

Krishnamurthy, V. and M. Sundararajan. 1987. "Studies on some Indian Nemaliales I. *Liagora doridis* Zeh, *L. orientalis* J. Ag. And L. *visakhapatnamensis* Umamaheswara Rao." *Seaweed Research and Utilization* 10: 71-80. 37 figs. In text & on suppl.

Krishnamurthy, V. and P. C. Thomas. 1977. "Some new or interesting marine algae from the Indian shores." *Seaweed Research and Utilization* 2: 42-51. 103: 27 figs.

Krishnamurthy, V. and K. Varadarajan. 1991a. "Studies on the *Dasyaceae* from India. I. The genera *Dasya*." *Seaweed Research and Utilization* 14: 45-52. 23 figs.

Krishnamurthy, V. and K. Varadarajan. 1991b. "Studies on Indian Delesseriaceae. III." *Seaweed Research and Utilization* 14: 61-71. 39 figs.

Krishnamurthy, V. and K. Varadarajan. 1992. "Studies on the Dasyaceae from India. II. The genera *Heterosiphonia* and *Dictyurus*." *Seaweed Research and Utilization* 14: 137-143. 25 figs.

Kumar, V. 1993a. "Biochemical constituents of marine algae from Tuticorin coast." *Indian Journal of Marine Sciences* 22: 138-140. 3 tables.

Kumar, V. 1993b. "Lipid content of some algae from Tuticorin coast." *Seaweed Research and Utilization* 16: 45-48. 1 table.

Kutzing, F. T. 1843. "Phycologia generalis…Leipzing." XXXII + 458 [459= Verbesserungen] pp, 80pls.

Kutzing, F. T. 1845. "Phycologia germinaca.......... Nordhausen." X + 340 Pp.

Kutzing, F. T. 1849. "Species algarum. Lipsiae [Leipzig]." VI + 922pp.

Kutzing, F. T. 1859. "Tabulae phycologia............ Vol. 10." Nordhausen. IV + 39pp., 100pls.

Kutzing, F. T. 1868. "Tabulae phycologia............ Vol. 18." Nordhausen. [III +] 35pp., 100pls.

Kylin, H. 1956. "Die Gattungen der Rhodophyceen." Lund: CWK Gleerups Forlag. XV + 673 pp. 458 figs.

Lawson, G, W. 1980. "A Check-list of East African Seaweeds (Djibouti to Tanzania)." Lagos, Nigeria: Department of Biological Sciences, University of Lagos. 65 pp. [Mimeographed].

Lewis, E. J. 1962. "Studies on the proteins, peptides and free amino-acid contents in some species of *Padina* from south-eastern coast of India." *Current Science* 31: 90-92. III tables.

Lewis, E. J. and E. A. Gonzales. 1962. "The proteins, peptides and free amino-acid contents of some species of marine algae from Bombay." *Annals of Botany* ser. 2. 26: 301-316. 3 tables.

Linneaus, C. 1753. "Species plantarum." Vol. 2. Holmiae [Stockholm]. Pp. 561-1200 [+ 1-31].

Lucas, A. H. S. and F. Perrin. 1947. "The seaweeds of South Australia." Part II. The red seaweeds. Adelaide. Pp. [107]–458, 202 figs.

Mairh, O. P., B. K. Ramavat, A. Tewari, R. M. Oza and H. V. Joshi. 1989. "Seasonal variation, bioaccumulation and prevention of loss of iodine in seaweeds." *Phytochemistry* 28: 3307-3310. 2 tables.

Mairh, O. P, K. Eswaran, R. G. Parekh, and G. Raja Krishna Kumar. 1997. "Studies on culture and carrageenan content of *Kappaphycus* and *Eucheuma* (Rhodophyta, Solieriaceae) grown under Indian condition." *In* National Symposium on Recent Trends in Algal Research. 1997. Programme and Abstracts, Andhra University, Visakhapatnam. P. 41.

Mairh, O. P., C. R. K. Reddy and G. Raja Krishna Kumar. 1998. "The resources of India. In A. T. Critchley & Ohno (eds.), *Seaweed resources of the World*." Japan International Cooperation Agency, Japan. Pp. 110–126.

Martens, G. von. 1968. "Die Tange. *In*: Die Preussische Expedition nach Ost-Asien. Nach amtlichen Quellen. Botanischer Theil." Berlin. 152 pp., VIII pls. [Title page dated 1866, but published in 1868; see F. A. Stafleu & R. S. Cowan, Taxonomic Literature, vol. 3, p. 317, 1981.]

Martens, G. von. 1869. "Beitrage zur Algen-Flora Indiens." Flora. 52: 233–238.

Martens, G. von. 1871. "List of algae collected by Mr. S. Kurz in Burma and adjacent islands." *Journal of the Asiatic Society of Bengal* 40, Part. 2: 461-469.

May, V. 1965. "A census and key to the species of Rhodophyta (red algae) recorded from Australia." Contributions from the New South Wales National Herbarium. 3: 349-429.

Maya, S. and N. B. Nair. 1992. "Ecology of tropical intertidal rock pools. II. Incidence and relative abundance of macroalgae of rock pools at Kovalam, south-west coast of India." *Seaweed Research and Utilization* 15: 21-38. 12 figs. 13 tables.

Menon, V. S., B. Leelambika, M. S. Krishnamurthy, T. G. B. Palode and R. I. Palode. 1993. "Some observations on the distribution of macrophytes in Tuticorin coast." *Seaweed Research and Utilization* 16: 57-61. 2 tables.

Misra, J. N. 1966b. "Phaeophyceae in India." New Delhi: Indian Council of Agricultural Research. [x +] 203 pp., including VI pls., 100 figs., V tables.

Misra, J. N. 1967. "The Phaeophyceae of the west coast of India." In V. Krishanmurthy ed. *Proceedings of the Seminar on Sea, Salt and Plants*. Bhavnagar. India. Pp. 227-233.

Mohammed Gulshad, A. K. V. Nasser and C. N. Hanifa Koya. 1998. "Domestic wastes and its impact on production of *Caulerpa racemosa* (Forssk.) Weber v. Bosse at Minicoy Island, Lakshadweep." *In* National Symposium on *Seaweeds and their Utilization* 1998. Programme and Abstracts, Madurai Kamaraj University, Madurai. P. 15.

Mohan Kumar, A., K. Rajendran and L. Kannan. 1991. "Copper toxicity to *Chaetomorpha linum* (Muell.) Kuetz. (Cladophorales, Chlorophyta)." *Seaweed Research and Utilization* 13: 115-121. 2 figs.

Monica Ram, M. R. Vijayaraghavan and Shashi B. Babbar. 2000. "Wound-response and regeneration in *Coelarthrum opuntia* (Endlicher) Boergesen." *Seaweed Research and Utilization* 22 (172): 45-50.

Montagne, C. 1842. "Botanique. Plantes cellulaires." *In* R. de la Sagra, Histoire physique, politique et naturelle de l'ile de Cuba. Pairs. [Vol. 11.] x + 549 pp. Atlas: pls. I–XX.

Montagne, C. 1845. "Plantes cellulaires." In J. B. Hombron and jacquinot, Voyage au Pole Sud Et dans l'Oceanie surles corvettes l'Astrolabe et la Zelee. Pendant les annees 1837–1838–1839–1840. sous le commandement de M. J. Dumont-d'Urville. Botanique. Pairs. Vol. 1. XIV + 349 pp. Atlas: Botanique: Cryptogamie. 20 pls.

Montagne, C. and P. M. A. Millardet. 1862. "Botanique, cryptogamie, Algues." *In* L. Millardet, Notes sur l'ile de Reunion (Bourbon). Annexe O. 25 pp., pls. XXIV–XXVII.

Murray, G. 1986. "On a new species of *Rhipila* (R. andersonii) from Merugui Archipelago." Transactions of the Linnean Society of London, Second Series, Botany, 2: 225–227, pl. 31.

Nageswara Rao, I., Ratna Kumari and G. Krupanidhi. 1999. "Biochemical composition of some marine algae along the Visakhapatnam coast, east coast of India." In G. Subba Rangaiah (ed.), Recent Trends in Algal Research, Visakhapatnam Pupl. Marine Algal Laboratory, Department of Botany, Andhra University, Waltair., Visakhapatnam Pp. 129-138.

Nair, N. B., V. Sobha and M. Arunachalam. 1982. "Algae from Southern Kerala coast." *Indian Journal of Marine Sciences* 11: 266-269. 1 fig. 1 table.

Nair, N. B., V. Sobha, R. Chandran, M. R. Ammal, S. Maya and H. Suryanarayanan. 1993. "Algal resources of the Kerala coast. VIII. Occurrence and relative abundance of Rhodophyta." *Seaweed Research and Utilization* 16: 183-197. 9 figs. 1 table.

Nair, N. B., V. Sobha, R. Chandran, P. A. Paul, P. I. Miranda and H. Suryanarayanan. 1986. "Nature and Distribution of the littoral algae and sea grasses of the south-west coast of India." Proceedings of the Indian National Science Academy, Part B. *Biological Sciences.* 52: 733-744. 5 figs. 3 tables.

Nair, N. B., V. Sobha, R. Chandran, M. Rathiamma, S. Maya and H. Suryanarayanan. 1990. "Algal resources of the Kerala coast. IX. Occurrence and relative abundance of Chlorophyceae. south-west coast of India." *Seaweed Research and Utilization* 12: 125-135, 9 figs. 2 tables.

Nashier, I., A. Jha and M. R. Vijayaraghavan. 1998. "Wound regeneration in *Solieria robusta* (Greville) Kylin (Solieriaceae, Gigartinales) from Port Okha, India." *Seaweed Research and Utilization* 20(1&2): 119-24.

Norris, R. E. 1987. "*Claudea elegans* (Delesseriaceae, Rhodophyta) in Natal, its first record in the western Indian Ocean and Africa." *South African Journal of Botany* 53: 311-315. 6 figs.

Ohno, M. and T. Critchley. 1993. "Seaweed cultivation and marine ranching." Japan International Cooperation Agency, Japan. Pp. 1-150.

Ohno, M. and O. P. Mairh. 1982. "Ecology of green alga *Ulvaceae* occurring on the coast of Okha, India." Reports of the USA Marine Biological Institute [Kochi University]. 3: 51-59. 6 figs. 4 tables.

Oza, R. M. 1997. "Culture of life history of *Bryopsis plumosa* 9Huds.) Ag." *In National Symposium on Recent Trends in Algal Research.* 1997. Programme and Abstracts, Andhra University, Visakhapatnam. P. 39.

Oza, R. M. and O. P. Mairh. 1997. "Experimental field Cultivation of *Ulva fasciata* Delile in the intertidal region on the coast of Diu." *In National Symposium on Recent Trends in Algal Research.* 1997. Programme and Abstracts, Andhra University, Visakhapatnam. P. 39.

Oza, R. M. and Supriya Gorasia nee Goswamy. 1998. "Life cycle of *Gracilaria corticata* J. Ag. (Gigartinales, Rhodophyta) *in vitro*." *In* National Symposium on *Seaweed and their Utilization* 1998. Programme and Abstracts, Madurai Kamaraj University, Madurai. P. 12.

Oza, R. M. and B. K. Ramavat. 2000. "New record of red alga *Digenea simplex* (Wulf.) C. Ag. (Ceramiales, Rhodomelaceae) from Indian coast." National Symposium on Seaweeds of India: *Biodiversity and Biotechnology* 2000. Programme and Abstracts, Central Salt and Marine Chemical Research Institute, Bhavnagar. P. 22.

Padmini Sreenivasa Rao, P. 1991. "Biological investigation of Indian marine algae 4. Screening of some green, red and brown seaweeds for their antimicrobial activity." *Seaweed Research and Utilization* 14: 37-43. 3 figs, 2 tables.

Padmini Sreenivasa Rao, P., P. Sreenivasa Rao and S. M. Karmakar. 1991. "Chemical Investigation of Indian *Sargassum." Seaweed Research and Utilization* 14: 15-19. 3.

Pandya, A. C., A. H. Bapodra and H. H. Parekh. 1990. "Brominated phenolic compounds in Indian marine algae." *Planta Medica* 56: 557.

Panikkar, M. V. N., P. Ampili and V. D. Chauhan, 1992. ('1991'). "Histochemical localisation of polysaccharides in the tissues of three Indian marine algae." *Phykos* 30: 61-66. 2 tables.

Pant, A. and A. G. Untawale. 1989. "Indian Seaweeds: resources and utilization. In A. C. Shukla (ed.)" *Advances in applied phycology* – II. New Delhi: Today and Tomorrow's. Pp. 35-41. 7 tables.

Parekh, R. G. and O. P. Mairh. 1997. "Studies on phycocolloids from *Chondrococcus hornemanii* (Cryptonemiales, Rhizophyllidaceae) from Mandapam, South East Coast of India." National Symposium on Recent Trends in Algal Research. 1997. Programme and Abstracts, Andhra University, Visakhapatnam. P. 40.

Parekh, R. G. and V. D. Chauhan. 1987. "Lipid content of some Indian seaweeds." *Indian Journal of Marine Sciences* 16: 272-273. 1 table.

Parekh, R. G. and V. D. Chauhan. 1990. "Chemical constituents of some red seaweeds of Okha coast." *Seaweed Research and Utilization* 21: 121-124. 1 table.

Parekh, R. G., M. J. Dave and E. J. Lewis. 1980. "Alginic acid content in common benthic brown algae of Saurashtra coast." *Journal of Marine Biological Association of India.* 17: 245-247. 1 table.

Parekh, R. G, L. V. Maru and M. J. Dave. 1977. "Chemical composition of green seaweeds of Saurashtra coast." *Botanica Marina* 20: 359-362. 2 tables.

Parekh, R. G. and O. P. Mairh. 1999. "Studies on phycocolloids from *Chondrococcus hornemanii* (Cryptonemiales, Rhizophyllidaceae) from Mandapam, South East Coast of India." In G. Subba Rangaiah ed. Recent Trends in Algal Research,Visakhapatnam Publ. Marine Algal Laboratary, Department of Botany, Andhra University, Waltair., Visakhapatnam. Pp. 99-105.

Pasricha, S, M. R. Vijayaraghavan and D. Sahoo. 1998. "Studies on the yield, nature and properties of carrageenan from *Laurencia pedicularioides* Boergs. (Rhodophyta)." *Seaweed Research and Utilization* 20 (1& 2): 19-22.

Prema, M., B. Subramanian and V. Krishnamurthy. 2000. "Scanning electron microscopic studies on some species of *Hypnea* Lamouroux (Gigartinales, Rhodophyta)." *Seaweed Research and Utilization* 20 (1& 2): 25-29.

Premnathan, M., K. Chauhan, S. K. Bajpai and K. Kathiresan. 1992. "A survey of some Indian marine plants of antiviral activity." *Botanica Marina* 35: 321-324. I table.

Quoy, J. R. C. and P. Gaimard. 1842. "Zoologie." In L. de Freycient, Voyage autour du monde. Execute sur les corvettes de S. M. I'Uraine et la Physicienne, pendant les annees 1817, 1818, 1819 et 1820. Pairs. [V +] 713pp., 96 pls

Radhakrishnan, S. 1976. "Some aspects of the distribution and seasonal abundance of macrophytic flora in the brackish Lake Pulicate, India." *In* C. K. Varshney & J. Rzoska (eds.). Aquatic weeds in South East Asia. The Hague: W. Junk. Pp. 107-117. 2 figs., 3 tables.

Raghukumar, C. S. Nagarkar and S. Raghukumar. 1992. "Association of thraustochytrids with living marine algae." *Mycological Research* 96: 543-546. 6 figs. 1 table.

Ragothaman, G. and M. V. Haranatha Babu. 1990. "Studies on the epiphytic diatoms of Devka coast." *In* V. N. Raja Rao (ed.). Perspectives in Phycology. New Delhi: Today and Tomorrow's. Pp. 305-308. 1pl.

Ragothaman, G. and M. V. Haranatha Babu. 1992. "Studies on the marine algae of Devka coast." *Seaweed Research and Utilization* 15: 91-97. 1 fig. 2 tables.

Rajendran, K., C. Govindasamy, R. Kannan, P. Sampathkumar, M. Ganesan and L. Kannan. 1991. "Seaweed resources of the Bay of Bengal on the northern part of the Tamil Nadu coast." *Seaweed Research and Utilization* 13:123-130. 4 tables.

Ramaiah, N. and D. Chandramohan. 1992. "Densities, cellulases, alginate and pectin lyases of luminous and other heterotrophic bacteria associated with marine algae." *Aquatic Botany* 44: 71-81. 3 figs. 6 tables.

Ramalakshmi. Y. and V. D. Chauhan. 1990. "Nitrate and ammonium uptake by algae *Sargassum swartzii* (Turn.) C. Ag. (Phaeophyta)." *Indian Journal of Marine Sciences* 19: 66-67. 1 fig. 3 tables.

Rama Rao, K., P. V. Subba Rao, P. C. Thomas, S. H. Zaidi, K. Subbaramaiah, N. Kaliaperumal, S. Kalimuthu, K. Muniyandi, J. R. Ramalingam, N. Najmuddin and V. S. K. Chennubhotla. 1993. "Seaweed resources of Tamil Nadu coast," sector – IV, Kilakkarai – Rameswaram Island (Dhanushkodi). *Seaweed Research and Utilization* 16: 103-110. 3 figs. 1table.

Ramavat, B. K., A. S. Mehta and O. P. Mairh. 2000. "Preparation chemical investigation and application of liquid seaweed fertilizer (L. S. F.) from *Sargassum tenerrimum* J. Ag. Of Saurashtra coast." *In* National Symposium on Seaweed of India: *Biodiversity and Biotechnology* 2000. Programme and Abstracts, Central Salt and Marine Chemical Research Institute, Bhavnagar. P. 76.

Ramavat, B. K., M. V. N. Panikkar and A. Tewari. 2000. "Mannitol content of some brown algae species of Indian coasts." In National Symposium on Seaweed of India: *Biodiversity and Biotechnology.* 2000. Programme and Abstracts, Central Salt and Marine Chemical Research Institute, Bhavnagar. P. 78.

Rao, D. S., K. G. Girijavallbhan, S. Muthusamy, V. Chandrika, C. P. Gopinathan, S. Kalimuthu and M. Najamuddin. 1991. "Bioactivity In marine algae." In M. F. Thompson, R. Sarojini & R. Nagabhushanam (eds.), Bioactive compounds from marine organisms, with emphasis on the Indian Ocean, New Delhi: Oxford & IBH. Pp. 373-377, 2 tables.

Rao, G. M. Narsimha. 1995. "Seasonal growth and biomass and reproductive behavior of three species of red algae in Godavari estuary, India." *Journal of Phycology* 31: 209-214. 7 figs. 5 tables.

Rao, G. M. Narasimha and M. Umamaheswara Rao. 1986. "Numerical study of intertidal zonation along Visakhapatnam coast, east coast of India." *Indian Journal of Marine Sciences* 15: 102-106. 4 figs.

Rao, G. M. Narasimha and M. Umamaheswara Rao. 1991. "Spore discharge in the red algae *Bostrichia tenella* and *caloglossa leprieurii* from the Godavari estuary, India." *Journal of Applied Phycology* 3: 153-158. 3 figs.

Rao, P. S. N. and Marcel Tigga. 2000. "Seaweed flora of Andaman and Nicobar Island with references to rare economically important species. *In* National Symposium on Seaweeds of India:" *Biodiversity and Biotechnology* 2000. Programme and Abstracts, Central Salt and Marine Chemical Research Institute, Bhavnagar. P. 14.

Rattan Kumar, V. and S. Naidu. 1997. "Culture of *Enteromorpha compressa* and *Ulva fasciata* in the laboratory conditions." *In* National Symposium on Recent Trends in Algal Research. 1997. Programme and Abstracts, Andhra University, Visakhapatnam. P. 43.

Redekar, P. D. and P. C. Raje. 2000. "Alginic acid and agar-agar content from the common seaweeds of Ratnagiri coast (west coast of India)." *Seaweed Research and Utilization* 22 (1&2): 41-43.

Reeta Jayasankar. 1997. "Fatty acid of *Gracilaria*." *In* National Symposium on Recent Trends in Algal Research. 1997. Programme and Abstracts, Andhra University, Visakhapatnam. P. 42.

Rengasamy, R. 1991. "Studies on *Padina distromatica* hauck." *Seaweed Research and Utilization* 13: 185-189. 18 figs.

Renuka Bai, N. Sr., M. R. Basil Rose and Sr. P. D. Mercy. 1998. "A naturally occurring agglutinin in the extract of the seaweed *Turbinaria ornata*." *Seaweed Research and Utilization* 20 (1&2): 63-96.

Roth, A. W. 1806. "Catalecta botanica." Fasc. 3 Lipsiae [Leipzig]. [VIII+] 350 [9] pp., XII pls.

Ruma Pal. 2000. "A brief report on the taxonomy and in vitro culture of *Enteromorpha* and *Ulva* from West Bengal coast." *Seaweed Research and Utilization* 20 (1&2): 51-60.

Sampathkumar, P., K. Rajendran, C. Govindasamy and L. Kannan. 1999. "Biochemical composition of *Chaetomorpha antennina* from southeast coast of India." *In* National Symposium on Recent Trends in Marine

Algal Research in India. 1999. Programme and Abstracts, Annamalai University, Parangipettai. P. 24.

Sathe-Pathak, V. S. Raghukumar, C. Raghukumar and S. Sharma. 1993. "Thraustochytrid and fungal component of marine detritus. I. Field studies on decomposition of the brown algae *Sargassum cinereum* J. Ag." *Indian Journal of Marine Sciences* 22: 159-167. 9 figs., I table.

Selvi, M., R. Selvaraj and S. Kabilan. 2001. "Phytochemistry of *Stoechospermum marginatum* (Huds.) Lamour." *In* National Symposium on Marine Algal Research in India: Prospect and Retrospect, 2001. Programme and Abstracts, Bharathidasan University, Tiruchirappalli. P. 23.

Setchell, W. A. 1933. "Some early algal confusions." II. University of California Publications in Botany. 17: 187-254, incl. Pls. 26-45.

Shanmugam, M., K. H. Mody, R. M. Oza and B. K. Ramavat. 2000. "Blood anticoaugulant activity of a marine green alga *Codium dwarkense* Borgs.: Influence of ecological parameters." *In* National Symposium on Seaweeds of India: *Biodiversity and Biotechnology* 2000. Programme and Abstracts, Central Salt and Marine Chemical Research Institute, Bhavnagar. P. 35.

Shanmugam, M, K. H. Mody, R. M. Oza and B. K. Ramavat. 2001. "Blood anticoaugulant activity of a marine green alga *Codium dwarkense* (Codiaceae, Chlorophyta) in relation to its growth stages." *Indian Journal of Marine Sciences* 300: 49-52.

Siddhanta, A. K., K. H. Mody, B. K. Ramavat, V. D. Chauhan, M. Sharma and H. S. Garg. 1995. "Characterization of sulphonoglycolipid from the red alga *Laurencia pedicularioides*." *Botanica Marina* 38: 329-331, 1 fig.

Siddhanta, A. K., A. M. Goswami, M. Shanmugam, K. H. Mody, B. K. Ramavat and O. P. Mairh 1997. "Studies on the sulphated polysaccharides of *Laurencia* species of the west coast of India." *In* National Symposium on Recent Trends in Algal Research 1997. Programme and Abstracts, Andhra University, Visakhapatnam. Pp. 33 - 34.

Silva, P. C. 1952. "*Codium* Stackhouse. In L. E. Egerod, An analysis of the siphonous Chlorophycophyta with special reference to the *Siphonocladales,* Siphonales and Dasycladales of Hawaii." University of California Publications in Botany. 25: 381-395, figs. 11-18. pls. 34b, 35, 36.

Silva, P. C. 1959. "The genus Codium (Chlorophyta) in South Africa." *Journal of South African Botany* 25: 101-165. 22 figs., pls. II – XVI.

Silva, P. C. 1962. "Comparison of algal floristic patterns in the Pacific with those in the Atlantic and Indian oceans, with special reference to *Codium*." *In*: Proceedings of the Ninth Pacific Science Congress [Bangkok, 1957]. Vol. 4. Bangkok. Pp. 201-216. 13 figs.

Silva, P. C., Philip, W. Basson and Richard L. Moe. 1996. Catalogue of the benthic marine Algae of the Indian Ocean." 1996. University of California in Botany. London, England. Pp. 1260.

Sindhu, P. and M. V. N. Panikkar. 1992. "Marine algae of Kerala. I: *Gelidiales*. *Seaweed Research and Utilization* 14: 119-127. 32 figs.

Sindhu, P. and M. V. N. Panikkar. 1995a. "On the Occurrence of new species of *Pterocladia* J. G. Agardh (Gelidiaceae, Rhodophyta) from Kanyakumari, India." *Seaweed Research and Utilization* 17 (1&2): 23-29.

Sindhu, P. and M. V. N. Panikkar. 1995b. "Observations on *Peyssonnelia obscura* Web. V. Bosse.Var. *bombayensis* Borgesen from Kerala, India." *Seaweed Research and Utilization* 17 (1&2): 147-150.

Sindhu, P. and M. V. N. Panikkar. 1995c. "Species of *Ulva* (Cholorophyceae, Ulvales) from Kerala, India." *Feddes Repertorium*. 106: 121-125. 2 figs.

Sindhu, P., C. Anilkumar and M. V. N. Panikkar. 1995. "Ecological studies on *Padina tetrastromatica* Hauck. (Phaeophyta, Dictyotales)." *Seaweed Research and Utilization* 17 (1&2): 135-149.

Singh, S., G. Jose and Y. B. K. Chowdhry. 1981. "Observations on the cytology of *Valonia ginnani*." *Phykos* 20: 65-69. 5 figs., 1 table.

Siraimeetan Pon and M. Selvaraj. 1998. "Experimental cultivations of *Gracilaria edulis* (Gmelin) Silva at Tuticorin in Gulf of Mannar." *In* National Symposium on *Seaweeds and their Utilization* 1998. Programme and Abstracts, Madurai Kamaraj University, Madurai. P. 20.

Sivakumar, K. and R. Rengasamy. 2000. "Culture studies on the effect of nitrate on growth, pigments, kappa-carrageenan content and ultrastructure of *Hypnea musiformis* and *H. valentiae*." *Seaweed Research and Utilization* 22 (1&2): 67-74.

Sivanand, N., and D. Subramanian. 1993. "Cytological studies in some species of *Caulerpa* from Tamil Nadu." *Seaweed Research and Utilization* 16: 29-35. 28 figs.

Sobha, V., and N. B. Nair. 1985. "Marine algae of the south-west coast of India. In: V. Krishnamurthy & A. G. Untawale (eds.). Marine Plants." Madras. *Seaweed Research and Utilization* Association Pp. 17-24. [2] maps, 1 table.

Sobha, V., M. Surendran and T. V. Nair. 1992. "Heavy metal, biochemical studies of different groups of algae from Cape Comorin and Kovalam." *Seaweed Research and Utilization* 15: 77-87. 6 figs, 5 tables.

Sonder, O. G. 1854. "Algae." *In* H. Zollinger, Systematisches Verzeichniss der im indischen Archipel in den Jahren 1842-1848 gesammelten sowie der aus Japan empfangenen pflanzen. Fasc. I. Zurich. Pp. 1-4.

Sonder, O. G. 1871. "Die Algen des tropischen Australiens." Abhandlungen aus dem Gebiete Der Naturwissenchaften herausgegeben von dem Naturwissenchaften Verein in Hamburg. 5 (2): 33-74, VI pls.

Srinivasa Rao, A. and M. Umamaheswara Rao. 1998. "Effect of environmental factors on the growth of *Gracilaria corticata*." *Phykos* 37 (1&2): 11-17.

Srinivasa Rao, A. and M. Umamaheswara Rao. 2000b. "Seasonal variation in biochemical content of *Sargassum polycystum* C. Agardh at Visakhapatnam, East coast of India." *In National Symposium on Seaweeds of India*: *Biodiversity and Biotechnology* 2000. Programme and Abstracts, Central Salt and Marine Chemical Research Institute, Bhavnagar. P. 27.

Srinivasa Rao, A. and M. Umamaheswara Rao. 2000c. "Influence of some physico-chemical condition on the callus induction rate in *Sargassum polycystum* C. Agardh at Visakhapatnam, East coast of India." *In National Symposium on Seaweeds of India: Biodiversity and Biotechnology.* 2000. Programme and Abstracts, Central Salt and Marine Chemical Research Institute, Bhavnagar. P. 39.

Srinivasa Rao, A. and M. Umamaheswara Rao. 2001. "Enhancemnt of callus induction rate in *Sargassum polycystum* C. Agardh by providing plant growth regulators." *In National Symposium on Seaweeds of India: Prospect and Retrospect,* 2001. Programme and Abstracts, Bharathidasan University, Tiruchirappalli. P. 33.

Srinivasa Rao, A., M. Umamaheswara Rao and V. Rattan Kumar 1998. "Effect of temperature, photon flux density, salinity and pH on growth of *Sargassum polycystum* under laboratory conditions". *In National Symposium on Seaweeds and their Utilization* 1998. Programme and Abstracts, Madurai Kamaraj University, Madurai. P. 21.

Srinivasa Rao, A., V. Rattan Kumar and M. Umamaheswara Rao. 1998. "Interaction of salinity-photon flux density on the growth of six marine algae." *In National Symposium on Seaweeds and their Utilization* 1998. Programme and Abstracts, Madurai Kamaraj University, Madurai. P. 22.

Srinivasa Rao, P. 1965. "*Polysiphonia* coacta Tseng, new to Indian Ocean." In: *Proceeding of the combined 51st and 52nd Indian Science Congress,* Calcutta, 1964-65. Part III. Abstracts. Calcutta. P. 301.

Srinivasa Rao, P. 1967. "The genus *Polysiphonia* from the Kathiawad coast of India." *In* V. Krishnamurthy ed. *Proceeding of the Seminar on Sea, Salt and Plants*. Bhavnagar. 1965. Pp. 169-177 + addendum, 23 figs.

Srinivasa Rao, P. and K. S. Parekh. 1981. Antibacterial activity of Indian seaweed extracts. *Botanica Marina* 24: 577-582. 2 figs., V tables.

Sridharan, R., J. Moses Babu, G. K. Trivedi and H. H. Mathur. 1991. "Fatty-acid composition of some marine algae from Indian waters." In M. F. Thompson, R. Sarojini & R. Nagabhushanam (eds.), *Bioactive compounds from marine organisms, with emphasis on the Indian Ocean.* New Delhi. Oxford & IBH. Pp. 201-204, 1 table.

Srinivasan, K. S. 1961. ('1960'). "Distribution patterns of marine algae in Indian seas." *In* P. Kachroo 9eds.), *Proceedings of the symposium on Algology,* New Delhi, 1959, New Delhi: Indian Council of Agricultural Research. Pp. 219-242. 13 figs., II table.

Srinivasan, K. S. 1965. "Algarum species ex India Oriundae." Bulletin of *Botanical Survey of India* 7: 88-266.

Srinivasan, K. S. 1967. "Conspectus of *Sargassum* species from Indian territorial waters." *Phykos* 5: 127-159. 73 figs.

Srinivasan, K. S. 1969. "Phycologia Indica (Icones of Indian marine algae)." Vol.1. Calcutta: *Bulletin of Botanical Survey of India* xix + 52 pp., LI pls., 7 photos, [24] group of figs.

Srinivasan, K. S. 1973. "Phycologia Indica" (Icones of Indian marine algae). Vol.2. Calcutta: *Bulletin of Botanical Survey of India* xvii + 60pp., LIV pls.

Srivastava, M. N. 1996. "On occurrence of *Struvea anastomosans* (Harvey) Piccone in Andaman Islands." *Phykos* 35 (1&2): 161-162.

Subbaramaiah, K. 1971. "Distribution of marine organisms on the Jalleshwar shore, Veraval." *Seaweed Research and Utilization* 1: 19-29. 13 figs.

Subbaramaiah, K. 1972. "Intertidal ecology of a rocky shore at Mandapam Camp in the Gulf of Mannar." *In*: Proceedings of the Fifty-ninth Indian Science Congress, Calcutta. 1972. Part III. Abstracts. Calcutta. P. 395.

Subbaramaiah, K. 1974. "Intertidal ecology of a rocky shore at Mandapam Camp, India." *Indian Journal of Marine Sciences* 3: 58-60. 3 tables.

Subba Rangaiah, G. and B. Apparao. 2000. "Seasonal growth, phenology and spore shedding capacities in three species of *Sargassum* growing along the Visakhapatnam coast." *In National Symposium on Seaweeds of India: Biodiversity and Biotechnology* 2000. Programme and Abstracts, Central Salt and Marine Chemical Research Institute, Bhavnagar. P. 18.

Subba Rangaiah, G. and C. Vanilla Kumari. 1999. "Seasonal growth, phenology and spore Shedding in *Amphiroa fragilissma* (L.) Lamouroux (Rhodophyceae, Cryptonemials) of the Visakhapatnam coast." *In* G. Subba Rangaiah (ed.), *Recent Trends in Algal Research*, Visakhapatnam. Pupl. Marine Algal Laboratory, Department of Botany, Andhra University, Waltair., Visakhapatnam. Pp. 255-269.

Subba Rao, P. V. 1997. "Effect of N-Methyl, N-Nitro N-Nitrosoguanidine on the growth and agar content of *Gelidiella acerosa* under field condition." *In National Symposium on Recent Trends in Algal Research.* 1997. Programme and Abstracts, Andhra University, Visakhapatnam. P. 43.

Subba Rao, P. V. and M. Ganesan. 1995. Survey of *Asparagopsis* (Delilei) Montagne (Nemalionales, Rhodophyta) resources for iodine content in the Gulf of Mannar Islands, South East coast of India." In *National Symposium on Algal Potential and its Exploitation*. 1995. Programme and Abstracts, American College, Madurai, India. P. 36.

Subba Rao, P. V., K. Rama Rao, T. K. Mal, K. Subbaramaiah, N. Kaliaperumal, S. Kalimuthu, K. Muniyandi, J. R. Ramalingam and V. S. K. Chennubhotla. 1992. "Seaweed resources of Tamil Nadu coast." Sector II. Alanthalai – Mandapam and Vembar – Nallathani Thivu. *Seaweed Research and Utilization* 15: 177-183. 2 figs. 2 tables.

Subba Rao, P. V., K. Rama Rao, M. R. P. Nair and K. Subbaramaiah. 1985. "A systematic list of marine algae from Andhra Pradesh and their distribution." In: V. Krishnamurthy & A. G. Untawale (eds.). Marine Plants *Seaweed Research and Utilization Association*. Pp. 29-34. 1 table.

Subramanian, A., L. Kannan, T. Pugalendhi and M.Muruganantham. 1998. "Heparin from a seaweed *Grateloupia filicina* (Rhodophyta)." *In National Symposium on Seaweeds and their Utilisation* 1998. Programme and Abstracts, Madurai Kamaraj University, Madurai. P. 17.

Subramanian, B. 1985a. "An annotated list of Ceramiaceous algae (Rhodophyta) of Tiruchendur coast, South India." *Seaweed Research and Utilization Association* 7: 71-81. 39 figs. 1 table.

Subramanian, B. 1985b. "On some Ceramiaceae (Rhodophyta) from the Indian shores." In V. Krishnamurthy & A. G. Untawale (eds.). Marine Plants. *Seaweed Research and Utilization* Association. Pp. 191-204. 51 figs.

Subramanian, D. 1991. "Chaemotaxonomical studies in South Indian species of *Caulerpa.*" *Journal of Phycology* 27 (suppl.): 70. [Abstract].

Subramanian, D. 1993. "Chaemotaxonomical studies in South Indian species of *Caulerpa.*" *Seaweed Research and Utilization* 16: 205-218. [6] figs., 6 tables.

Sundararajan, M. and K. Annakkili. 2000. "Three species of *Codium* from South India." *In National Symposium on Seaweeds of India: Biodiversity and Biotechnology* 2000. Programme and Abstracts, Central Salt and Marine Chemical Research Institute, Bhavnagar. P. 73.

Suresh, K., M. S. Ahamed, G. Durairaj and V. K. Nair. 1993. "Impact of power plant heated effluent on the abundance of sedentary organism, off Kalpakkam, east coast of India." *Hydrobiologia* 268: 109-114. 3 fig.

Taylor, W. R. 1966. "Records of Asian and western Pacific and marine algae particularly algae from Indonesia and the Philippines." *Pacific Science* 20: 342-359. 2 figs.

Tewari, A. 1998. "Studies on preservation of *Sargassum*." Iv. The role of alginate degrading bacteria at different stage of processing of *Sargassum*. *Seaweed Research and Utilization* 20 (1&2): 1-8.

Tewari, A. and H. V. Joshi. 1988. "Effect of domestic sewage and industrial effluents on biomass and species diversity of seaweeds." *Botanica Marina* 31: 389-397. 1 fig., XI tables.

Tewari, A., S. Thampan and H. V. Joshi. 1990. "Effect of chlor-alkali industry effluent on the growth and biochemical composition of two marine macroalgae." *Marine Pollution Bulletin* 21: 33-38. 6 figs, 2 tables.

Thomas, P. C. and K. Subbaramaiah. 1991. "Growth of marine algae in a *Sargassum wightii* dominated population at Mandapam." *Seaweed Research and Utilization* 13: 201-206. 1 fig., 2 tables.

Thomas, P. C. and K. Subbaramaiah. 1994. "Seasonality in growth and Phycocolloid content in *Hypnea pannosa* J. Agardh (Gigartinales, Rhodophyta) from Mandapam region." *Proceeding of the Indian National Science Academy, Part B. Biological Science*. 60: 93 – 98, 3 tables.

Umamaheswara Rao, M. 1969. "Catalogue of marine algae in the reference collections of the Central Marine Fisheries Research Institute." *Bulletin of the Central Marine Fisheries Research Institute.* [Mandapam Camp, India]. 9: [i +] 37-48.

Umamaheswara Rao, M. 1970a. ('1968'). "Additions of the flora of the Gulf of Mannar and Palk Bay from Mandapam area." *Journal of the Marine Biological Association of India* 10: 366-369.

Umamaheswara Rao, M. 1970b. ('1968'). "On two new records of Codiaceae from India." *Journal of the Marine Biological Association of India* 10: 407-409. 8 figs.

Umamaheswara Rao, M. 1972. "Coral reef flora of the Gulf of Mannar and Palk Bay." *In*: *Proceeding of the First International Symposium on Coral and Coral Reefs*. Mandapam Camp. 1969, Cochin: Marine Biological Association of India. Pp. 217-230. 6 figs., 1 table.

Umamaheswara Rao, M. 1974a. ('1972'). "Ecological observations on some intertidal algae of Mandapam coast." *Proceeding of the Indian National Science Academy*. Part B, *Biological Science* 38: 298-307. 3 figs. [Reprinted in M. R. N. Prasad 9ed.), 1974].

Umamaheswara Rao, M. 1974b. ('1972'). "On the Gracilariaceae of the seas around India. *Journal of Marine Biological Association of India* 14: 671-696. 5 figs., IV pls., 1 table.

Umamaheswara Rao, M. 1974c. "Additions to the algal flora of the Gulf of Mannar and Palk Bay from Mandapam area – II." *Phykos* 13 (2): 56-69. including I pl., 33 figs.

Umamaheswara Rao, M. 1987. "Key for the identification of economically important seaweeds." *Bulletin of the Central Marine Fisheries Research Institute* [Cochin, India]. *Seaweed Research and Utilization in India* 41: 19-25.

Umamaheswara Rao, M. 1990. "Autecological and eco-physiological studies on marine algae of Visakhapatnam and Mandapam." In V. N. Raja Rao (ed.), *Perspectives in Phycology*. New Delhi: Today and Tomorrow's. Pp. 323-335. 3 figs., 8 tables.

Umamaheswara Rao, M. 1991. "On a new species of *Helminthocladia* from India." *Seaweed Research and Utilization* 13: 169-174. 18 figs.

Umamaheswara Rao, M. and N. Kaliaperumal. 1987. "Diurnal periodicity of spore-shedding in some red algae of Visakhapatnam coast." *Journal of Experimental Marine Biology and Ecology* 106: 193-199. 2 figs., III tables.

Umamaheswara Rao, M. and T. Sreeramulu. 1970. "An annotated list of the marine algae of Visakhapatnam (India)." *Botanical Journal of the Linnean* Society [London]. 63: 23-45. 65 figs.

Untawale, A. G., V. K. Dhargalkar and V. V. Agadi. 1983. "List of marine algae from Goa,India." National Institute of Oceanography. [iii+] [42] pp. + 36 figs.

Untawale, A. G., and T. G. Jagtap. 1984. "Marine macrophytes of Minicoy (Lakshadweep) coral atoll of the Arabian sea." *Aquatic Botany* 19: 97-103. 3 figs., 1 table.

Untawale, A. G., and T. G. Jagtap. 1989. "Observation on marine macrophytes of the Republic of Seychelles." *Botanica Marina* 32: 115-119. 3 figs., 1 table.

Untawale, A. G., C. R. K. Reddy and V. D. Ambiye. 1989. "Marine algal flora of submerged Angria Bank. (Arabian Sea)." *Indian Journal of Marine Sciences* 18: 207-209. 2 figs., 1 table.

Untawale, A. G., C. R. K. Reddy and G. V. Deshmukhe. 1989. "Ecology of intertidal benthic algal of northern Karnataka coast." *Indian Journal of Marine Sciences* 18: 73-81, 3 figs., 2 tables.

Vanilla Kumari, E. and G. Subba Rangaiah. 1997a. "Seasonal growth, reproduction and spore shedding in *Amphiroa fragilissima* (L.) Lamouroux (Rhodophyceae, Cryptonemiales) of the Visakhapatnam coast." *In National Symposium on Recent Trends in Algal Research.* 1997. Programme and Abstracts, Andhra University, Visakhapatnam. P. 22.

Vanilla Kumari, E. and G. Subba Rangaiah. 1997b. "Effect of environment factors on spore shedding in *Jania rubens* (L.) Lamouroux (Rhodophyceae, Cryptonemiales) of the Visakhapatnam coast." *In* National Symposium on Recent Trends in Algal Research. 1997. Programme and Abstracts, Andhra University, Visakhapatnam. P. 22 - 23.

Vellaiyan, R., S. Balakrishnan and R. Thenmozhi. 2001. "Biological screening of antibacterial activity of *Sargassum wightii*." *In National Symposium on Marine Algal Research in India: Prospect and Retrospect.* 2001. Programme and Abstracts, Bharathidasan University, Tiruchirappalli. P. 25.

Venkateswarlu, Y. and M. A. F. Biabani. 1995. "A diterpene from the brown alga *Stoechospermum marginatum*." *Phytochemistry* 40: 331-333.

Vidyavathi, N. and K. R. Sridhar. 1991. "Seasonal and geographical variations in the antimicrobial activity of seaweeds from the Mangalore coast of India." *Botanica Marina* 34: 279-284. 2 figs., II table.

Vijaya Parthasarathy, M. D. and V. Krishnamurthy 2000. "*Udotea caribaea* Littler & Littler – A new for the Indian shores." *Seaweed Research and Utilization* 22 (1&2): 15-17.

Wahidulla, S. and L. D' Souza. 1995. "Phthalate esters from the brown alga *Stoechospermum marignatum* (C. Agardh) Kuetzing." *Botanica Marina* 38: 333-334.

Weber-van Bosse, A. 1898. "Monographic des Caulerpes." Annales du Jardin Botanique de Buitenzorg. 15: 243-401. Pls. XX – XXXIV.

Weber-van Bosse, A. 1913. "Marine algae, Rhodophyceae, of the "Sealark" Expedition,"collected by Mr. J. Stanley Gardiner, M. A. Transaction of the Linneaen Society of London, Second Series, Botany. 8: 105-142. 1 fig., pls. 12-14.

Weber-van Bosse, A. 1914. [The Percy Sladen Trust Expedition to the Indian Ocean in 1905, under the leadership of Mr. J. Stanley Gardiner, M. A. Reports:] No. XIV. "Marine algae, Rhodophyceae." Transactions of the Linnean Society of London, Second Series, Zoology. 16: 269 – 306, 1 fig., pls. 16-18. [Reprinting of weber-van Bosse 1913b].

Wynne, M. J. 1991. "A change in the name of the type of *Chondria* C. Agardh. (Rhodomelaceae, Rhodophyta)." *Taxon* 40: 316-318.

Zaidi, S. H. and O. P. Mairh. 1997a. "Biomass of promising alginophytes of the Indian coasts." In V. P. Agarwal & V. P. Kamboj (eds.), *Chemistry and Biology of Herbal Medicine* Publ. Society of Biosciences, Muzaffarnagar. Pp. 113-130.

Zaidi, S. H. and O. P. Mairh. 1997b. "Biomass of promising carrageenophytes on the Indian coast." *In National Symposium on Recent Trends in Algal Research*. 1997. Programme and Abstracts, Andhra University, Visakhapatnam. P. 39.

Zaidi, S. H. and O. P. Mairh. 1999a. "Impact of varying salinities on the biomass increase of *Gelidiella acerosa* (Forsskal) Felman et Hamel." *In* National *Symposium on Recent Trends in Marine Algal Research in India.* 1999. Programme and Abstracts, Annamalai University, Parangipettai. P. 28.

Zaidi, S. H. and O. P. Mairh. 1999b. "Biomass of promising carrageenophytes on the Indian coast." *In* G. Subba Rangaiah ed. *Recent Trends in Algal Research* Visakhapatnam, Publ. Marine Algal Loboratory, Department of Botany, Andhra University, waltair., Visakhapatnam. Pp. 66-82.

Zaidi, S. H., R. M. Oza and O. P. Mairh. 1998. "Effect of different temperatures on the biomass increase of *Gracilaria verrucosa* (Huds.)." Papenfuss *In National Symposium* on *Seaweeds and their Utilization* 1998. Programme and Abstracts, Madurai Kamaraj University, Madurai. P. 15.

Zaidi, S. H, R. M., Oza, and O. P. Mairh. 1999. "Current status of phycocolloids of the Indian seaweeds, their use and market potentials." *In* G. Subba Rangaiah ed. *Recent Trends in Algal Research.* Visakhapatnam, Publ. Marine Algal Loboratory, Department of Botany, Andhra University, Waltair. Visakhapatnam. Pp. 66-82.

Zanardini, G. 1858. "Plantarum in mari Rubro hucusque collectarum enumeration (juvante A. Figari)." Memorie del Reale Istituto Veneto di Scienze, Lettere ed Arti. 7: 209-309. pls. III-XIV.